Extreme Weather

Published by
Black Dog & Leventhal Publishers, Inc.
151 West 19th Street
New York, NY 10011

Distributed by
Workman Publishing Company
225 Varick Street
New York, NY 10014

Manufactured in China

Cover design by Becky Terhune
Interior design by Jon Morgan

ISBN-13: 978-1-57912-834-0

h g f e d c b a

Library of Congress Cataloging-in-Publication Data available on file.

Extreme Weather

Understanding the Science of Hurricanes, Tornadoes, Floods, Heat Waves, Snow Storms, Global Warming and Other Atmospheric Disturbances

H MICHAEL MOGIL

BLACK DOG
& LEVENTHAL
PUBLISHERS
NEW YORK

CONTENTS

FOREWORD

I have had a 55-year love affair with the weather across the US and the world which was ignited when, as a nine year old, I was fascinated by a series of hurricanes that struck New York City. It led to learning and sharing information about weather that has involved teaching courses in Earth and Physical Science as well as satellite image interpretation, climate change and meteorology.

I have focused on understanding the natural world, how we fit into it, how we affect it and how, ultimately, it affects us. I am continually amazed at how organized the world really is. Patterns are everywhere – in the clouds, in weather maps and satellite images and even in oceans. This quest for knowledge and understanding has enabled me to develop a panorama which has caused me to question and rethink our knowledge base. Nowhere do I question it more than in the debate that has developed on global warming and climate change.

This book looks at what we know about short- and long-term variations in weather and climate and explains the difference between the two while raising the questions we should be asking. It looks at all aspects of weather and climate and attempts to work out what we do, and don't know. It also shows that the natural world needs a balance and I would suggest that climate change is an extremely complicated issue where simplistic conclusions can damage our understanding of the issues and the solutions we should seek.

Am I correct to question the conventional beliefs on global warming? I don't know, but I hope that this book will get people talking and assessing what has happened and what still is happening to our planet – past, present and future. Even though I can be skeptical about how much we humans have contributed to global warming, I still believe that we should conserve our fossil fuels, minimize pollutants and work to save all endangered species. Both individually and collectively we can keep our planet alive.

I should like to thank the many people who have helped to make this book a reality: my wife Barbara Levine and our dog, Pepper; Gene Rhoden and Ron Holle; and the many meteorologists and organizations that are too numerous to mention. I should also like to thank Dr William Bonner, Herb Lieb, Earl Estelle and Jo Bryant and her associates. Should you wish to contact me the address is: extreme-weather@weatherworks.com

H Michael Mogil

Certified Consulting Meteorologist

1 INTRODUCTION

> "The snow squall may have only lasted 15 minutes, but at last snow finally fell in New York City on January 10 2007. This marked the latest-arriving snowfall in the City's history since 1878, when snow arrived on January 4."
>
> *H Michael Mogil*

When tornadoes move over agricultural areas, they often pick up large amounts of dust. If the dust completely engulfs the tornado, it makes the tornado appear even larger.

While writing this book, extreme weather made the news almost every day. First, it was unseasonably warm across the entire United States and Europe. Snow droughts were a concern to skiers and the trend was, according to some, "clear testimony to continued global warming".

A tornado ravaged parts of London in winter and in late fall/early winter twisters struck the southern US. Snow fell in parts of Australia in September, but New York City remained snowless far beyond the date of its usual first snow. Then extreme cold struck California and oranges froze…

Setting the stage

A national television network contacted me regarding a possible interview about the extremely warm weather across the United States. They wanted to address climate change, El Niño and the northward shift of the jet stream as possible causes.

Fortunately a few days earlier I had looked carefully at the upper level wind pattern (30,000 feet and higher) in the eastern Pacific Ocean. It suggested the onset of a major pattern shift, one that typically involves "Siberian air moving into Canada and the United States." I told my wife (you always have to tell your wife first) and then I called the major news network. I explained that pattern shifts like this occur often and I sent the network representative a monthly temperature graph that I had produced as part of a classroom weather activity package. The graph was for Washington, DC (January 2005).

Within hours the representative called me back to advise that the story would not run (they had more important news). While I was glad, I was also saddened. I was hoping that the story would be used to highlight upcoming pattern shifts and that such shifts were to be expected. I also suspected that, as bitterly cold weather, snow, and ice arrived, many people might be wishing that global warming would stay. Unfortunately, such explanations are not what drive news today. It has to be the extreme or the most awful. Or as they say in media, "if it bleeds, it leads."

This approach has ensured that people from around the world believe the worst when it comes to weather and climate (climate can be viewed as long-term weather conditions). They are sure that every hurricane will be a Katrina; that every year will be warmer than the last; and that we are handcuffed by the forces of nature that are only getting more powerful and vengeful.

Notice that, in the graph above, the (far) above average temperature pattern in the first half of the month was replaced by an equally below average pattern for the rest of the month. In fact, when all the highs and lows for the month were averaged, the monthly temperature was just about average. Yet, on not a single day did the daily average temperature match the monthly daily average temperature. In short, the daily weather this month was never average!

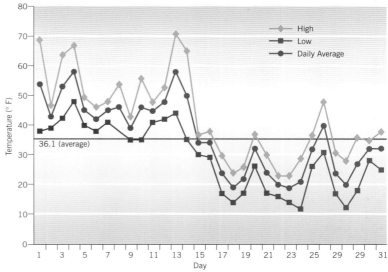

Washington DC (DCA) Temperatures for January 2005

Temperature variation for Washington, DC in January 2005. Notice how the very warm temperatures at the start of the month were replaced by comparable very cold temperatures later in the month. The average temperature for the month matched the 30-year "average" even with such extremes.

Tending toward average

However, nothing could be further from the truth and, in writing this book, I have attempted to bring some sense of scientific understanding and reasonableness to the fields of weather and climate, in particular extreme weather and climate change. I want each reader to take away knowledge about how the weather works and realize that almost everything in nature seeks balance. In much the same way that the temperature graph transitioned, from much above average to much below average, over the month (and the actual January temperatures were just "average"), so, too, do other weather and climate variables tend toward a middle ground.

Severe storms battered the UK on January 18, 2007. Many died and transportation was brought to a standstill while 70-80 mph winds ravaged the coastline at Blackpool, sending waves crashing over the sea barriers.

You probably know this yourself from playing in your bathtub as a child that, try as you might, you could never get all the water to collect on one side of the tub. As soon as the water level on one side was higher than the level on the other, the water flowed back to balance out the water level.

While finishing work on this book, the weather took on a similar pattern. The above average temperature regime across much of the United States was quickly replaced by one of bitter chill. Ice storms crippled parts of the central US and the northeast; it snowed in Malibu (southern coastal) California for the first time in recorded history; and California citrus fruits froze. But the citrus freeze was not something new. It has happened many times before. In fact, according to Jim Bagnall, a National Weather Service (NWS) meteorologist based in Hanford, California, "…the unusual weather pattern bringing such low temperatures to California… comes about once every eight years." The most recent situations occurred in 1990, 1998, and now in 2007. Hence, there have been many more non-freeze years punctuated with only a few that seriously affect the crops. This "one-in-eight" cycle is therefore the expected pattern or "average".

Icicles created by a spray irrigation system (designed to prevent the freezing of the fruit itself) hang from an orange tree in Orange Grove, California. But the toll in lost agricultural produce from an extreme freeze in January 2007 exceeded $1.4 billion. Citrus and vegetable crops were both hard hit. Governor Arnold Schwarzenegger immediately declared a state of emergency.

Sometimes, the averages take longer or shorter periods to balance. For the Atlantic Ocean area hurricanes, 2004 and 2005 were much above average in number and impact to the United States. We heard dire warnings that global warming was at work, yet, in 2006, with alleged "global warming" (and I will use quotes around the term until we've had a chance to explore what this really means), in full swing, not a single hurricane struck the US. So, there were two years of record-breaking high US strike numbers followed by a year with much below numbers. El Niño was blamed for tempering "global warming" in 2006; could something in addition to "global warming" have contributed to the active hurricane seasons in 2004 and 2005?

In mid-January, the worst winter storm to strike Europe in nine years raced across much of the continent. Transportation was disrupted, power was lost, and dozens were killed. But, after all, it was the worst storm in only nine years! In between, there were many lesser storms, each with lesser impacts. Still, there were storms. That meant the average was lesser events. I don't think anyone (save a few meteorologists who thrive on stormy weather) would have wanted the average to be more extreme storminess.

Perspective

Put into context, this means that everyday, somewhere around the earth, extreme weather is occurring. It can be a blizzard, a tornado outbreak, a hurricane, typhoon or cyclone, or just a plain, old-fashioned winter-type storm (with wind, rain, snow, and/or waves). If storms aren't extreme enough, there is the prolonged lack of storms or rain (drought), high winds, heat waves and cold waves. And then there are the weather-caused or weather-related events like forest and grass fires and pollution episodes. If weather isn't the culprit, it may be the accomplice of other earth hazards including acid rainfall and volcanic ash deposition.

In addition, we are continually impacting the weather (short-term conditions) and climate (longer-term ones) around us. This can involve so-called "global warming" (based in part on the fossil fuels we burn), the creation of local urban heat islands by building up urban centers, and the changes we make to the shape and character of land surfaces that in turn affect daily solar heating and rainfall runoff. Trees and grasslands

Fog is one reason that lighthouses were built along coastlines. Bright lights like these alerted mariners that they were close to the shore.

absorb water run-off where concrete and asphalt surfaces, roofs and other impervious surfaces do not. Even indoors, we control our weather with heaters, air conditioners, humidifiers, dehumidifiers and oxygen generators.

However, it is extreme weather (climate) that makes the news and catches our attention, and that is what this book is about. This book is designed to let you understand both how extreme conditions fit into the overall global weather and climate pattern, and that such extreme events really are to be expected. In the process, you should recognize that "extreme" can mean different things to different people. A one-inch snowfall to people who live near Buffalo can be a nuisance; to people living in places like London, England, or Atlanta, Georgia, a similar snowfall can almost cripple the area.

I hope that this book makes you a better global citizen and a more science-oriented person. I anticipate that it will help put the "news" into perspective so you can better use weather and climate information to advantage.

Information at your fingertips

A good place to start looking at extreme weather is on the National Weather Service's (NWS) main web page – www.nws.noaa.gov. Real-time hazards, shown on a county-by-county basis, are continually updated. Similar individual country real-time hazard summaries and forecasts are available at other national weather service sites, including Canada, Australia and the UK. For a more a more global view, check out NOAA's National Climatic Data Center "Monthly Hazard" at http://www.ncdc.noaa.gov /oa /reports/ weatherevents.html.

The National Weather Service's website provides real-time information about all types of hazardous weather conditions.

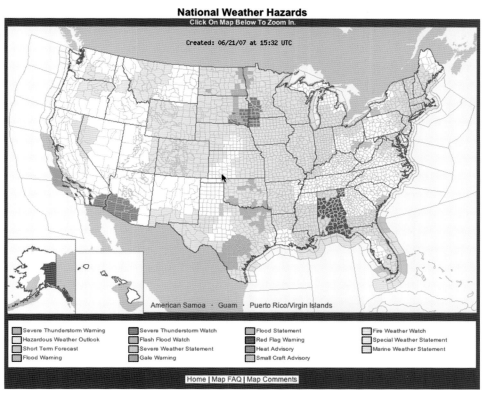

Locator #	Continent	Temperature Extremes (°F)	Location	Elevation (feet)	Date
1	Africa	136	El Azizia, Libya	367	9/13/1922
2	North America	134	Death Valley, CA Greenland Ranch	-178	7/10/1913
3	Asia	129	Tirat Tsvi, Israel	-722	6/22/1942
4	Australia	128	Cloncurry, Queensland	622	1/16/1889
5	Europe	122	Seville, Spain	26	8/4/1881
6	South America	120	Rivadavia, Argentina	676	12/11/1905
7	Oceania	108	Tuguegarao, Philippines	72	4/29/1912
8	Antarctica	59	Vanda Staton, Scott Coast	49	1/5/1974
9	Antarctica	-129	Vostok	11,220	7/21/1983
10a	Asia	-90	Oimekon, Russia	2,625	2/6/1933
10b	Asia	-90	Verkhoyansk, Russia	350	2/7/1892
11	Greenland	-87	Northice	7,687	1/9/1954
12	North America	-81.4	Snag, Yukon, Canada	2,130	2/3/1947
13	Europe	-67	Ust'Shchugor, Russia	279	date unknown
14	South America	-27	Sarmiento, Argentina	879	6/1/1907
15	Africa	-11	Ifrane, Morocco	5,364	2/11/1935
16	Australia	-9.4	Charlotte Pass, NSW	5,758	6/29/1994
17	Oceania	12	Mauna Kea Observatory, HI	13,773	5/17/1979

Locator #	Continent	Rainfall Extremes (inches)	Location	Elevation (feet)	Years of Record
18a	South America	523.6	Lloro, Colombia	520	29
19	Asia	467.4	Mawynram, India	4,597	38
20	Oceania	460.0	Mt Waialeale, Kauai, HI	5,148	30
21	Africa	405.0	Debundscha, Cameroon	30	32
18b	South America	354.0	Quibdo, Columbia	120	16
22	Australia	340.0	Bellenden Ker, Queensland	5,102	9
23	North America	256.0	Henderson Lake, British Colombia	12	14
24	Europe	183.0	Crkvica, Bosnia-Herzegovina	3,337	22
25	South America	0.03	Arica, Chile	95	59
26	Africa	0.1	Wadi Halfa, Sudan	410	39
27	Antarctica	0.8	Amundsen-Scott South Pole Station	9,186	10
28	North America	1.2	Bataguse, Mexico	16	14
29	Asia	1.8	Aden, Yemen	22	50
30	Australia	4.05	Mulka (Troudaninna), South Australia	160	42
31	Europe	6.4	Astrakhan, Russia	45	25
32	Oceania	8.93	Puako, Hawaii, HI	5	13

Highest Temp Extremes

Lowest Temp Extremes

Highest Average Annual Precipitation Extremes

Lowest Average Annual Precipitation Extremes

Earth's global extremes

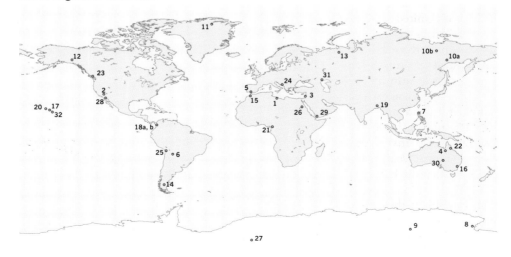

Meteorologists, much like sports enthusiasts, thrive on records. This map (and the accompanying table) show global annual temperature and rainfall extremes. These records, however, only span about a 100-year period.

Earth's extremes (short- and long-term)

The US Geological Survey (USGS) has compiled a list of global extremes in temperature and precipitation (with data from about the past 100 years).

Even with the specter of "global warming", it is interesting to note that many of these global extremes occurred before 1950. The latest extreme was at Charlotte Pass, New South Wales, Australia (June 29, 1994) and it was a record Low (-9 °F (-22 °C)) continental temperature. However, high altitude locations (5,758 feet (1.75km)) are used to determine lowest continental extreme temperatures.

Yet within this extreme framework, Earth enjoys a fairly stable overall climate. There are ups and downs over geologic time and very short time periods. However, the daily range of temperatures around the globe is fairly narrow given the ranges experienced on other planets. Further, Earth's average temperature is more uniform and hospitable than that of other planets.

Global and large-scale temperature ranges

On any given day, the range in temperature (difference between the highest and lowest reported) across our planet far exceeds 100 °F (55.6°C). And the range across the contiguous United States alone is often 70-109 °F (21-43°C). In fact, just look to the graph (page 12) for the year 2005 to see the range of temperatures across the contiguous United States. Although not shown, on several days – especially in summer – California reported both the national US high and low temperatures. National extremes hinge on many factors, mostly latitude, altitude, snow cover (or lack thereof), and land-sea variations. California extremes depend mostly on altitude and land surface. Truckee (located in a valley in the Sierra Nevada Mountains) tends to be the coldest site in the summer; Death Valley (282 feet below sea level) is often the hottest spot.

The atmosphere that is responsible for such perceived large temperature extremes and home to "global warming", is the same atmosphere that

Temperature changes through time

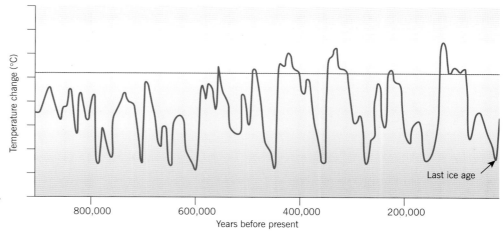

Representation of long-term temperature change in earth's history. Notice the wide swings in the temperature. Dips in temperature pattern are cold periods (some being glacial); rises represent warmer periods (during which glacier and ice sheets retreated poleward). The horizontal line represents the present global average temperature of about 15 °C (59 °F). Thus the solid curves show small changes from this average; note that the temperature drops only about 5 °C during a glaciation.

TEMPERATURE OF THE PLANETS

The temperature observed on the planets in our solar system is based on several factors. With the exception of Venus, almost all planets get colder as distance from the Sun increases. The amount and type of atmosphere is important, too. Earth has an atmosphere composed of many gases (including carbon dioxide); Venus's atmosphere is almost exclusively carbon dioxide; planets far from the sun (and the Earth's moon) have little atmosphere. Although not shown, the four most distant planets have temperature ranges, too, but they hinge on whether the planet's surface is facing toward or way from the sun. Of all the planets in our solar system, only Earth has a temperature range favorable for life as we know it.

Mean Temperature at Surface (° F)

Mercury	Venus	Earth	Mars	Jupiter	Saturn	Uranus	Neptune
-292 to 806	869	-128 to 136	-115 to 32	-238	-274	-328	-346

Daily Temperature Range in 2005 – US

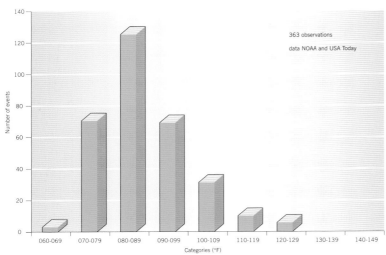

363 observations

data NOAA and USA Today

Summary of daily temperature ranges across the contiguous United States for 2005 (Alaska and Hawaii excluded). Typically, the variation between the highest and lowest temperature reported is in the 70s to 90s.

tempers our planet's overall weather/climate (average temperature around 59 °F (15 °C)) and allows human and other life forms to exist. In short, our atmosphere with its "greenhouse gases" is both an alleged culprit and saviour.

Yet, when we talk about extreme weather, events like tornadoes, hurricanes and blizzards come immediately to mind. But, extreme weather goes far beyond the storms that make the news headlines.

Temperature variations at one place

Even on a given day, in one place, the daily temperature variations often range between 20 and 40 °F (11 and 22 °C) or more. Some of this is linked to latitude, proximity to mountains, amount of cloudiness, presence of snow cover, length of day and the movement of transitory weather systems (including highs, lows and fronts). In highly vegetated regions during the growing season, the range is tempered because of evaporation from plants. The daily range is also small along some coastal regions (e.g. US west coast and northwest Europe). In dry, desert-like regions, especially those with sandy land surfaces or in drought-affected regions, the daily range can be large.

Similar variations occur over the seasons. Yakutsk, Russia has an average monthly high temperature of 77 °F (25 °C) in July (almost 24 hours of sunlight) and an average monthly high temperature of -34 °F (-37 °C) in January (almost no sunlight). The range in average high monthly temperature at Yakutsk is more than 110 °F (43.1 °C)! At Palu, with around 12 hours of sunlight year round, the annual range is only a few degrees.

All of these variations, and more, are either directly or indirectly related to latitude, which is linked to the number of sunlight hours and sun angle.

Frontal temperature variations

In January 2007, an arctic air mass invaded the United States. As the cold air moved slowly southward (displacing the warmer air), a large temperature gradient developed across the advancing cold front. During the late afternoon of January 13,

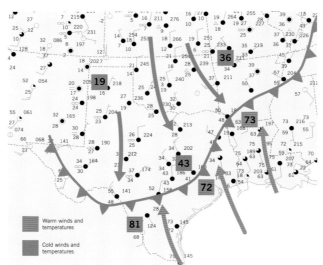

the temperature range across the front, in the state of Texas alone, was 62 °F (34.5 °C) Across the much smaller state of Arkansas, the temperature range was 37 °F (20.6 °C) Immediately across the cold front in eastern Texas, the variation was 29 °F (16.1 °C).

A Tale of Two Air Masses

A cold front lies across Texas and Arkansas. Northerly winds are transporting cold air toward the Equator; southerly winds are transporting warm air toward the North Pole. This north-south temperature transportation helps keep the Earth's temperatures in balance.

Temperature difference across Texas and Arkansas on January 13, 2007

	Temperature difference (°F)	Distance (miles)	Temperature gradient (°F per 100 miles)
Laredo - Amarillo, Texas	81 – 19 = 62	547	11
Eldorado, - Harrison, Arkansas	73 – 36 = 37	211	18
Houston, - College Station, Texas	72 – 43 = 29	91	31

Some weather fronts have very sharp temperature variations (known as temperature gradients). Here are some measures of this gradient for the cold front shown in the figure on page 12. When this front passed, people in the area knew immediately that colder air had arrived.

Considering airline distance between the city pairs (see table above), the temperature gradient borders on incredible! The Houston-College Station gradient translates to almost one degree F per three miles (0.345 °C per kilometer).

Extreme temperature variations

Sometimes, strong (even extreme) localized winds, known as Chinooks, can bring dramatic temperature variations to locations just to the east of the Rocky Mountains in both the US and Canada. Similar events, (known as Foehns) occur in the European Alps. As high winds force air down from the mountains, the air is heated as it comes under higher air pressure. This warmer air, driven by high winds, tries to displace colder air below. Depending upon the strength of the warm downslope wind and the coldness of the air it is trying to displace, an ongoing battleground can exist. When the Chinook wind arrives, the warming can be dramatic – they can elevate winter temperature dramatically. Folklore has given the wind the name "snow eater" because it can melt and evaporate snow cover very quickly.

One of the most dramatic examples of these winds occurred in Loma, Montana on January 15, 1972. The temperature rose from -54 °F to 49 °F (-47 to 9.5 °C), the greatest temperature change ever recorded in the United States during a 24-hour period.

Cold air (much denser than warm air) can also pool near the ground with warmer air atop. This is called an inversion because the temperature profile is inverted from what it normally is. The air temperatures of the lowest 8-10 miles (13-16km) of the atmosphere are usually warmer closest to the ground. If an oscillation (or wave pattern) develops on the interface between the warm air (top) and cold air (bottom) there can be dramatic, almost instantaneous, fluctuations. One such event occurred at Spearfish, South Dakota (just east of the Black Hills) on January 22, 1943. In a two-minute period, the temperature rose from -4 °F (-20 °C) at 7:30 am to 45 °F (7 °C) at 7:32 am, a 49 °F (27 °C) change. The temperature then climbed to 54 °F (12 °C) by 9:00 am before plunging back to -4 °F (-20 °C) at 9:27 am, a change of 58 °F (32 °C) in 27 minutes.

Temperature variations in your neighborhood

Being curious about neighborhood temperature variations, I once assessed how the temperature varied in my Rockville, Maryland sub-division (located just to the northwest of Washington, DC). Using a rapid response thermometer, I walked three city blocks down my street, crossed a drainage ditch, walked back up a slight hill and then crossed the drainage area back to my house. The area had relatively new homes and small trees. The elevation difference from my home down to the bottom of the hill was 15 feet (4.5 meters); the horizontal distance I walked was around 1,500 feet (457 meters). It took me less than 15 minutes to complete the entire circuit, so the transit time from top to bottom of the hill was around seven to eight minutes.

The weather for the day of this November "experiment" involved mostly clear skies and light winds. I purposely carried out the study around sunset to maximize local variations. I could have also done this around sunrise.

Some locations reported rainfall of more than a foot; nearby locations had rainfall of less than two inches. Rainfall gradients like these are common in heavy rainfall events.

What I found startled me. The temperature variation from the top to the bottom of the gentle slope was 9 °F (5 °C). During my transit, I discovered cooler, localized drainage winds, so the temperature variation up and down the slope was not uniform. This local temperature gradient on this day was 31 °F (17 °C) per mile (three times the gradient shown in the figure on the previous page).

This variation, just within my neighborhood, also turned out to be far greater than that which was being reported across the entire Washington, DC metropolitan area that evening on television weather reports.

More than temperature

Similar extremes can be found in most weather variables. They can involve changes over short distances or changes over short time periods at a particular location. When you drive on a foggy morning, you may drive in and out of the fog several times; if you live on the east or south sides of the Great Lakes – where "lake effect" snows occur – you can also drive in and out of intense snow bands in a matter of seconds. ("Lake effect" snows occur when cold air blows across large, warm lakes in late fall and early winter. The Great Lakes in the United States are one the most active "lake effect" regions in the world.)

Tropical rain showers have been known to produce rain on one side of the street but not on the other. And winter storms often bring rain to the southeast suburbs of major US east coast cities, while northern and western areas are covered with snow or freezing rain.

Rainfall and snowfall

Rainfall extremes can make temperature extremes pall by comparison. Consider what happened in Washington, DC in mid-2006. Overall, Washington National Airport (just southwest of the downtown area) reported an above average rainfall year. Yet the year was punctuated by significant dry periods. During the last week of June, during a three-day period, some 10.34 inches (26 cm) of rain (more than 25 percent of the annual average) fell. Nearby locations received as much as 15 inches (38 cm) of rain during the overall six-day rainfall event. So while vegetation suffered for parts of the year, serious, property-damaging flooding also occurred.

Locally, to the south and east of the Great Lakes, the winter of 2006-07 was not a favorable one for lake effect snows. Although the Great Lakes were warm well into the winter, the air crossing them was also warm. This was true at least through mid-January.

On June 29, 2006, Delaware River flood waters surge past these three Trenton, NJ bridges. Image view is looking northeast toward downtown Trenton.

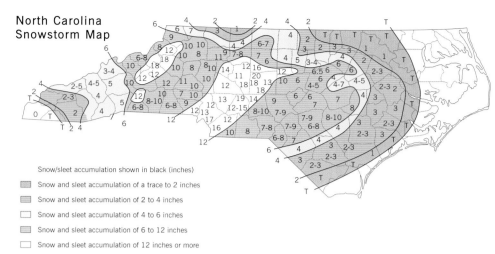

North Carolina Snowstorm Map

Snow/sleet accumulation shown in black (inches)

- Snow and sleet accumulation of a trace to 2 inches
- Snow and sleet accumulation of 2 to 4 inches
- Snow and sleet accumulation of 4 to 6 inches
- Snow and sleet accumulation of 6 to 12 inches
- Snow and sleet accumulation of 12 inches or more

Just as rainfall patterns have large gradients, so, too, can snowfall events. For this February 26-27, 2004 storm, measurable snow fell across much of North Carolina. Yet, some areas received more than a foot of snow and sleet while nearby areas (sometimes in the same county) had significantly less.

Then the lake effect began with a vengeance, as bitterly cold air blew across the region. Within two weeks, many locations had wiped out their two-month snowfall deficits, and week-long snowfall totals measured five to eight feet (4.5-2.5 meters) in places. Still, while lake effect snows had gotten mostly back to "normal" by early February, the overall snowfall in the northeast part of United States remained much below average.

To the west, however, snowfall was much above average in parts of west Texas and Colorado, and Alaska – which had also had suffered a "snow drought" in recent years – was reporting much above average snowfall. In early 2007, the Australian Bureau of Meteorology issued its Annual Australian Climate Statement for 2006. The report noted, "…Rainfall was well below normal in the southeast and far southwest, but close to normal when averaged over the whole country."

The map on page 16 shows where rainfall was above to much above average across Australia. Even by just looking quickly at the blue and red areas, it is easy to see how rainfall did balance itself across the continent.

Driving on snow-covered roads is always treacherous. It's even more so in mountainous areas due to curvy and sometimes steeply-sloped roadways. Here, drivers slowly navigate a mountain roadway in the western US.

Australian Rainfall 2006

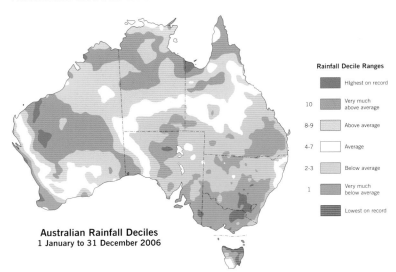

Australian Rainfall Deciles
1 January to 31 December 2006

Rainfall Decile Ranges

	Highest on record
10	Very much above average
8-9	Above average
4-7	Average
2-3	Below average
1	Very much below average
	Lowest on record

Rainfall across Australia for 2006. Values in deciles (which are similar to percentiles, only expressed without the right-most zero), brightest blue, for example, indicates those places with rainfall values in the top 10-20% of rainfall years.

Average temperatures over the week

	Cloud cover	Temperature	Precipitation
Sunday	0%	80	0.00"
Monday	0%	80	0.00"
Tuesday	30%	78	0.00"
Wednesday	100%	64	1.35"
Thursday	50%	66	0.05"
Friday	0%	78	0.00"
Saturday	0%	79	0.00"
Actual Average daily	20%	75	0.20"
"normal" daily	20%	75	0.20"

Representation of a week's worth of weather to demonstrate how "average" may not adequately represent daily weather conditions.

Globally, snowfall for the winter period (December 2005-February 2006) was slightly above average for the Northern Hemisphere. Yet, for North America, the snow cover was below average by about the same amount. According to NOAA's National Climatic Data Center (NCDC), much of the global surplus, "…was due to anomalously cold and snowy conditions across Asia and Europe." Again, when viewed globally, the above and below values tended to average out.

Definitions

In studying anything, vocabulary is everything. We have to be "on the same page", so to speak. We could delve into a lengthy dictionary style discourse here, but instead, let's look at just a few of the key definitions and see how they've been used (intentionally or otherwise) to confuse our perception of weather.

Is "average" really normal?

For years, meteorologists and climatologists referred to the 30-year average weather conditions as "climatological normals". Mathematically, the two terms are interchangeable; meteorologically, they convey very different conditions.

If you vacationed in a new locale and someone there told you that the weather conditions were expected to be "normal" today, you'd probably think these were what happened every day. This would be the typical weather here, and you'd likely expect the same conditions for the duration of your vacation.

Imagine that during the course of your stay, say one week, there were also some cool and rainy days along with several sunny days with very warm temperatures. In short, there were some more extreme conditions.

If we averaged things out (as shown in the box left), you can see the average sky condition, precipitation and temperature matched what was indicated as "normal".

In weather, the so-called "normal", is really the average of extremes and, unless you are in a place with almost unchanging weather conditions, the odds of having the weather on any particular day exactly match the longer-term average are extremely small. Most meteorologists now recognize this and go out of their way to refer to the climatic record as "average".

"Average" can refer to almost any aspect of weather – winter snowfall, number of hurricanes, number of sunny or cloudy days. In fact, weather statistics can rival sports statistics if we let them. Listen to the weather reports and eventually you'll hear something like "the record low high temperature today was…" Translated, this means that of all the high temperatures recorded on a particular day, since

weather record-keeping began at that place, the lowest high temperature recorded has always been higher than the high temperature recorded today.

Partly sunny or partly cloudy?

There are some specific meteorological conditions that have to be met for the sky condition to be labelled "partly sunny" versus "partly cloudy". But our perception of these terms can be quite different.

Consider two distant locations during the winter. One location is Cleveland, Ohio. Here, due to the lake effect (see page 14) and other factors, skies tend to be cloudy for a large part of the winter. If the sun peeks out for a short period one afternoon, chances are good that many people will declare the weather "partly sunny". Similarly, Phoenix, Arizona has many cloud-free winter days. Locals might call the weather "partly cloudy" even when only a few puffy cumulus clouds dot the sky. Regardless, the scenarios are testimony to very extreme differences in cloud cover between one place and another.

Windy versus breezy

In 2005, I testified at a trial. The trial centered on the definition of a not-so-extreme weather event – wind conditions that were "breezy" versus those that were "windy". "Breezy" is defined as having slightly lower wind speeds than "windy".

But the plaintiff's attorney and the other weather expert contended that the accepted definition of breezy was different than the one used in official National Weather Service weather forecasts.

The weather forecast in the newspaper for Pennsylvania that day used "windy" in the forecast, but also included a description of wind speeds that were in the "breezy" category. There was also a suggestion that wind gusts satisfied the wind speed forecast categorization.

The result was that several definitions of wind were being applied. In this case, the jury had to decide which one was correct? Breezy won because those were the conditions that were also observed.

Sometimes, the definition of stormy weather is blurred through error. For example, I once heard of a report that noted winds gusting to 81 mph in New Jersey. The meteorologist then noted something such as "you know that hurricane force is only 74 mph." But, the accepted definition of hurricane force winds is based on

When skies have been cloudy for a while, just the slightest hint at blue skies or thinning clouds can sometimes prompt a forecast that reads "becoming partly sunny". Conversely, following days of cloudless skies, even a few cumulus skies can make skies appear "partly cloudy".

sustained winds, not gusts. This expansion of the definition actually lowers the bar for reporting such events, making them more common. Viewed from another angle, it takes non-storm events and elevates them to storm status, always keeping us at the edge of our seats as the storminess worsens.

Does it have to be a tornado?

When it comes to thunderstorm-related wind damage, there's a tendency for those suffering a loss to want a tornado to be involved. From the insurance company's perspective, there's nothing different between wind damage from a downburst (straight-line outflow winds) and a tornado (rotating winds).

After a tornado or wind damage event, the National Weather Service (and others) may conduct a damage survey from the ground and/or the air. The intent is to assess the damage and estimate wind speed for the individual event and the climatological record. There are criteria for carrying out this survey and for categorizing the data according to the Fujita Scale including the new Enhanced Scale (Chapter 7).

Sometimes, however, even meteorologists may be jaded toward calling an event a tornado or applying a slightly higher damage estimate to the classification. The LaPlata, Maryland tornado of April 28, 2002 was initially classified as an F5; a damage survey after the event lowered the classification to an F4 when some of the F5 damage was attributed to issues in building construction. Yet assessment by others, some of who surveyed the damage, was that the worst damage was actually a category F3.

A significant wind damage event occurred in Reisterstown, MD in October 1979. The event was described as tornadic. Yet, in viewing the damage scene, the only rotational component was that linked to wind flow around buildings. The event still stands in the climatological database as a tornado.

In the case of hurricanes, the reverse can occur. Some 10 years after Hurricane Andrew struck southeast Florida, and after numerous engineering studies, Andrew's classification was raised from a Saffir-Simpson category 4 to a 5.

Each year, dedicated men and women assess hundreds of storm events and apply much-needed metadata to ensure that there is as complete a record as is possible. However, they are human and few have the resources to always assess and reassess what goes into the record. We'll talk more about weather records as we look at specific extreme weather events throughout the book.

Smoke rises straight up. Less than 1 mile per hour. CALM (0)	Large branches in motion; umbrellas hard to hold; telephone wires whistle. 25-31 miles per hour. STRONG BREEZE (6)
Smoke drifts; weather vanes still. 1-3 miles per hour. LIGHT AIR (1)	Whole trees in motion; walking against wind difficult. 32-38 miles per hour. MODERATE GALE (7)
Leaves rustle and weather vanes move. 4-7 miles per hour. SLIGHT BREEZE (2)	Twigs break off the trees. 39-46 miles per hour. FRESH GALE (8)
Leaves and small twigs in constant motion; light flag extended. 8-12 miles per hour. GENTLE BREEZE (3)	Slight building damage. 47-54 miles per hour. STRONG GALE (9)
Dust, dry leaves, loose papers raised; small branches move. 13-18 miles per hour. MODERATE BREEZE (4)	Seldom happens inland; trees uprooted; much damage. 55-63 miles per hour. WHOLE GALE (10)
Small trees in leaf start to sway; crested wavelets form on inland waters. 19-24 miles per hour. FRESH BREEZE (5)	Very rare; much damage. 64-72 miles per hour. STORM (11)
	Anything over 73 miles per hour is a HURRICANE (12)

Using the Beaufort Wind Scale on land
Using the Beaufort Wind Scale (designed for estimating winds over water bodies) on land.

> "Heavy rainfall in January 2007 slightly alleviated the ongoing drought conditions in eastern and southern Australia. But serious, short-term deficiencies – with rainfall amounts in the lowest 10% of historical (1900-2005) totals – have developed in southeast Queensland and parts of the northwest territories."
>
> *H Michael Mogil*

Meteorologists can be likened to sports aficionados. They like to keep track of records and they often create new categories to document even more records.

Take sports for a minute. In Super Bowl XLI (41) played in early 2007, sportscasters were quick to pick up that during the past half a dozen or so championship games no touchdowns were scored in the first quarter. That was after one player scored a record-breaking kick-off return. In tennis, we are always looking to see who wins the Grand Slam (holding the title in all four major events – Australian, French and US Opens, and Wimbledon – at the same time). In horse racing, there is always interest in the Triple Crown winner. In the Olympic Games, there are almost an infinite number of comparisons made to past Olympic events (including scores, medals per country, winning times, and so on).

Weather records

In the weather arena, the United States had a Fujita Tornado Scale (now replaced by an Enhanced Fujita Scale); in Europe they use the TORRO Scale for tornadoes; the US also uses a new Northeast Snow Impact Scale, a Saffir-Simpson Hurricane Scale, a drought Index and many more. The National Weather Service Office in Buffalo, New York, classifies lake effect events using its informal "Lake Flake Scale". This scale, "…subjectively ranks the events from one flake (*) for minor (wimpy) events to five flakes (*****) for epic mega-storms, based not only on snowfall amounts but the impact on population centers as well." Each season's events are named following a theme (e.g. cats, trees, etc.).

Then there is the spate of daily, monthly and annual statistics and records for official weather-reporting sites. One can add daily maxima and minima temperatures by country and around the world, significant events as they occur and any other statistic that can be generated. The list is endless!

A SELECTION OF MEASUREMENT SCALES

Snow index: http://www.ncdc.noaa.gov/oa/climate/research/snow-nesis/

Lake Flake: http://www.erh.noaa.gov/buf/lakeeffect/indexlk.html

Fujita: http://www.tornadoproject.com/fscale/fscale.htm

TORRO: http://www.torro.org.uk

Saffir Simpson: http://www.hwn.org/home/saffir-simpson-scale.html

Drought: http://drought.unl.edu/dm/monitor.html

During the cold spell of winter 2007 in the United States, television meteorologists kept track of the number of consecutive snowy days in lake effect areas, the number of consecutive hours the temperature remained below zero in Minnesota, and when the last time such low wind chill values had been recorded in Chicago. They also concentrated on the percentage of seasonal average snowfall in lake effect regions, how many days were above and below average temperature in New York City, the number of consecutive days that snow was on the ground in Denver, and when the sun last appeared in various locations. In hurricane and tornado seasons, there are comparably long lists of records.

Comparing and contrasting

In addition to specific records, there are also comparisons from place to place. One such comparison involved the amount of snow that fell during an early February day in a heavy lake effect snow band and the seasonal total at New York City. Seasonal lake effect snows are normally three to five times or more than the New York City average (New York City snows often come later in February and March). The comparison sounds impressive (tens of inches versus a few inches), but is much like comparing apples to oranges.

In all of these weather scenarios (as with events like the Super Bowl), the period of record or comparison is small. In most cases, it is far less than 100 years. This makes breaking records of all types, or being far from an average condition, easier. In weather, it also says that whatever is happening now may bear little resemblance to what may have happened hundreds of thousands of years ago.

Periods of record

Consider the daily graph of any stock market average. For a moment, assume that it represents the long-term condition of any weather variable over the Earth, over a part of the Earth's 4.5 billion year history. Then look at the graph at the end of the stock trading day. Assume that is the current value and the current trend of the market variable (and that it represents the Earth today). How well does that instantaneous measure compare to the myriad of ups and downs that occurred throughout the day?

Many forces govern the Earth; some physical, many even external to the Earth, and some human-caused. Such instances also affect stock markets. If there is a shortage of fossil fuels, fruits, vegetables or other

LAKE EFFECT SNOWS

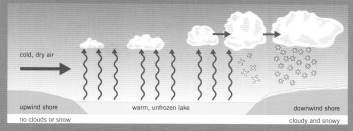

cold, dry air   upwind shore   no clouds or snow   warm, unfrozen lake   downwind shore   cloudy and snowy

Lake effect snows develop as cold, dry air blows across relatively warm, unfrozen lakes. During its passage across the lake, the air is moistened and heated from below making the air unstable. Thus, the warmed air wants to rise, eventually rising enough for clouds to form. If the process continues long enough, clouds can become tall enough to produce rain or snow showers.

When winds blow along the length of a large lake, and when the winds remain in a more or less constant direction, a long line of clouds can develop. As each cloud arrives at the downwind shore of the lake, it can deposit many inches of snow. As successive clouds arrive, the accumulated snowfall can be measured in feet. The process is referred to as "training" in which the long line of snow-producing clouds resembles a railroad train.

Radar and visible satellite images (late morning to early afternoon on February 4, 2007) show a long lake effect band. The band formed as southwesterly winds blew along the length of Lake Ontario. Notice how the clouds and precipitation increase as the band approaches the downwind shoreline.

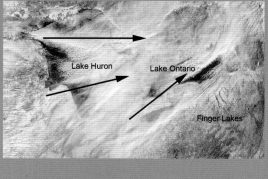

Lake Huron   Lake Ontario   Finger Lakes

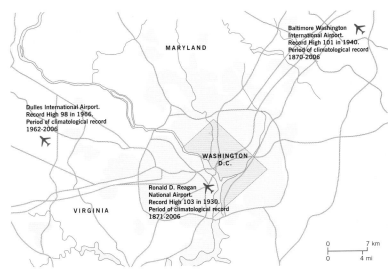

There are three airports in the greater Washington, DC area. All lie within a 50-mile radius of downtown Washington, DC. Plotted here are record high temperature data for July 26. Notice that all three airports have reported their record for this date in different years. One reason for this centers on varied periods of climatological records.

View looking west through the road cut at Sideling Hill. Before the road cut, the mountain ridge blocked westerly winds and cold air from infiltrating the valley to the east.

commodities (for example, by weather extremes), then if political or economic forces are at work, or if traders buy up or sell off certain stocks, the market can react. These impacts can occur over a few minutes or span longer periods.

Now compare the instantaneous trend and value to the annual stock market graph. Again, how representative is it to the longer-term?

The Washington, DC metropolitan area is unique in that it has three major airports all within a 50-mile circle, centered just to the northwest of the downtown area. If an extreme weather event occurred (e.g. very high temperatures) you would expect that all three sites would report a new record. Yet, often only Washington Dulles Airport (25 miles west of Washington, DC) makes the record books.

This happens because the period of record at Dulles Airport is only about 35 years while Washington Reagan National Airport (adjacent to the US capital) and Baltimore-Washington Airport have records dating back around 100 years. The longer the period of statistics, the harder it is to break records.

Location, location, location

However, the weather reports at these two long-term weather reporting sites were not always taken at the airport location. Washington's record-keeping, for example, began in downtown Washington and wound up being relocated several times until it finally arrived at the airport location.

Even when the weather reporting site doesn't move, populations do. Many urban centers have expanded significantly during the past 50 years. Thus, a site once located in a rural or suburban locale may now be included in the "urban heat island" region. Even if the site hasn't moved or been taken over by a metropolitan area, nearby construction, other changes in land use, the creation of a water storage or recreational reservoir, or the growth, or removal, of trees, can all effect how local weather changes.

In west-central Maryland, construction crews had to make a cut through a large gap in one of the Appalachian Mountain ridges to build Interstate Highway 68. The gap at Sideling Hill now allows colder air, driven by westerly winds, to move eastward more easily, affecting the nearby valley. Local residents have explained to me, in great detail, the changes they have experienced since the road cut was made.

A hurricane in 1933 opened up an inlet along Maryland's Atlantic shoreline. To capitalize on this "positive" benefit of a hurricane, it was decided to keep the inlet open by building bulkheads and carrying out periodic dredging. An unintended effect was to change the transport of sand along the coastline, which in turn affected how sand was collected or moved along beaches. Assateague Island (just to the south of the inlet) has now lost so much sand deposition that the island has migrated almost a mile toward the coast in the last 70

2,000 year temperature comparison

This graph shows that temperatures in the early years of the 21st Century are indeed warmer than at any time in the past 2,000 years.

■ **(1-1979):** *A Moberg, DM Sonechkin, K Holmgren, NM Datsenko and W Karlén (2005). Highly Variable Northern Hemisphere Temperatures Reconstructed From Low- And High-Resolution Proxy Data*

▬ **(200-1980):** *ME Mann and PD Jones (2003). Global Surface Temperatures Over The Past Two Millennia*

■ **(200-1995):** *PD Jones and ME Mann (2004). Climate Over Past Millennia*

years. Ocean City, just to the north of the inlet, requires periodic beach sand replenishment programs to maintain its beach.

The New Iberia, Louisiana, area has a tornado frequency slightly below the Louisiana state average. However it is 65 percent greater than the US average. On January 5, 2007 a tornado struck the city, killing two people. News reports concentrated on the fact that the tornado was the first to strike the city. Given the size of a tornado and the area within a defined city location, such strikes would be rare indeed. Yet, on October 29, 1974, an F3 tornado occurred 16 miles away from the New Iberia city center, injuring one person and causing between $50,000 and $500,000 in damage. On April 21, 1977, a F3 tornado touched down 22.4 miles (36 km) away from the city center. That twister killed one person, injured 11 people and caused between $500,000 and $5,000,000 in damage.

As you'll read about in the chapter on tornadoes, the relatively recent relocation of Denver's Airport to the east of the city has affected the number of tornado sightings in that area.

So what's extreme?

Given that our fully-documented historical weather data record dates back around 100 years or less in many places (slightly longer for parts of Europe), most of our comparisons to extreme events as the 21st Century begins are limited. However, news stories often manage to take advantage of this (e.g. "worst storm in nine years").

The last Little Ice Age (LIA), which began around 700 years ago, reached minimum temperatures about 350 years ago. Thus, most of our current extreme

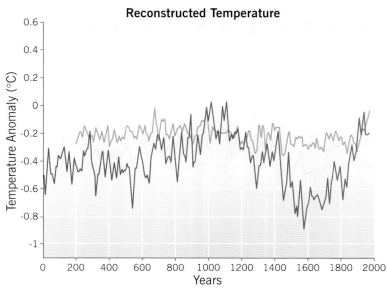

Reconstructed Temperature

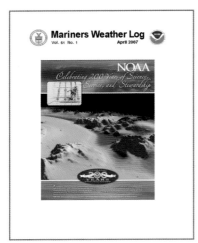

The Mariners Weather Log *is published by the National Weather Service (NWS) and contains articles, news and information about marine weather events, storms at sea and weather forecasting.*

weather comparisons are almost completely based on the period of warming following this LIA.

However, how does this compare to conditions over the longer-term? And, let's also recall that a small amount of snow in one place can be a nuisance; in another (not used to it) it can be an extreme and disruptive event.

National and international record keeping

Most weather records are kept by governmental weather agencies. Yet, such agencies do not have a very long history. For example, the United Kingdom's weather services began in 1854 and were developed to serve the marine community; Australia's individual state weather services were consolidated into a national service in 1908; Japan's weather services began in 1875; and the Canadian and United States weather services both formed around 1870. In many less developed countries, national weather services have much shorter histories.

Just about every national weather service had humble beginnings and it took decades for each to grow into a viable weather observing, forecasting and data archiving entity. In fact, according to Morley Thomas who chronicled the evolution of the Canadian Weather Services, "…very little statistical data, delineating the climate of the country, were available prior to 1900."

Morley was talking about the basic climatic information, not extreme weather and storm climatologies. In fact, specialized services to forecast and document severe thunderstorms and hurricanes did not come into existence in the United States until the mid-20th Century. Most snowfall records in the United States date back to the same time, although at least one record category, greatest daily snowfall, dates back to 1893. Data for assessing lake effect snow events in New York State only dates back to the late 1990s (and, for most locales, the data start in the early 2000s).

The International Meteorological Organization (IMO) was founded in 1873 and was one way for fledgling governmental organizations to develop global interactions. The IMO became the foundation for the World Meteorological Organization (WMO) established in 1950. In 1951, the WMO became the specialized agency of the United Nations for meteorology (weather and climate), operational hydrology and related geophysical sciences.

In recent years, interested individuals (e.g. David Ludlum), some organizations (e.g. the Tornado Project in the United States and TORRO in the UK) and various weather publications (*Weather* – Royal Meteorological Society, *Weatherwise* – Heldref Publications and *Mariners Weather Log* – NOAA) have gone far beyond routine governmental weather services to reconstruct and/or document the longer-term stormy weather history. Many people and organizations are also documenting weather and climate history on the Internet.

Steve Newman, creator of the Earthweek™ column only has a web-based history dating back to 2002. However, his information dates back to around 1987 – still a very short period.

Sources of extreme weather/climate information

Clearly, any governmental weather organization will have some data that defines average climate and summarizes extreme events. However, completeness of the record, the period of record, and the format vary widely. In some nations (e.g. the United States and Australia), there is a separate climatological agency. In writing this book, I found

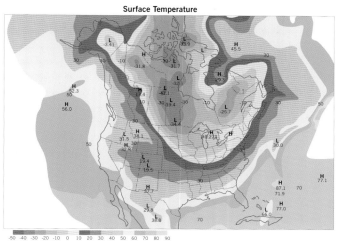

it extremely difficult to find required data sets. Instead of being able to find, access, and navigate a database using criteria-based searches, many databases simply provided summaries or showed pages that provided details for specific events. I had to become an even more creative web miner to find what I needed.

For these reasons, I have provided many links to national and global web sites. I hope these make your search for more information easier. I have also provided a listing of books about extreme weather. While this book explores the entire range of such events, many of the books I have listed key on one specific type.

Although every nation that collects weather and climate data does so under an umbrella of international standards, these standards are always subject to interpretation and availability of local resources. As a result, a uniform global database does not exist.

Looking at the weather

Meteorologists employ many types of displays to visualize and compare weather. These include larger-scale satellite images, weather maps that can span full hemispheres, and local or national radar displays. Some weather maps can provide very useful information about extreme weather, for example, here is a pair of isotherm maps for North America. Take a moment to compare the two maps.

The first map shows overnight temperatures in early January 2007; the second map shows overnight temperatures a month later. The colored bands of temperature, called isotherms (red is usually the warm color and blues and magentas the cold colors) clearly show a very different temperature pattern across North America. On January 5, the isotherms show a warm pattern, with cold air located near the North Pole. On February 5, bitterly cold air covered nearly all of Canada and a large part of the northern United States.

What is especially interesting is that the Inter-Governmental Panel on Climate Change (IPCC) issued an updated report about global warming while this extreme cold event (second map) was in the process of seizing control over much of North America.

Looking at Climate

Climate is harder to visualize than weather. Weather affects us every day; climate effects creep up on us over a much longer time. Weather is measured in minutes to days; climate in expressed in 30-year periods (known as "averages" or "normals") to periods spanning thousands to hundreds of thousands of years.

Mountains covered with snow allow for formation and maintenance of glaciers. When glaciers reach oceans and other water bodies, pieces of ice (icebergs) can break off.

Blue ice is caused by light being refracted as it enters a dense ice mass. While this iceberg seems massive, note that about 90% of its mass is hidden underwater.

We can also see how weather happens and changes. Not so for climate change. Glaciers have advanced and retreated over time. Glaciers and ice packs calve (pieces break off and drop into the ocean) each year, creating icebergs; yet news reports show the calving and attribute it to global warming. One such iceberg was responsible for the catastrophic sinking of the Titanic almost 100 years ago. Forest fires occur in marginally dry areas

almost every year. At least one tree ring analysis shows burn scar marks on trees dated back to 1696. Warm and cold periods have both short- and long-term fluctuations that dwarf what we are experiencing now (see also, Chapter 19). These indicate that what we are experiencing now is nothing new. It has happened before many times.

Understanding weather and climate and its changes

The 2007 IPCC report stated strongly that humans are *the* cause of global warming (emphasis mine). However, it might also be considered that, looking at research, it might not all be human-based activity that is the cause. For example, before cars and factories took over, we were already warming from the recent LIA (Little Ice Age). I am hard pressed to arrive at the same conclusion, although there is no doubt that we humans have contributed to the recent warming. This is based, in part, on the fact that Earth was already warming from the recent LIA; this trend started centuries before cars and factories took over and entered the picture.

Humans have affected our planet (in ways too numerous to reiterate here), but longer-term weather and climate shifts have occurred and continue to occur. As good global citizens, it is important that we fully understand the myriad of physical and other forces at work and integrate scientific, anecdotal and other information to arrive at informed choices. These choices can and should include public policy (e.g. emission controls, alternative fuels, tax credits) and individual choice (e.g. I will replace windows on my home with energy efficient windows). They will clearly require global support in order to have the desired effects.

After reading this book, I hope you will look at the reporting of extreme weather in a new light and will be better able to recognize short-term variability as distinguished from long-term climate change. I also hope that you'll become involved in the debate and decisions that will surely keep weather and climate in the forefront of the news.

Forest fires often produce large amounts of smoke as they burn. The skies over France were filled with smoke from this large blaze.

> "It's another one for the record books as 2006 will go down as a year of record warmth (second warmest in the US). It will also include a water-logged Pacific Northwest, a parched farming belt and an unbelievable fire season in the southwest US. Of course, not a single hurricane landfalled in the US either…"
>
> *H Michael Mogil*

The panel of scientists who drafted the 2007 IPCC report on global warming clearly believe that the weather is getting warmer, weather extremes are getting worse and significant, possibly irreversible, changes are happening to our planet's weather and climate.

Time and space distribution of costly weather disasters in the US.

The National Oceanic and Atmosphere Administration (NOAA) statistics indicate that costly weather-related disasters are on the rise. Since 1980, for example, there have been 70 weather-based disasters, resulting in $560 billion in losses. Although 1988 was the year with the most events, fiscal extremes have been largest in recent years.

Billion Dollar Weather Disasters 1980-2006

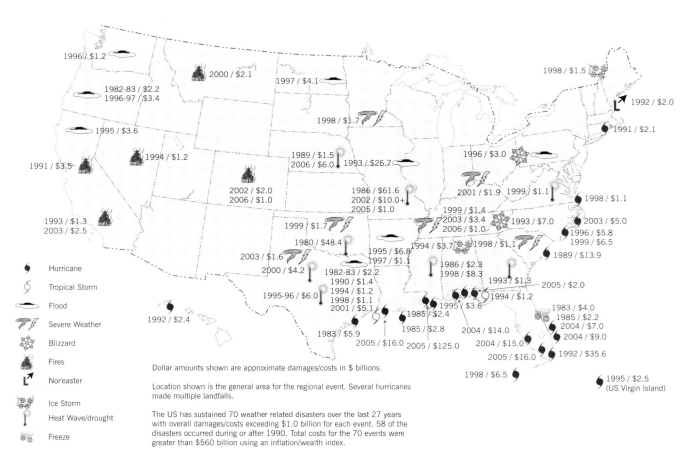

Dollar amounts shown are approximate damages/costs in $ billions.

Location shown is the general area for the regional event. Several hurricanes made multiple landfalls.

The US has sustained 70 weather related disasters over the last 27 years with overall damages/costs exceeding $1.0 billion for each event. 58 of the disasters occurred during or after 1990. Total costs for the 70 events were greater than $560 billion using an inflation/wealth index.

Legend:
- Hurricane
- Tropical Storm
- Flood
- Severe Weather
- Blizzard
- Fires
- Nor'easter
- Ice Storm
- Heat Wave/drought
- Freeze

The figure opposite shows the geographic distribution and types of these disasters.

Given these, and other, statistics it is possible to jump to the conclusion that the weather is getting worse, that it is getting warmer, and that storms are becoming more frequent and more severe. But is this really true?

In the previous chapter we saw that record keeping was quite limited. Aside from longer-term climatic proxy data (such as, ice core samples, tree rings, and carbon and oxygen dating), which address overall conditions, not site-specific records, our period of records is so small as to make it almost meaningless. This has led some to question what we are really looking at. Is it long-term climate change or short-term weather and climate variability?

I have talked with many people who note that the "weather isn't what it used to be." They recall growing up with major snowfalls or "real winters". Aside from just looking at records, what else helps or hinders us in determining if the weather and/or climate are changing?

Long-term versus short-term memory

Part of explaining weather and climate change involves our memories. In general, we remember the most recent events. If you grew up more than 30 years ago, notwithstanding any variations in the weather patterns, you probably just remember last year better than when we were young. Our mind can play tricks on us (hence the often heard phrase "back in the good old days…").

The last time the River Thames completely froze over was in 1814 and this drawing depicts that year's Frost Fair. Note the swings and horseracing, along with the two tents labeled Wellington and Moscow.

In 1063 the River Thames, running through London, England, froze over for 14 weeks. In 1076 it did so again. Some 350 years later, from November 24 to February 10, 1434, the waters beneath the famous London Bridge froze solid. One hundred years after that, the ice was so hard that carriages were able to cross the river.

This was all recorded before the words "climate change" had entered the vocabulary, indeed, before climate changes were even measured. The river has frozen only briefly since. This may be due to the design of the old London Bridge that more easily allowed ice to collect at its piers. It can also be linked to the ending of the Little Ice Age (see temperature graph in Chapter 2)."

Besides, where you grew up may not be where you live now. Over the past three to four decades, in the US at least, there has been a major population shift to warmer climes. We moved from Maryland to Florida in mid-2005. It took us about two months to acclimatize to the warmer and more humid weather. During the cold wave in February 2007, I visited Maryland. Here, temperatures were in the teens with sub-zero wind chills.

When I returned to Florida, a few hours later, I alighted from the airplane into temperatures that were in the mid-50s (10-15 °C). People stepping off the plane said, "ahh, feels nice." People awaiting family and friends were noting how cold it was. The human thermometer is relative and we can forget this over time.

The role of the media and the Internet

The media (including the Internet) plays a role, too. When we were younger, it took days (sometimes weeks) to fully visualize how an extreme weather event was unfolding in some other part of the world. Today, there is instantaneous coverage. It can even involve "special reports" that interrupt regular television broadcasting. In the past, newspapers (the mainstay of weather reports) were gradually superseded by television, and now by the Internet. Once-a-day news coverage has now been replaced by reports

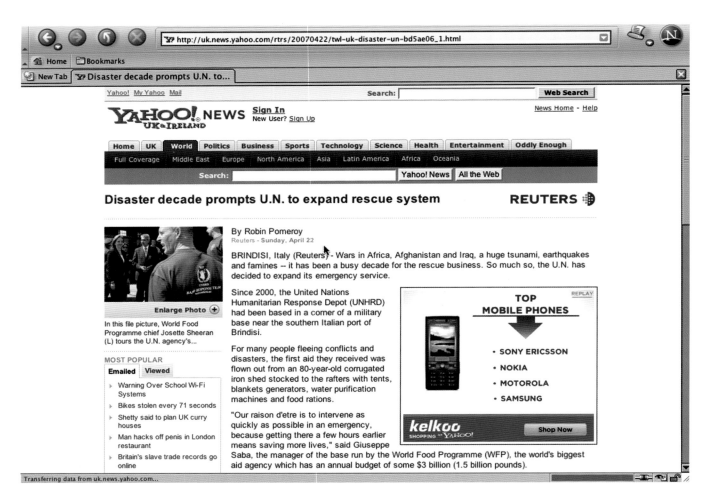

An example of how instantaneous news reporting has become in the 21st century.

broadcast several times a day and almost an infinite number of times on the Internet. Furthermore, once those reports make it to the web they remain there and almost all of the disaster stories feature immediate images and/or video feeds, making the story even more impressionable.

Aside from routine news reports, the United States is home to a cable television network dedicated solely to reporting the weather, storms, and climate change. It's been documented that the worse the weather, the more people visit the television station and its website; thus, making the weather or climate appear worse becomes good for business.

Reports about extreme weather often highlight that it "struck without warning." Often, the worst is reported, and headlines may often be worded in ways that don't convey the whole story. In short, there is no easy way of escaping news stories about extreme weather/climate events and our awareness of weather and climate has increased, whether we realize this or not.

Covering Hurricane Wilma

Consider, for example, Hurricane Wilma that stalled over the Yucatan Peninsula of Mexico then, several days later, accelerated eastward and struck southern Florida. The Yucatan was devastated by the storm, with damage estimates of around $17 billion (US). Stories about this were limited in the United States though because the storm was en route to affect Florida. Once the storm struck Florida all attention in the United States became internally focused.

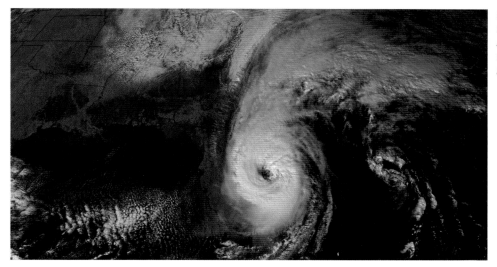

Hurricane Wilma passes over Florida on its way back out to the Atlantic. This image was taken by the GOES-12 satellite at 1km resolution.

My wife and I were in Washington, DC as Wilma approached and we decided to return to our home in Naples, Florida, before the storm struck. It was a conscious and well-thought out decision. Our home was built to withstand a category 2 storm; we have shatterproof windows and I had installed our hurricane protection coverings before we left. We are also seven miles (11 km) from the coast and outside most flood-prone areas.

We weathered the storm. Yes, we lost some roof tiles. A few of our trees blew over and a few others snapped. We lost power for 12 hours and cable for four days. There was a lot of wind.

When the storm passed, my thought was, "So, it was a windy day!" Winds at the Naples Airport never reached hurricane strength, and although some nearby areas suffered significant damage, most did not. It was a combination of distance from the eye of the storm along with quality and type of home construction. Florida's east coast, however, actually did suffer more significant damage than anticipated because Wilma strengthened as she crossed the Florida peninsula.

My son called from the Washington, DC area wanting to know if we were okay. He had heard news reports that, "Naples was destroyed." Exaggeration can, sometimes, make stories seem more interesting than they really are!

Wilma's Fury: compare the damage that Hurricane Wilma caused along Florida's East Coast on October 26, 2005. Wilma passed across the area as a Category 3 storm. Damage closest to waterways and the coast (left) is often worse due to the combined effects of wind and waves.

These apartments near Paisley (Lake County), Florida, were no match for an early February 2007 tornado.

Being led by subjective viewpoints

During the great January-February 2007 "Deep Freeze" that enveloped much of the north-central and northeast US, people interviewed on television news programs complained about the cold. Weeks earlier, when abnormally warm weather ruled, similar reports noted that people were complaining about the warmth and were "begging" for a return to winter.

Following the early-morning tornadoes that ravaged parts of north-central Florida in early February 2007, volunteers cleaning up the area were interviewed. One worker noted that the damage from the tornadoes was far worse than the damage from Hurricane Katrina. There was no mention of her role in the Katrina clean-up (if she was, indeed, there) and whether she had served in a wind or flood ravaged area of Florida or Louisiana.

But everything's relative...

For various reasons, some media support environmental causes. Thus, environmental stories (especially about global warming) often showcase the "worst case scenario" and don't provide alternative interpretations to the climate record.

For example, it rained hard during Super Bowl XLI (February 4, 2007). The crowd even sang "Purple Rain" at half time. The day after the football game a sports reporter commented that the weather during Super Bowl XLI was, "...the worst ever!" That may be true in the precipitation department, but Super Bowl VI in New Orleans was a lot colder (kick-off temperature was 39 °F (4 °C)).

Many will recall the "Freezer Bowl", a football championship play-off game played in Cincinnati's Riverfront Stadium on January 10, 1982. The game was played in the coldest wind chill temperature in National Football League history. Although the air temperature was -9°F (-23 °C), the wind chill was -59 °F (-50 °C). Let's not forget that, since that event, many play-off and Super Bowl events are played in domed or Southern area stadiums to avoid such extreme weather events.

The tornado damage at this Tennessee garage is plain to see. The automobiles that had been displayed on the forecourt now lie in a heap of scrap metal.

As you can see, everything is relative – the "worst ever" could simply be redefined (so, rain instead of cold) and, say the rain at the Super Bowl that game had been slightly worse than the heaviest downpour, then that would be the worst ever weather. "Worst ever" or "worst in 20 years" are among the terms often used to report extreme events. Just recall that "ever" is usually not much longer than 20 years!

Mixed messages

As noted in Chapter 1, the definitions of climatic events become blurred. For example, some have attributed hurricane force to wind gusts rather than sustained winds. Others have misused terms for cloudiness (e.g. "partly sunny"). However, the difference between weather and climate (and changes therein) may be the greatest problem. Today, even the most recent weather event is attributed to climate change or global warming, and if not these, then El Niño is the culprit.

Sometimes, El Niño and La Niña are called upon to explain all types of events. As a result, people think these eastern Pacific Ocean temperature patterns control climate. Several years ago, my wife and I discovered that the same resulting weather pattern had been attributed to both El Niño and La Niña. Talk about muddying the meteorological waters!

Typically, the most recent extreme weather event is compared to the closest such historical event. Thus, we hear terms like, "worst storm in the last nine years," "coldest winter since 1995" or "this is the coldest morning low since last winter." In this scenario, the context is weather, but the implication becomes climate. Hence, people think climate change can be determined by recent extreme events.

For the El Niño event, following the media's proclamation, scientists noted that the event was actually relatively weak. Since El Niño was not at work in early 2007, the flooding in Indonesia needed to be blamed on something. Climate change was a good choice!

El Niño and La Niña don't actually change climate patterns; rather they are part of it. When these periodic transitions in sea surface temperatures (above average and below average, respectively) occur off the northwest coast of South America westward to the

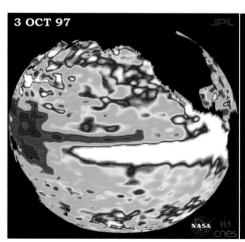 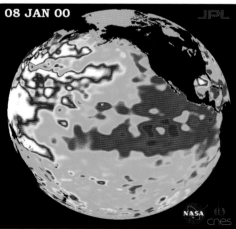

Specially-processed satellite images showing sea surface temperatures across the eastern Pacific Ocean. The left image shows an El Niño pattern with very warm temperatures (white) across Equatorial regions. The right image contrasts with very cool sea surface temperatures. Each pattern is associated with very different weather conditions, both in the local area and around the world (see Chapter 19).

El Niño causing worldwide weather chaos

"…El Niño is at it again. Scientists attribute the latest warming in the eastern Pacific Ocean for freakish weather conditions around the world. More than 140 people have died in storms across Europe and Asia in the past few days, while parts of southeast Asia and the US are seeing the worst droughts in decades…".

Two examples of how the internet has come to play a big part in reporting on extreme weather conditions around the world.

Indonesian Flooding Linked to Climate Change Official Reports

"…Climate change has contributed to the worst flooding to strike the Indonesian capital in years, an environmental official noted yesterday.

Floodwaters have submerged large parts of Jakarta and its surroundings since last week and caused the deaths of more than 50 people. Hundreds of thousands have been displaced.

The official, the head of the Environmental Affairs, claimed that although flooding is a natural event in the area, that it has been made worse by human actions – global warming.

Rescue efforts continue…"

ENSO (El Niño/Southern Oscillation) Index values since 1950. Red values indicate El Niño events; blue values La Niña's. Prior to 1980 there were many more dramatic La Niña events; since 2000, both events have shown a subdued character.

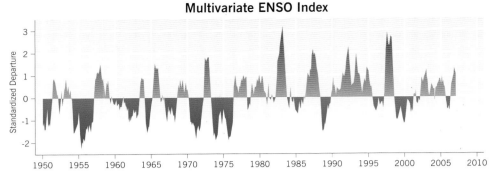

Multivariate ENSO Index

International Dateline. In the process, they influence atmospheric wind and pressure patterns. These, in turn, translate into periodic weather patterns that have happened for years before any of us were born. Alternating droughts and floods are something that most parts of the world experience on a routine basis.

But, if it is a change in wind pattern that creates or destroys El Niño and La Niña and the phenomena change the wind patterns themselves, then what causes the wind and pressure pattern shift that reverses the process? And why does the process reverse itself so quickly (1997-98 El Niño to 1998-2000 La Niña) sometimes and take so long at other times? Right now, the cause-effect picture is quite unclear.

Regardless, whatever weather patterns El Niño and La Niña bring globally or regionally also involve oscillations. The only thing that can be safely said about these phenomena in terms of climate is that they recur and that over time, they tend to average out both in sea surface temperature values and in global weather patterns.

Although there are anecdotal records about such events dating back several centuries, formal documentation of sea surface temperatures began only around 1950.

Documenting the weather

Governmental weather stations provide the best record of recent weather and climate change. However, many have histories that don't even span a full century. Others have either relocated several times and/or have been overrun by urban sprawl (and the effects of the associated urban heat island). Their documented weather records can be used to help define short-term weather and climate variability, but makes looking at long-term climate change questionable.

Most weather records (in the past 30-50 years) are also based on newer technologies (e.g. satellites, radar, snow gauges and automatic weather sensors). With acceleration of technology, newer instruments surpass older ones. Yet, each instrument and/or observing system comes with its unique set of capabilities, limitations, and quirks. Sometimes instrument upgrades or implementations result is discontinuities in the data record. Sometimes, the rules for recording the weather change.

For example, early satellite systems showed cloud and temperature patterns. Newer systems (and improved computer algorithms) now paint a far more detailed picture of the earth-atmosphere-ocean system. However, satellite science is new (the first weather satellite was launched in 1960) and still evolving.

In the United States, severe thunderstorm and tornado statistics have only been kept for about half a century. In other parts of the world, such information has a mere 10-20 year lifespan. In early 2007, an Enhanced Fujita Scale replaced the long-standing Fujita

EXAGGERATION

One of the easiest things to exaggerate is the size of hailstones. "The size of golf balls" or "the size of tennis balls" are phrases often heard. In reality, their ragged shape – with uneven edges – make them hard to quantify and they tend to be much smaller than you think. One of the best ways to measure the real size of a hailstone is to take it home with you, pop it in the freezer while you find a ruler (it shouldn't get any bigger but make sure you don't accidentally freeze it onto something else), then, ruler in hand, take it out and measure it.

The largest ever hailstone to be measured in the US was seven inches (17.5 cm) in diameter, and fell in Nebraska. The previous largest hailstone really was the size...of a grapefruit!

Scale. In the process, wind speeds associated with the various tornado categories were changed. Comparing the two data sets (and the information attributed to each) will not be an easy task.

Supplementing record keeping

Co-operative observers (people who thrive on watching and recording the weather) have provided much detail to US weather history. In other locales, such supplementary data is limited, at best. Storm spotters have become a mainstay in helping governmental agencies gather information during stormy or extreme weather, as well as afterwards. Yet, such spotters don't cover the landscape. Lacking so-called "ground-truth", warnings are not issued on some extreme events.

Meteorology students monitor an isolated supercell thunderstorm in south-central Kansas on June 5, 2004.

In Naples, Florida, in the summer of 2006, torrential thunderstorm rains flooded storm drains and streets in and near my community. Some streets were flooded and impassable. My wife was unable to drive her car into our neighborhood – I had to retrieve her in my SUV. I was able to get a storm report into the National Weather Service (I was not yet an official storm spotter). The meteorologist at the NWS office told me that they had been trying to get information about the storm's impacts but had no contact in the immediate area.

However, I am only one site. There are many gaps in the overall ground-based network. If these types of gap exist in the United States, think about the situation in lesser-developed countries and in regions of low population around the world. Observers are not "official" and accurate information might not always be put forward.

Following a heavy (approximately six-inch) afternoon thunderstorm, plant debris clogged storm drains in this Florida community. As a result, water was unable to drain away and ponded in low spots on roadways

The climate arena suffers many of the same types of data shortfalls as weather reporting. The most significant difference is that climate records have been derived by proxies, not real observations. Since we can't go back and measure weather variables like temperature and rainfall, we can only refer to what they were based on; things like tree rings, fossils, sedimentary rocks, and other information. Plate tectonics (the slow movement of continental masses), solar energy variations, mountain formation and erosion, and meteor impacts, are even harder to integrate.

The effects of population drift

As local and global populations grow, and we migrate in ever increasing numbers to coastal locales, the risk of extreme weather affecting humans increases. For example, the US coastal population has swelled in the past 50 years, making some 54 percent of the total US population at risk to coastal storms and hurricanes. Internationally, the numbers are similar.

Fort Lauderdale is but one of many cities that lie on the coast. When hurricanes and other major storms strike, there may almost be no buffer between land and water.

According to Don Hinrichsen, United Nations consultant and author, "Population distribution is increasingly skewed. Recent studies have shown that the overwhelming bulk of humanity is concentrated along or near coasts on just 10 percent of the earth's land surface. As of 1998, over half the population of the planet – over three billion people – live and work in a coastal strip just 120 miles (200 km) wide, while a full two-thirds, four billion, are found within 240 miles (400 km) of a coast."

Thus, any storm (or related geologic/oceanographic hazard) that does strike will have a much greater effect. Even if the storm doesn't strike a particular population center, evacuations of millions of people on clogged highways will dominate news reports. The figure on page 28 (Billion Dollar Disasters) provides graphic testimony to how vulnerable coastal areas are to weather hazards (especially hurricanes). With more people crowded into cities and urban centers, and with associated changes in the land's ability to handle runoff, it should come as no surprise that some cities are

These views of Cape Town, South Africa (right) and Freemantle, Australia (left) show clearly how people are crammed into coastal communities.

experiencing more flooding. In early 2007, flooding displaced more than 300,000 people in Jakarta, Indonesia.

With our growing population and its reliance on electricity (especially for computer applications), any disruption to power makes news. I estimate that storms in Europe and the United States in the winter of 2006-07 left some three million people without power (some for lengthy periods). Even without a high population density, population shifts can affect weather reporting and/or weather impacts. Consider Denver, Colorado, where a new airport was built in the 1990s. This produced a population shift east of the metropolitan center and a corresponding increase in tornado sightings (see Chapter 7 for more on this phenomenon).

Hence, the more people who live in places with unusual or extreme weather, the more we hear about it as their lives are affected. If nobody lived in a region where a F5 tornado occurred, would we hear about it at all?

The newsworthy factor

In Florida, it is interesting to hear when the northeast United States has a blizzard and when California's oranges freeze, and equally, they are interested when Florida experiences a big hurricane season. Change the weather variable and location, and the storyline will be same. This is true even when the weather impact affects an overpopulated nation and the numbers of people killed, injured, or displaced soars into five or six figures.

Following a disaster somewhere, there is often a call for aid and people want to help. It may involve simply providing cash donations or helping a community rebuild. Sometimes, great ideas surface; three sisters from Bethesda, Maryland wanted to help students affected by Hurricane Katrina and started Project Backpack with a goal of

February 2007 brought devastating floods to Jakarta, Indonesia. It was only after flood waters like these started to recede that residents could sift through mud and debris to salvage personal belongings. This flood event killed 50 and displaced hundreds of thousands.

A Florida home is shuttered up for the hurricane season.

distributing 1,000 backpacks filled with children's playthings on January 9, 2005. In one month, they had collected and sent more than 25,000 backpacks to the kids. After two months, they had sent 50,000 backpacks from more than 100 communities.

Climatic disaster or poor building materials?

Following the storms of the past few decades, there have been major changes in building codes. Use of new materials, new structural connectors, and new storm shields and windows have all lessened damage from storms like hurricanes. Shatterproof windows for homes (much like their counterparts in cars), easy to install or activate storm coverings for windows, improved connector systems that tie the roof to the home, and even new cap anchoring systems for tile roofs are all becoming the mainstay in hurricane-prone zones in the United States.

Insurers are responding favorably to these because such homes suffer less damage than homes built to lesser standards. Florida is one of several states that are mandated to offer insurance discounts to homes built to withstand some of a hurricane's fury.

However, following many storm events, damage surveys discover hidden construction flaws (and sometimes lax building code enforcement). Some of the most significant damage caused by Hurricane Andrew in southeast Florida in 1992 may not have happened if builders had adhered to proper building codes and inspectors had enforced them.

Changes of land use

Flooding is a different matter, however. More and more people are living in or near flood zones, and with growing changes in land use – that may increase impervious surfaces and lessen tree and grass coverage – areas not previously considered flood zones are now being flooded. While many coastal zones fall into this scenario, many urban and suburban areas are where the most significant increase in flooding is taking place.

According to statistics released by the Center for Research on the Epidemiology of Disasters (CRED) in January 2007, there was a total of 395 disasters recorded in 2006; 226 (57.2 percent) caused by floods. This is a sharp increase when compared with an average of 162 floods (out of 398 disasters) over the 2000-05 period.

Flooding (including those associated with hurricanes) accounted for most of the 26 disasters that occurred in the United States in 2006. China, affected by 35 disasters during the same period, also suffered a large number of flooding events. In fact, of the top 10 natural disasters in 2006, five of the top eight weather-related events were either floods or had flooding associated with them.

During the summer of 1993, recurring periods of heavy thunderstorms caused major flooding along parts of the central Mississippi River Valley. Prior to this event, flood-weary residents of bottomland homes had no choice but to return to the floodplain, repair their homes, and wait for the next disaster to strike. Recognizing that alternative measures were needed, several national political leaders helped craft legislation which required that 15 percent of all disaster relief funds be set aside for relocation, land acquisition and other forms of hazard mitigation.

The largest voluntary relocation of floodplain homes and businesses followed. More than 10,000 Midwest structures were moved from the river bottoms to higher ground nearby.

In Illinois and Missouri alone, 5,100 homes and businesses were relocated at a cost of $66 million – structures that previously had received $191 million in flood insurance payments for rebuilding.

In other cases, entire communities were relocated to higher ground – forever reducing the risk of flooding. In Grafton, near the confluence of the Illinois and Mississippi Rivers, more than 100 homes and businesses have been relocated to a 235-acre site on the bluff. Replacing them in the floodplain, a park, marina and bike trail support recreation and tourism. As expected, floodwaters returned in 1995, 1996 and 1998. But Grafton residents remained high and dry.

A flood marker on the Mississippi River showing the level of the record 1993 flood. St Genevieve is located south of St Louis, Missouri.

Climate change or just bad weather?

While there is no doubt that temperatures have become warmer in recent times, these may be partially due to urbanization, instrument changes, and even reporting procedures. Flooding is becoming more of a problem as more and more people are living in or near flood zones. And with growing changes in land use – that may increase impervious surfaces as well as lessen tree and grass coverage – areas not previously considered flood zones are now being flooded. While many coastal zones fall into this scenario, many urban and suburban areas are where the most significant increase in flooding is taking place. Other records, such as those involving hurricanes and tornadoes, are being set, in large part, because our historical basis for these is so limited.

4 WINTER STORMS

Because it is a low-pressure system, the lowest air pressure is at the center of the storm. Thus, as a storm approaches, barometric pressure usually falls. Not surprisingly, the words on the face of classic barometers reflect that low-pressure readings indicate rainy or stormy weather.

Air mass source regions and their typical tracks.

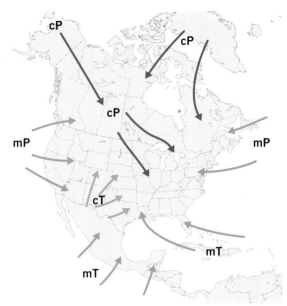

> *"Snowflakes are one of nature's most fragile things, but just look at what they can do when they stick together."*
> *Vesta Kelly*

Middle-latitude storms are known by many names. Typically, they are referred to as low-pressure systems or cyclones. Near the US east coast they are called "nor'easters". To the Norwegians, who originally studied them in great deal because of their impact on fishing fleets, they are "wave cyclones".

Although these storms occur year-round, winter storms provide the greatest impact because of their strength and the wide range of weather that they can bring. While all lows can bring heavy rain and high winds, winter storms add a frozen potpourri and possibly an arctic chill.

Cyclones = low pressure systems

Weather fronts are attached to most middle-latitude low-pressure systems. Fronts are actually transition zones along which air masses with different characteristics clash. These clashes help storms develop and mature. When air mass contrasts weaken (in the storm's old age), the storm also weakens and eventually dies.

If the terminology used to describe the cyclogenesis process (e.g. air mass clashes and fronts) seems militaristic, that's because it is. The "wave cyclone model" earned its name around the time of World War I and the process itself does resemble a military battle between air mass armies.

Classic cyclogenesis over middle latitude oceans often resembles a side view of a wave crashing on a beach (see box, page 42). The "wave" starts out as a small bump on an otherwise flat frontal zone, develops a wave-like appearance and finally curls itself into oblivion. At the end of its life cycle, the low fills in or "dies" and leaves behind a flat frontal zone. This sets the stage for the process to repeat itself.

Although the Norwegian model, as shown, is a good one to follow, each storm system develops and goes through a unique lifetime. So look for both similarities and differences as you monitor storms that affect your area.

Air mass

An air mass is a large (tens of thousands of square miles in area) volume of air with relatively similar horizontal characteristics of temperature and humidity. Each air mass develops over a particular type of land or water surface, drawing its characteristics from that underlying surface. For example, an air mass that forms over warm, ocean waters would become "maritime Tropical". The map and table

	Tropical (warm)	Polar (cold)
maritime (moist)	maritime Tropical (mT)	maritime Polar (mP)
continental (dry)	continental Tropical (cT)	continental Polar (cP)

Snowstorms can bring both suburban and urban areas to a standstill. The scene on top is from a residential area in Canada; above is a snow-covered (and almost car-hidden) street in New York City.

on page 40 summarizes the different types of air masses and how they affect North America. A classic wave cyclone starts to develop at the interface between two (and sometimes three) air masses.

Along the east coast of continents, there is a warm season air mass shifting poleward. Thus, some places that have maritime Polar (mP) air in winter will have more maritime Tropical (mT) intrusions in the warmer months, especially on the east coast of continents. This includes the New York City area southward to Washington, DC and in Japan.

In most places, the clash is between Tropical and Polar air masses, but sometimes, it can be between continental and maritime air masses. Off the east coast of Asia, storms often develop as continental Polar air masses move offshore and battle with Maritime air masses. In the United States, three air masses – continental Polar, maritime Tropical, and continental Tropical – frequently provide the battleground for storminess across the Central Plains. The same three air masses (on a more localized scale) interact over Bangladesh. In and near Australia, maritime Polar, maritime Tropical and continental Tropical air masses

THE COIL OF A SNAKE

In the 1939 movie *The Wizard Of Oz*, Dorothy urged her dog, Toto, to run because there was a "cyclone" coming. Since then, "tornado", has replaced the term "cyclone" for small-scale violently spinning winds (see Chapter 7). But the word "cyclone" has come to mean any swirling wind pattern that spins according to the sense of the Earth's rotation in its hemisphere.

Cyclones spin counterclockwise (in the same sense as the Earth looking down from above the North Pole) in the Northern Hemisphere. They spin clockwise in the Southern Hemisphere (the same sense as the spin of the Earth looking "up" onto the South Pole). High-pressure systems (known as "anticyclones"), spin in the opposite sense from their stormy counterparts (clockwise in the Northern Hemisphere and counter-clockwise in the Southern Hemisphere).

In 1848, British East India Company official Henry Piddington first used the word "cyclone" to describe the devastating tropical storm of December 1789 in Coringa, India. The word stems from the Greek word kyklon – moving in a circle or "the coil of a snake". Piddington's focus was on warning ship captains that the storms in the Bay of Bengal spun counterclockwise north of the Equator and clockwise south of the Equator. By understanding their circulation, Piddington was convinced that mariners could more safely use the circulations for navigation and avoid the strongest winds. Much of his basic knowledge of these storms is still used today.

NORWEGIAN CYCLONE MODEL
(NORTHERN HEMISPHERE VERSION*)

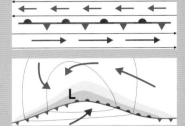

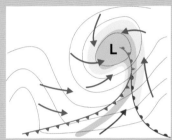

Vilhelm Bjerknes founded a geophysical institute at the University of Bergen, Norway in 1918 at which he and other scientists studied storm evolution and structure. In 1919, his son Jacob introduced the conceptual model of the "extratropical cyclone", which became a foundation for the long-range weather forecasting service envisioned by his father. Defining the front on which traveling surface lows moved as the "Polar Front" (borrowed from World War I terminology), the so-called "Bergen School" named the two active convergence lines of the cyclone model the "warm front" and the "cold front". Today, the evolutionary process is referred to as the "wave cyclone model".

A Cyclogenesis (cyclone formation) begins with a boundary, or front, separating warm air to the south from cold air to the northern hemisphere. The front is often stationary or almost stationary. The points on cold fronts and half moons on warm fronts (as above) indicate the direction toward which the front will move. For stationary fronts, the symbols indicate which way the front would move if one of the air masses "got pushy".

B A wave (or a kink) can develop on the front. Sometimes this occurs in conjunction with an upper-level disturbance embedded in the jet stream pattern. Sometimes it forms due to weather conditions at the Earth's surface or to favorable land-sea interactions. As the wave develops, precipitation will develop and increase with the heaviest occurring along the front (dark green).

C As the wave intensifies further, both cold and warm fronts become better organized and the circulation around the low-pressure center strengthens. Thunderstorms may develop ahead of the cold front at this time (shown as a dark green line in an otherwise rain-free area).

D The wave becomes a mature low-pressure system, when the cold front - moving faster than the warm front - "catches up" with the warm front. As the cold front overtakes the warm front, an "occluded" front forms. The occlusion process starts the complete capture of the low-pressure system within the colder air mass. The low-pressure system typically reaches its most intense stage at this time. The heaviest precipitation often occurs along the warm front, back towards and to the north and west of the low-pressure center.

Stationary front (neither air mass moves)

Warm front (warm air displaces colder air)

Cold front (cold air displaces warmer air)

Occluded front (cold air on both sides of occlusion; one cold air mass is colder than the other)

Isobars (lines of equal pressure)

Precipitation (darker green is for heavier amounts); there is no distinction made between frozen and non-frozen types

Warm wind

Cold wind

E As the low-pressure center weakens, pressure values increase, winds start to weaken and precipitation typically decreases. Depending upon location and time of year, the low can disappear within a few days or remain for several weeks. Meanwhile, the trailing cold front becomes stationary, setting the stage for the next "wave" to develop. A new low can also form at the "triple point" near where the three fronts meet.

*The same pattern exists in the Southern Hemisphere except that the sense of rotation is reversed and the fronts associated with the storm system extend northward from the low.

wage war. In northwestern Europe, maritime Polar air tends to dominate. Most of the time, cyclones bring stormy weather; high winds, heavy rain and snow, severe thunderstorms and – along coasts – dangerous and erosive waves. Some ocean storms can generate waves 30-50 feet (9-15 meters) high or higher!

The purpose of cyclones

With a stormy character, it may be hard to accept that cyclones have a global purpose – they actually help to balance the world's temperature, moisture and wind patterns. Consider, for a moment, a world without storminess. Due to solar heating and radiational cooling (every object not at absolute zero (-453 °F) gives off energy in the form of infrared waves), the poles would get colder and the tropics hotter. Tropical latitudes would have moisture-laden air and polar regions would have very dry air. Over time, the imbalance between high and low latitudes would continue to grow.

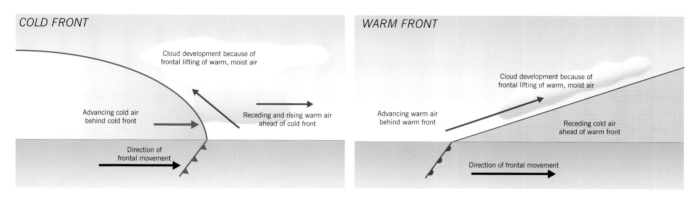

COLD FRONT

Cloud development because of frontal lifting of warm, moist air

Advancing cold air behind cold front

Receding and rising warm air ahead of cold front

Direction of frontal movement

WARM FRONT

Cloud development because of frontal lifting of warm, moist air

Advancing warm air behind warm front

Receding cold air ahead of warm front

Direction of frontal movement

Jet streams, high-speed ribbons of air roughly 6-12 miles (9-19 km) above the Earth's surface, are driven by temperature variations. The greater the temperature contrast at the Earth's surface and through the lower atmosphere (known as the troposphere), the stronger the jet stream winds.

Unless something happened to interrupt the process, some locales on Earth would become too hot, humid, cold and/or dry to support life as we know it. The precipitation pattern would be equally imbalanced. Of course, jet stream winds would strengthen, as well. The "something" that gets things back into balance is the "wave cyclone".

As the cyclone starts to develop, winds ahead of the storm, blowing from lower latitudes, bring warmth and moisture poleward. Behind the storm, winds bring colder (and often

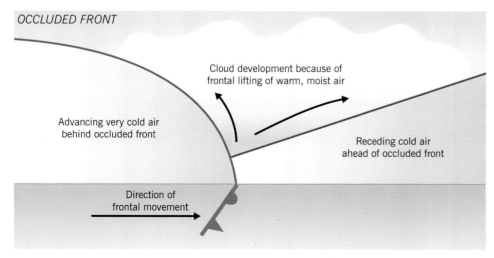

OCCLUDED FRONT

Cloud development because of frontal lifting of warm, moist air

Advancing very cold air behind occluded front

Receding cold air ahead of occluded front

Direction of frontal movement

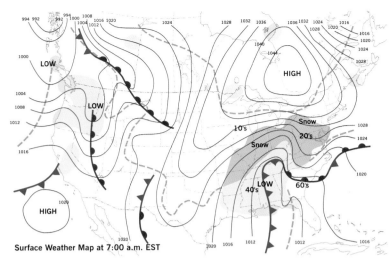

Although most surface low pressure systems are not exact replicas of the classical storm system, they contain many of the expected attributes. This surface low pressure system, in its developing stages at 7 am EST on February 16, 2003, paralyzed the middle Atlantic region with some two feet of snow. Temperatures to the southeast of the low (Florida and Georgia) were in the 60s. while to the north of the low (Washington, DC to Illinois) they were in the teens. Notice how the precipitation was focused along the frontal zone.

drier) air Equatorward. This energy transfer also fuels the storm. Often, the greater the difference in air mass characteristics, and/or the stronger the jet stream winds, the more intense the ensuing storm.

Each cyclone (large or small, intense or weak, newly formed or in old age) helps in this global balancing process. Consider the number of storms present at any time around the globe (see page 46) and you can see that the process is never-ending.

Looking closely at fronts

Colder air, because it is denser, is better able to displace warmer air. It undercuts warmer air more easily by plowing under it, while warmer air typically only slowly displaces colder air by sliding over it. Over time, the colder air eventually encircles the cyclone, cutting off the supply of warmer air, creating an occluded front. See box on page 42.

The top two images on page 43 portray classical cross sectional views (vertical slices) through cold and warm fronts. Notice the steepness of the cold-frontal slope compared with the slope of the warm front. Upward motion is stronger and more concentrated ahead of the cold front. This leads to the formation of thunderstorms. The more gentle slope of the warm front typically leads to a larger area of clouds and more gentle precipitation.

As the advancing cold air overtakes the retreating cold air, the warmest air is lifted from the Earth's surface. The newly-formed occluded front (lower image page 43) can exhibit the weather of both fronts or just one of them depending upon their strength. The formation of the occluded front signifies that the wave cyclone will soon be in its dying stages.

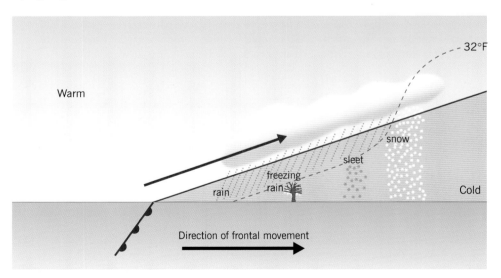

During the cyclone's life cycle, warm air moves aloft over large regions in which cold air is trapped near the ground. If the cold air near the ground is both deep vertically and cold enough, snow will fall. However, if the cold air is shallower (as it is closer to the warm front), then rain (or melted snow) can fall through the cold air and refreeze.

In this situation, small ice pellets, known as sleet, can fall. If the cold air is very shallow, rain will reach the ground and freeze upon striking ground based objects (trees, powerlines, cars, roadways, etc.). This freezing rain situation can manifest itself as an ice storm.

The figure on the bottom of Page 44 shows where these different types of wintry precipitation may fall relative to a low-pressure area and its associated fronts.

Cyclone redevelopment

Although storms over ocean waters best typify this wave cyclone process, storms over land areas exhibit similar characteristics. It's just that mountains, land-water boundaries, snow cover, and other factors can sometimes "confuse" or modify the evolution. This often happens along the east coast of the United States where cold air becomes trapped between the Appalachian Mountains and the Atlantic Ocean. This wedge of cold air often stops a developing cyclone from moving directly from the central United States into the Atlantic Ocean. Instead, the older storm becomes more quickly surrounded by cold air and a new storm develops just offshore where again, temperature and moisture characteristics are most favorable. Meteorologists often say that the low-pressure center has "jumped" to the coast when it is actually the development of a new low-pressure system.

Left: Two favored regions of cyclogenesis based on land-ocean boundaries and associated temperature variations.

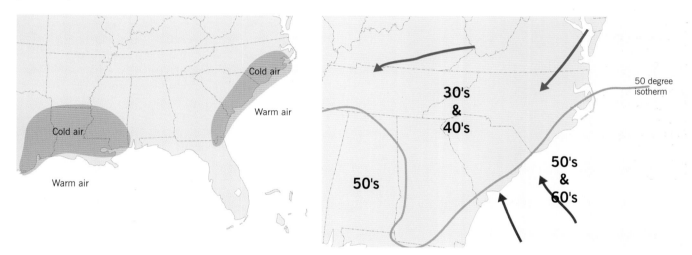

Geographical considerations

This leads logically to looking at favored geographical regions for the development of mid-latitude storms. Any place in which cold air lies to the west or poleward of warmer air becomes a potential breeding ground for these storms. Areas along the east coast of continents are perhaps the most favorable. Here (near Japan and the east coast of the United States) cold continental air masses interact with air masses that lie over very warm ocean currents (Gulf Stream, Kuroshio Current, respectively). The Greenland – Atlantic Ocean interface is another.

If one adds mountains to the mix (air masses are often colder over elevated terrain), the region near Nepal, the Andes (South America), the Rocky Mountains (United States) and the Alps (Europe) are other favored cyclogenesis regions.

Sometimes the shape of the coastline resembles the wave pattern of a newly formed cyclone. Imagine that a developing low-pressure system lies on a stationary front along the coastline of either the Texas-Louisiana area or the coastline of the Carolinas southward to north Florida. In winter, the land is often much colder than the adjacent water bodies; in turn, the air masses over these surfaces have similarly distinct

Above: On this surface weather map from just before sunrise on February 13, 2007, notice how the 50-degree isotherm "hugs" the US east coast. With cold air to the northwest and warmer, more humid air to the east, this area is primed for the formation of a mid-latitude cyclone.

temperature and moisture characteristics. In fact, there is almost a semi-permanent frontal zone along these coastal areas.

When snow cover blankets only part of a continental region, the resulting contrast in ground surface temperatures can also lead to the formation of a stationary front. Once the front is formed, can cyclogenesis be far behind?

"Triple Point" cyclogenesis

New low-pressure centers can also form at the so-called "triple-point" where a warm air mass interacts with two colder air masses (see page 42). This happens most often over Europe and the eastern United States. As a storm approaches Europe, the coldest and driest air mass lies to the east over land, while a cold and moist air mass lies to the west. To the southwest, warmer and more humid air comes into play. Cover up the original low its occluded front (Fig on page 42) and you'll see that the resulting map pattern looks just like the wave cyclone in its infancy.

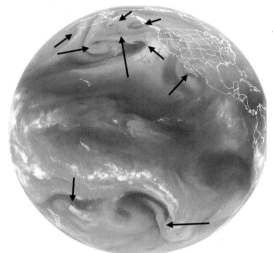

There are seven easy-to-see low-pressure circulations in the Northern Hemisphere (and two in the Southern Hemisphere) in the water vapor/infrared satellite image from February 8, 2007. Five of the lows are spinning within one storm circulation in the northeast Pacific Ocean.

Over the eastern United States, the coldest air is often to the west of the low-pressure center with cool and moist air offshore and between the Atlantic Ocean and the Appalachians. Again, warmer and more humid air lies to the south and east. Formation of a new low at the triple-point is not uncommon.

Similar triple-point situations exist off the east coast of South America. Because Australia lacks the very cold source regions that other places have, triple-point cyclogenesis is not as common there.

Upper level storms

Sometimes, low-pressure centers develop as upper level storm systems build downward. In this situation, a large upper-level, low-pressure system may spawn one or more surface low-pressure systems. Hawaii's wintertime Kona storm is one of these. The Icelandic or Greenland low provides a similar setting. In November 2006, an upper-level low developed over the southeastern United States, eventually spawning a strong surface low.

Sometimes, a large upper-level low will develop a series of low-pressure systems that spin around it. This is similar to a hurricane with multiple swirls in its eye and a tornado with multiple vortices. These "swirls inside swirls" situations will be described in more detail in their appropriate chapters. One such pattern with at least five lows within one circulation pattern can be seen in this satellite view of the Pacific Ocean (see figure above left).

Still, each storm system is unique and can have many different characteristics or development scenarios. In many ways, storm systems can be likened to people; each one slightly different than its neighbor. So, when you watch television weather reports (or check out storms on the Internet), be sure to keep this in mind.

Although most large-scale cyclones move fairly quickly (due to a high speed jet stream nearby), some get cut off from high-speed winds aloft. When this happens, the storm can sit almost stationary and spin for days. Hawaii's Kona storms, the perennial Hudson Bay low and the Icelandic-Greenland low are prime examples. In 1962, a winter storm sat nearly stationary off the US east coast for five days and brought high wind and high surf to areas from the northeast to the Carolinas. Known as the Ash Wednesday Storm, it caused record-breaking coastal flooding and erosion.

Highs and lows together

Sometimes, the central pressure inside the low is not very low and yet the storm brings incredibly strong winds. In these situations, a nearby high-pressure system may be the cause. It is the difference in pressure across a specific distance that defines the wind, not the value of the pressure itself. Drawn as isobars (lines of equal pressure on weather maps), pressure maps can be likened to topographic maps. The closer the lines, the steeper the slope (or change in altitude or pressure). In weather, a steep pressure gradient or slope means stronger winds.

Uncertainty

From experience, I can see the difficulty in forecasting winter storms; that's because these storms bring so many different kinds of weather, often across large areas, as shown above. A typical winter storm can bring snow, sleet, freezing rain and rain. It can also bring thunderstorms (sometimes severe), blowing snow and fog. Sometimes all of these weather types occur during the storm's passage in say a 24-hour period. For example, Cincinnati, OH, experienced freezing rain, sleet, snow, rain and haze during the 24-hour period ending at sunrise on February 14, 2007. Occasionally three, four or even five weather types can occur at the same time.

Hurricanes (see Chapter 5) are always accompanied by a forecast "zone of uncertainty". This tells people in, and near, the path of the storm that the center of the storm may be 50 miles (80 km), 100 miles (160 km) or even more than 200 miles (320 km) distant from the forecast location. This can be used to determine the types of actions that people and emergency management officials may have to consider. Tornado and severe thunderstorms watches (see Chapter 6) also come with uncertain information; "…tornadoes are possible in and near the watch area…"

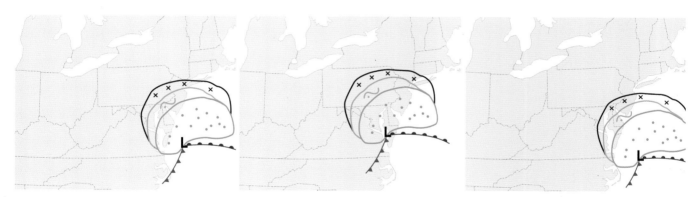

Winter storms come with no such geographical uncertainty portrayals. In fact, forecasts are often displayed with certainty information: a series of maps may be presented that shows the timeline of the snow's arrival; when and where snow is expected to change to sleet or freezing rain; and when the heaviest snow is expected to fall. Typically there is a map showing total expected storm snowfall. What happens when a similar type of forecast error, as can accompany hurricanes, manifests itself in a winter storm?

Three possible storm positions associated with a storm moving toward the northeast. A slight shift in position will greatly change the forecasted and observed weather in the highly populated northeast and Middle Atlantic region.

The figure above shows mostly rain moving across the mid-Atlantic region of the US east coast. Based on the track of the storm (with the center passing to the southeast of Washington, DC and Philadelphia, PA), these two cities should remain far in the cold air mass and experience a frozen precipitation event. Locations to their south and east would have a mix of precipitation, while coastal areas would have mostly rain.

Now, shift the storm track 75 miles (120 km) to the left of its expected path (center) and compare the weather in Philadelphia. Instead of heavy snow, Philadelphia now experiences a transition from snow to rain. Instead of a heavy snow event, a slushy wet, melting snow results.

Finally, shift the storm track 75 miles (120 km) to the right of the original path (right). Philadelphia may experience a few snow flurries, but otherwise miss any significant snowfall accumulation. Hence, within a reasonable 150-mile (240-km) error band, the weather that actually happens can range from flurries through heavy snow to slushy rain.

Is it any wonder that meteorologists get bad press when it comes to forecasting winter storms? Sometimes winter storms intensify at a different rate. Some storms intensify faster while others may by-pass an area before intensifying. The combination of these and other potential forecast uncertainties explain why parents and teachers alike always tell their charges to do their homework, just in case the storm doesn't happen as planned. Sometimes snow isn't forecast, but several inches fall. This is when meteorologists have to talk about "shoveling six inches (15 cm) of "partly cloudy" from the driveway."

Observed snowfall pattern for February 15, 2004 storm passage in southern Tennessee. The snow band itself is mostly a straight line, but is quite narrow (only about 50 miles wide) and contains large variations in snowfall across small distances. The distance between Nashville and Huntsville is only 104 miles (167 km).

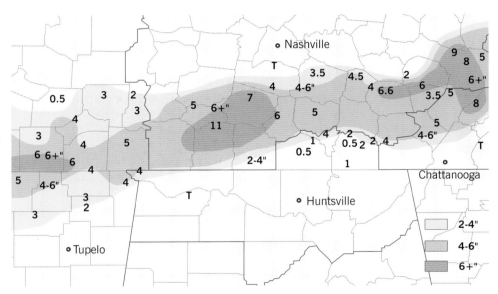

Snowstorms

While light snowfalls occur often, these are typically mostly a concern for travel (e.g. car, plane or school bus). These events do not usually disrupt whole communities and create lengthy travel delays. Once snowfall totals reach (or are expected to reach) about four inches (10 cm) in 12 hours or six inches (15cm) in 24 hours, heavy snow terminology comes into play. Criteria for heavy snow can be locally adjusted for mountain and "lake effect" areas where these snowfall amounts are more commonly observed. Similar limits for reaching "heavy snow" exist in countries around the world.

Heavy snow most often occurs along, and to the left of, the track traced by traveling cyclones. This is the "cold side" of the cyclone. Snowfall amounts depend upon the track of the cyclone, its speed of movement, and the availability of moist air. In mountainous areas, the cyclone may not be as evident on surface weather maps, but is still present in weather patterns in the middle and upper troposphere.

Typically, the band of heaviest snow is long and relatively narrow. Often, not far from the band of heaviest snow, precipitation transitions to sleet, freezing rain and/or rain. In larger counties or even over large cities, it is not unusual for heavy snow to fall over northwestern areas and rain to fall over southeastern areas. In the figure above, snowfall in Fayette County (second one in from lower left corner of state with the "11" in it) varied from 11 inches (28 cm) in the northern part of the county to less than two inches (5 cm) in the southeast part of the county (across a distance of about 30 miles (48 km)).

Ground temperatures leading up to, and during, the snow event and the rate at which snow falls are key to whether the snow provides picturesque scenes or whether roadways become snow-covered and slippery. Often secondary and side roads become snow-covered most quickly with main highways mostly free of snowpack. This is because the weight of many cars moving over snow compresses and warms the snow, partially melting it. Also, main roads are plowed and treated with chemicals earlier than side roads. While snowstorms can create incredible beauty, including sculpting of the snow by winds, they can also be deadly, economically disastrous and socially disruptive.

In addition to winter's snow, ice and high winds, fog can also ground airplanes and cause airports to close. Here, passengers at Beijing's Capital Airport are grounded by fog on February 21, 2007.

Snowstorm impacts

Airports are often the first to feel the pinch, with fewer flights allowed to land and take off due to spacing and ground control factors. If the snow falls too quickly or the winds are too strong, some runways may be shut down. In the worst case, airports may cancel large numbers of flights or even shut down.

In the winter of 2007, Denver, Washington, DC, and several other major airports closed for lengthy periods. Several major European airports were closed during a major winterstorm in January 2007.

In some regions of the US, schools, businesses, government offices and mass transit systems may close or have delayed openings or limited schedules. Many US school systems in snow-prone areas build "snow days" into their school calendars.

Although the US Postal Service once claimed that it would deliver the mail in all types of weather, that is no longer the case. If the snow is bad enough, delivery is curtailed for safety reasons.

During a major northeast blizzard in 1978, people could not mail mortgage and other payments and deposit paychecks. Thus, there was a significant disruption to the economics of the entire region. With today's electronic banking, such impacts are lessened but not completely eliminated.

Medical, police, fire, utility and related services are also affected when roads become ice- and snow-covered. Just because the main roads are clear doesn't mean that these service providers can reach your home on icy or snow-covered side streets. The Valentine's Day snowstorm (2007) in the northeast US probably delayed more than a few of Cupid's packages. Of course, people who shovel and plow driveways, sidewalks and parking lots benefit greatly from these events.

At home or at work

Following a snow event, look up to gutters and down to downspouts. These normally carry water away from your roof and your home. However, as heat from your home melts

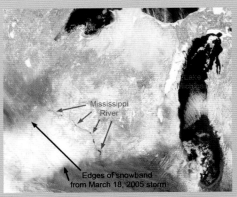

Mississippi River

Lake Michigan

Edges of snowband from March 18, 2005 storm

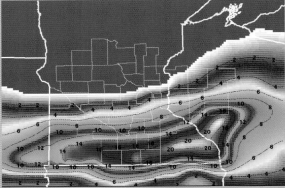

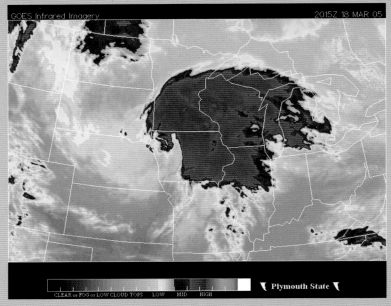

GOES Infrared Imagery

2015Z 18 MAR 05

CLEAR or FOG or LOW CLOUD TOPS   LOW   MID   HIGH

Plymouth State

A WINTER STORM CHASE

Most storm chasers head out looking for severe thunderstorms and tornadoes. On March 18, 2005, I had little choice but to conduct a chase-intercept of a middle-latitude winter storm across the northern Plains. Such transects are rare.

Upon landing in Minneapolis, Minnesota, en route to a severe weather/tornado conference in Lincoln, Nebraska, I learned that my connecting flight would leave a day late because heavy snowfall in the area had forced the airport to close. I knew it would be a close call, but I had hoped to beat the snow. I missed my window of opportunity by about two hours.

Determined to get to the conference on time, I rented a car and began my eight-hour drive. Little did I realize as I started my journey that it would turn into an educational venture.

I hit the highway and started photographing the scenery. It quickly became apparent that I was going to pass through the warm front and the cold front of the storm and that I would also be passing through the zone of heaviest snowfall. My trek did not disappoint. While most of us stay in one location as storms pass us by, I passed the storm in my "chase".

I have provided a visual documentation of my journey here. Notice how the weather I experienced matches the model of the wave cyclone presented earlier in the chapter, including where the heaviest snow fell. Notice also how the swirl pattern seen in satellite imagery resembles the wave cyclone. Coming so late in the winter, this storm event broke at least two snowfall records:

- The Twin Cities (Minneapolis-St. Paul, Minnesota) broke their daily snowfall record for March 18 with 4.6 inches (11.7 cm) of snow. The old record was 3.4 inches (8.6 cm) that occurred in 1998.

- Mason City, Iowa recorded an 11.8 inch (30 cm) snowfall on March 18. The previous record of 11.0 inches (28 cm) was established in 1971.

Look out below! Snow slides off metal roofs most easily. This is due to the smoothness of the metal and the metal's ability to warm more quickly than other types of roofs.

the snow on the roof slowly, the water refreezes in gutters, creating an ice dam. Since water expands upon freezing, it overflows the gutter, creating beautiful icicle displays. However, if these pointed missiles fall onto someone, they can cause serious injury. Meanwhile, the ice is also backing up under the roof line and into the attic area. There, heat from the house can melt it, creating water damage. In some locales, where ice dams are common, electrically-heated wires are installed in gutters and downspouts to maintain a water flow.

Roofs used to have snow catchers that kept ice and snow from sliding off. Some newer roof designs use smooth metal surfaces that foster roof avalanches.

Newer sidewalk and road chemicals are more environmentally friendly than the old mainstay "salt" (or sodium chloride). Less toxic chemicals that don't harm plants, waterway life, pet's paws and cars are now being used.

In many communities, there are potential legal implications and fines if sidewalks and driveways are not plowed or cars are parked in "snow emergency" zones.

Lake effect snows

Any place downwind from a large, relatively warm body of water is prone to "lake effect" snows. These localized heavy snow events are linked to snow showers that develop over nearby warmer waters and move ashore. As colder air moves across the warmer waters, it is heated and moistened from below, making it unstable. Unstable air can rise freely, allowing cumulus clouds to quickly form. If the clouds build sufficiently, snow can develop and fall from them. When the clouds make landfall, elevated terrain, frictional wind convergence, and other factors cause the air to rise further, enhancing the snowfall.

Snow showers often develop in bands parallel to low-level winds. When the wind blows persistently along the length of a lake (rather than across it), snow showers can become even more intense. In this situation, known as a "train echo effect" because it is like a series of train cars moving past a station, the snow showers move across an area in succession. This on-going effect, which can sometimes last a day or more, has cumulatively yielded snowfall totals measured in feet (not inches). Such was the case in February 2007 when locations east of Lake Ontario recorded around 12 feet (3.6 meters) of snow in a week. When winds blow across the lakes, lake effect snows can still occur, but snowfall totals are typically much less.

Lake effect snows are most likely in the earlier part of the colder season when the air temperatures are much colder than the relatively warmer lake waters. Also, the lakes may freeze later in the season, blocking evaporation and heating.

Because prevailing winds, following the passage of a large-scale storm system, blow across the Great Lakes primarily from a westerly or northerly direction, downwind often means the eastern, southeastern or southern shores of the Great Lakes (see below). Places such as Gary, IN; Buffalo, NY; and Watertown, NY (on the Tug Hill Plateau, eastern shore of Lake Ontario) are among the places known for their lake effect snows. The figure below shows the "Diemos" lake effect event in New York State lake that occurred New Year 1999. Buffalo's greatest snow event (known as "Bald Eagle") occurred between Christmas 2001 and New Year 2002 where 81.6 inches (207 cm) of snow fell on Buffalo and 127 inches (323 cm) fell on Montague, east of Lake Ontario. In areas affected by lake effect snows, they can contribute to 30-60 percent of total winter precipitation.

Such lake effect snows can also occur downwind from the Great Salt Lake. In continent-island settings, cold air can move offshore from a continent, develop snow showers over warmer bay or ocean waters and deposit snow on a downwind island. This happens in Japan, Korea and Scandinavia. Sometimes thunder and lightning can accompany the snow. This unusual phenomenon is known as "thundersnow".

Lake effect snows can wreak havoc in some areas, even though they are a common occurrence. That's because the events are so localized. One part of a county (or even a larger city) may experience several feet of snow while another may have mostly sunny skies. Drivers may be traveling along highways at high speed and suddenly enter a snow shower where visibility is nearly non-existent. If the heavy snow showers affect a local airport, it may be impossible to keep runways free from snow falling at a rate of four inches (10 cm) an hour, sometimes accompanied by high winds. If the snow falls early enough, as it did in Buffalo on October 12 and 13, 2006 when an early season record of 22.4 inches (56.9 cm) fell at Buffalo airport, trees may still be filled with leaves. Under these conditions, snow can easily collect and overload the trees, causing them to fall onto power lines. Almost half a million people in the greater Buffalo area lost power from the October 2006 lake effect snowstorm.

The National Weather Service in Buffalo, NY has become the premier lake effect authority. Forecaster Tom Niziol has led the effort to document, study and forecast these events. The office has also begun naming past lake effect snow events and categorizing them on a "Lake Flake" scale. For the lake flake scale, "1" describes a "wimpy" event and "5" describes an "epic-mega storm". The scale is linked to both snowfall amounts and the snowfall's impact on population centers.

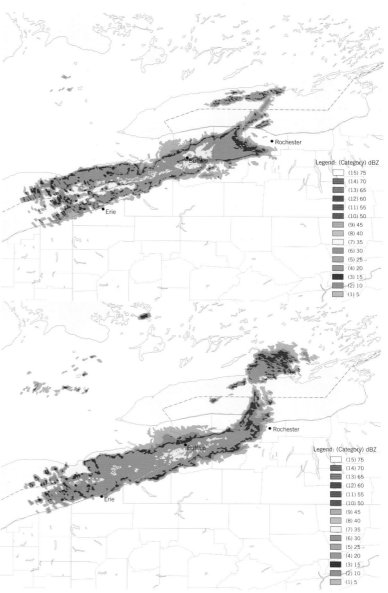

This radar image pair (October 13, 2006) shows a lake effect snow band during a three-hour period. Notice that the band has remained nearly stationary. Areas affected by this band received heavy snow, while nearby areas received none.

A northeast US snowfall index

The National Weather Service has recently adopted a similar categorization framework for larger-scale winter storms. At present, only the northeast part of the United States is covered by the Northeast Snowfall Index Scale (NESIS). NESIS is designed to integrate not only the amount of snowfall (the weather variable), but also where it fell (urban versus. suburban locales). As long as the storm produces at least 10 inches (25cm) of snow, it is possible to arrive at a rating - Extreme, Crippling, Major, Significant, and Notable. Hence, while other indices (e.g. the Saffir-Simpson Scale for hurricanes) address only weather, this index includes societal impacts.

Blizzards

Perhaps the best way to describe a "blizzard" is to say that it is a snowstorm gone wild. Although snowfall amounts do not have to be extreme, high winds, cold temperatures

"Bald eagle" lake effect event with "bulls-eye" snowfall patterns.

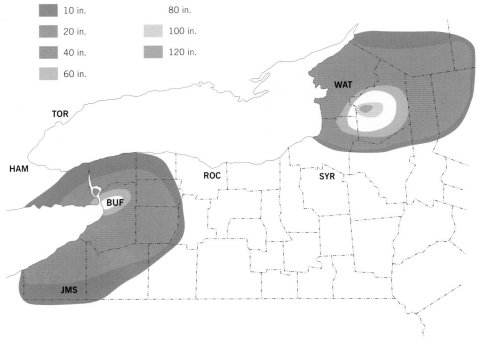

Lake Storm "Bald Eagle"
December 24, 2001 - January 1, 2002

10 in.	80 in.
20 in.	100 in.
40 in.	120 in.
60 in.	

WAT

TOR

HAM

ROC

SYR

BUF

JMS

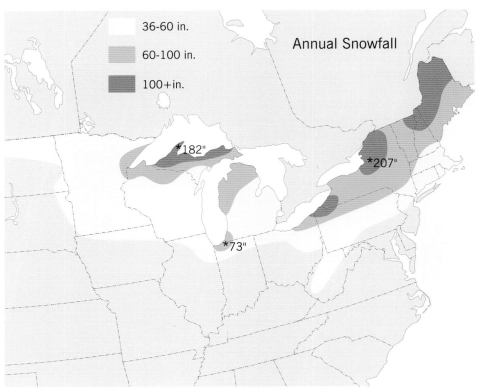

36-60 in.
60-100 in.
100+ in.

Annual Snowfall

*182"

*207"

*73"

Places with significant lake effect snows dwarf snowfall statistics for many of the larger east coast cities. Montague, NY in the Tug Hill Plateau receives an annual snowfall of 207 inches (526 cm), while New York City, Philadelphia, and Washington, DC are all below 25 inches (63 cm).

NESIS SCALE

Category	NESIS Value	Description
1	1-2.499	Notable
2	2.5-3.99	Significant
3	4-5.99	Major
4	6-9.99	Crippling
5	10.0+	Extreme

Northeast Snow Index Scale (NESIS) NESIS can range from 1 to more than 10. The higher the NESIS, the greater the snowfall and/or the greater its impact on larger urban areas. Adjectives are applied to the NESIS values to make the index more meaningful.

and near-zero visibility due to falling and/or blowing snow can create havoc. There are some very specific criteria that have to be met.

"Winds of at least 35 mph (frequent gusts are generally considered to qualify) and visibility reduced by falling and/or blowing snow to less than 1/4 mile. The strict application is for the wind and visibility criteria to be met for at least three consecutive hours." American Meteorological Society's *Glossary of Meteorology*

Blizzards are relatively rare in the scope of winter storms in the United States, except in mountainous areas in the western parts of the country and in the higher elevations of mountains. Mount Washington, New Hampshire, experiences blizzard conditions for a large part of the winter. According to Michael Branick, a National Weather Service (NWS) forecaster, some 10 percent of winter storm events can be described as having blizzard conditions at some point during the event. The region extending from Idaho across to Minnesota, and southward to Colorado and Nebraska, is the US' "blizzard alley".

Blizzards are more common in Canada and Alaska where cold, snow and windy conditions are more common. While the region of greatest blizzard occurrence in the US is in the interior region, Canada and Alaska know no such boundary. Europe, Russia, Scandinavia and parts of the European Alps have frequent blizzards. One such event affected large parts of northern Europe, including northeast France, parts of Germany and the Scandinavian Peninsula in late December 2005.

TOP 33 US STORM EVENTS SPANNING THE PERIOD 1956-2006

Rank	Date	NESIS	Category	Description
1	1993 Mar 12 - 14	13.20	5	Extreme
2	1996 Jan 06 - 08	11.78	5	Extreme
3	2003 Feb 15 - 18	8.91	4	Crippling
4	1960 Mar 02 - 05	8.77	4	Crippling
5	1961 Feb 02 - 05	7.06	4	Crippling
6	1964 Jan 11 - 14	6.91	4	Crippling
7	2005 Jan 21 - 24	6.80	4	Crippling
8	1978 Jan 19 - 21	6.53	4	Crippling
9	1969 Dec 25 - 28	6.29	4	Crippling
10	1983 Feb 10 - 12	6.25	4	Crippling
11	1958 Feb 14 - 17	6.25	4	Crippling
12	1966 Jan 29 - 31	5.93	3	Major
13	1978 Feb 05 - 07	5.78	3	Major
14	1987 Jan 21 - 23	5.40	3	Major
15	1994 Feb 08 - 12	5.39	3	Major
16	1972 Feb 18 - 20	4.77	3	Major
17	1979 Feb 17 - 19	4.77	3	Major
18	1960 Dec 11 - 13	4.53	3	Major
19	1969 Feb 22 - 28	4.29	3	Major
20	2006 Feb 12 - 13	4.10	3	Major
21	1961 Jan 18 - 21	4.04	3	Major
22	1966 Dec 23 - 25	3.81	2	Significant
23	1958 Mar 18 - 21	3.51	2	Significant
24	1969 Feb 08 - 10	3.51	2	Significant
25	1967 Feb 05 - 07	3.50	2	Significant
26	1982 Apr 06 - 07	3.35	2	Significant
27	2000 Jan 24 - 26	2.52	2	Significant
28	2000 Dec 30 - 31	2.37	1	Notable
29	1997 Mar 31 - 01	2.29	1	Notable
30	1956 Mar 18 - 19	1.87	1	Notable
31	1987 Feb 22 - 23	1.46	1	Notable
32	1995 Feb 02 - 04	1.43	1	Notable
33	1987 Jan 25 - 26	1.19	1	Notable

HISTORIC BLIZZARDS

Here are a few of the most memorable US blizzards:

• Blizzard of 1888, known as The "Great White Hurricane"

As with other recorded great blizzards, the preceding weather was unseasonably mild. As temperatures tumbled, heavy rain turned into heavy snow that continued for a day and a half. Snowfall estimates included 50 inches (127 cm) in Massachusetts and Connecticut and 40 inches over New Jersey and New York. Saratoga Springs, NY reported 58 inches (101 cm) of snow.

The storm paralyzed the east coast of the United States and Atlantic Canada from the Chesapeake Bay to the Maritime provinces of eastern Canada. With the telegraph system crippled, major cities such as New York and Washington, DC were isolated for several days. Property loss from fire alone was estimated at $25 million (in 1888). One hundred people were killed in New York City alone and it is estimated 400 people died from the storm.

• November 11, 1940, "Armistice Day Blizzard"

Unseasonably high temperatures preceded this event across Nebraska, South Dakota, Iowa, Minnesota, Wisconsin and Michigan. As the afternoon wore on temperatures dropped sharply, winds picked up and precipitation began. Shortly afterwards, the precipitation transformed into sleet and then snow. The storm continued into the next day, with snowfalls of up to 27 inches (68 cm), winds up to 80 mph (129 km/h) and 20-foot (6-meter) snowdrifts. Deaths associated with the storm totaled 154; many were duck hunters who had taken time off from work to take advantage of the good weather. More than 65 sailors died following ship sinkings on Lake Michigan.

• January 25-27, 1978, "Blizzard of 1978"

A tremendous blizzard struck Ohio, Michigan, Kentucky, Illinois, western Pennsylvania and southeast Wisconsin. One to three feet of snow was common throughout this area with 50-70 mph (80-128 km/h) winds whipping snow into drifts as high as 10-15 feet (3-4.5 meters). Ohio was hardest hit with 100 mph (161 km/h) winds and 25 foot (7.5 meter) drifts! Much of the affected area was paralyzed for several days. Over 70 deaths were blamed on this storm.

• February 6-7, 1978, "Blizzard of 1978" Part II

This was the second blizzard to strike the US in two weeks. Heavy snow fell in the mid-Atlantic region, into New England, at rates of 3-4 inches (7-10cm) per hour. Boston recorded 27 inches (68 cm) of snow with winds gusting to 79 mph (127 km/h); the city was shut down for almost a full week. Other locations in interior New England reported as much as 55 inches (140 cm) of snow.

A downed tree and snow-covered automobile attest to the magnitude of snowfall in the Asheville, North Carolina area on March 14, 1993. Some of the snowfall totals were: Mt Mitchell, North Carolina, 50 inches; Grantsville, Maryland, 47 inches; Syracuse, New York, 43 inches.

• March 12-15, 1993, "Storm of the Century", also "Superstorm '93"

This huge storm affected almost the entire eastern third of the United States. At its height, the storm system stretched from Canada to Central America, but its main wintry impact was on the eastern United States. Areas as far south as central Alabama and Georgia received 4-6 inches (10-15 cm) of snow and areas as far south as Birmingham, Alabama, received up to 12 inches (30 cm). Locally, up to two inches (5 cm) of snow was reported in the Florida panhandle. Strong winds, with gusts above 70 mph (112 km/h), helped to create a storm surge that rivaled that of hurricanes. Significant flooding occurred along some parts of Florida's Gulf Coast. For the first time on record, every major airport on the US east coast was closed sometime during the storm event. Hundreds of hikers had to be rescued from the Carolina and Tennessee mountains.

A curtain of angled icicles graces a railing along the shore of Lake Erie. Persistent high winds, associated wind-driven spray and below freezing temperatures created the unusual display.

Ice storms

While much of the stormiest wintry weather comes with low-pressure systems, sometimes it is the high-pressure system that dominates. The most insidious situation involves the arrival of a very cold and shallow arctic air mass. Often, ahead of this air mass, a warm and fairly humid air mass is in place. As the cold air undercuts the warm air, the very warm air aloft remains over it. Even with extensive cloud cover, sometimes the air near the ground can be close to zero, while the temperature of the air above the cold front can be 40 °F (4.5 °C) or higher.

Lacking a well-defined surface low, cold and high pressure dominates the news, until sleet and freezing rain starts to fall. If the front stalls or only moves slowly, there can be an extended period of sleet and/or freezing rain. A freezing rain event can be far more devastating than a major snowstorm. That's because ice is so much heavier than snow and because snow will often fall off of trees and power lines while ice builds up.

Situations are made worse when trees fall onto power lines. Hundreds, thousands and even tens of thousands of power line sections can be felled. Unsurprisingly, even with the help of power restoration crews from other parts of the country, people can still be without power for a week or more.

During the January 12-13, 2007 ice storm in Oklahoma, meteorologists – recognizing the potential for a major ice storm – forecast up to two-inch (5 cm) ice accumulations and there were pronouncements that the storm could leave millions powerless. Fearing the worst, Oklahoma's governor declared a state of emergency before the event began. Schools let out early, road crews went into action, and airlines cancelled flights.

Flooding events

Winter cyclones can bring flooding events, especially in lower latitudes where warmer and moist air is present: when a series of storms crosses the same area over a short period of time; and when heavy rain falls on melting snow and/or frozen ground. Along the west coast of continents, storms may be associated with long tropical-laden moisture plumes. If these plumes remain nearly stationary, and the rain pattern moves into favorable mountain ranges, heavy rainfall and flooding can result. Most of the time such plumes intersect north-south mountain ranges, but moisture bands can also affect east-

ICE STORM

Ice caked on trees in Missouri during an ice storm in early 2007. This is only about 1/2 inch (1.5 cm) of ice; imagine how much ice would have accumulated with a two inch (5 cm) deposit.

Ice on power lines and trees in Missouri in January 2007, making walking dangerous. As the ice accumulates, trees and powerlines become too heavy and collapse.

Ice storms can be more devastating than snowstorms because they typically cause massive power outages. Although brief periods of freezing rain or drizzle can create inconvenience and lead to traffic accidents, and slips and falls, more significant ice accumulations (ice storms) make the headlines.

For an ice storm to occur, there usually needs to be a region of sub-freezing air overlain by warmer, moist air. This happens most frequently in North America due to several reasons:

- Cold air source can feed polar air far to the south (there are no east-west blocking mountains).

- North-south mountain ranges act to "trap" cold air; this is known as "cold air damming". When the air is trapped between the mountains and a warmer air mass and in the western United States and Canada, valleys act to trap the cold air.

- A readily-available source of warm and humid air that can overrun the colder air mass.

Freezing rain and drizzle events are fairly common elsewhere in the world, especially in parts of large middle-latitude continents. However, ice storms, of the magnitude seen in North America, are much rarer. The UK, for example, has suffered major ice storms about four times since 1940. One very significant event occurred January 23-24, 1996, when across parts of Wales and central England there was an increase in the incidence of automobile accidents and electric power disruptions.

Most ice storms occur in a fairly narrow region, but that region can also be very long depending upon meteorological conditions.

west ranges (e.g. the Pyrenees on the Iberian Peninsula and the Santa Ynez Mountains to the northwest of Los Angeles, California). Coastal flooding and seiche flooding along the shores of large lakes can also occur. Look at Chapter 10 for a complete discussion of all types of heavy rainfall and flooding events.

Climatic Implications

There is little doubt that recent winter storms have had much more of a societal and economic impact than storms in the past 100 years. However, in terms of strength and frequency, there is little to suggest that storms are any more intense than their older cousins.

Instead, it appears that the location and timing of the storms is not meshing with what people recall from the past. Some events, such as the February 2007 lake effect in New York State, have lasted longer than similar events in the recent past. This can be partially related to a later than usual arrival of cold air masses and that when the cold air arrived, lakes were unusually warm. Some described Europe's big winter storm in early February 2007 as the worst in the last nine years. This involves a comparison to recent weather incidents, not a relationship with historical climatic information.

The weight of ice was too much for this tree in Springfield, Missouri. The trunk split down the middle during the ice storm in January 2007.

A PARTIAL CHRONOLOGICAL LISTING OF SIGNIFICANT NORTH AMERICAN ICE STORMS

January 31-February 1, 1951 Major ice storm from Louisiana to Ohio; worst impact in Tennessee; damage estimates of $100 million (in 1951). Some define this as the "worst" in US history.

March 4, 1976 Up to 600,000 homes without power north and west of Milwaukee as up to five inches (13 cm) of ice coated trees and power lines. High winds (up to 60 mph (96 km/h)) compounded the effect of the ice. Some were left without power for up to 10 days.

February 10, 1994 Severe ice storm occurred over portions of Arkansas, Louisiana, Mississippi, and Alabama. Ice thicknesses reached six inches (15 cm) in Mississippi. Some 3.7 million acres of commercial forestland, valued at an estimated $1.3 billion, was damaged. Some residents of Mississippi were without power for up to a month. Damage and cleanup costs exceeded $50 million in Arkansas alone.

January 5-10, 1998 Freezing rain coated trees and power lines in Ontario, Quebec, and New Brunswick (Canada) and parts of interior New England (US) with 3-4 inches (7.5-10 cm) of ice. There were massive power outages affecting up to four million people, some lasting as long as a month. Several dozen died of hypothermia. About 1,000 steel transmission line poles were felled by the weight of the ice. According to Environment Canada, this storm was the most expensive natural disaster in Canadian history and directly affected more people than any other previous Canadian weather event.

January 15-16, 2007 Ice coated areas from north Texas to New England. Some parts of Oklahoma had ice that was up to four inches (10 cm) thick. Massive power outages reported (more than 500,000 homes) and remained without power for weeks after the event ended.

"If a hurricane doesn't kill you dead, it will make you strong…"
Jimmy Buffett, "Breathe In, Breathe Out, Move On"

While middle-latitude cyclones (see Chapter 4) help transfer heat and moisture poleward and bring colder and drier air Equatorward, hurricanes and their counterparts in other locales – all within a broad class of storms known as tropical cyclones, move incredible amounts of tropical heat and moisture toward the poles. That's because these storms form much closer to the Equator, over warm ocean waters, and typically do not have fronts associated with them until they reach higher latitudes and are in their waning phases. Hurricanes are noted for their circular, banded pattern, low-level inflow, upper-level outflow and especially their eye (below and right). They typically form and move westward within the trade wind easterlies near the Equator, and recurve to the east only when they reach middle latitudes (opposite below).

The transformation from a warm and humid air mass into one filled with towering cumulus clouds, generates an incredible amount of energy. In fact, according to the Atlantic Oceanographic and Meteorological Laboratory (AOML) of the National Oceanic and Atmospheric Association (NOAA), an average hurricane (one with a radius of about 400 miles (644 km)) can produce the equivalent of 200 times the world's electrical generation capacity each day. That these storms have multi-day lifetimes, and that some 70-80 occur each year around the world, truly make the hurricane the "greatest storm on Earth".

A cross-section or slice through a Northern Hemisphere hurricane. Winds spiral into the storm counter-clockwise at low levels and are most intense within the eye wall and "rain bands". Winds spiral out from the storm clockwise at high altitude. Although air spirals upward within the eyewall and outer edges of the eye, air actually sinks within the eye.

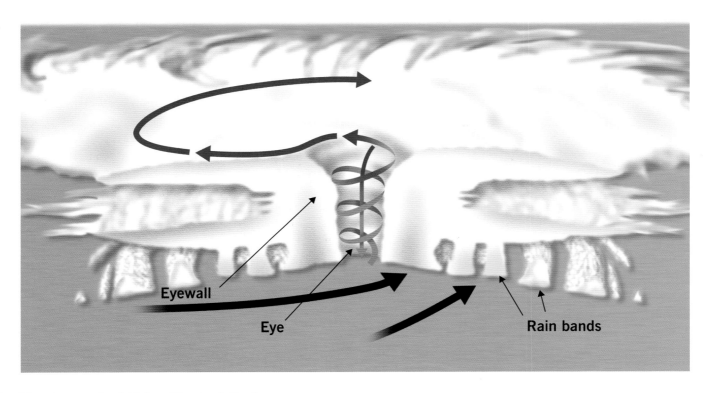

Eyewall

Eye

Rain bands

A satellite image taken by the US National Oceanic and Atmospheric Administration (NOAA) released on August 31, 2005 shows intense Hurricane Katrina approaching the Louisiana and Mississippi Gulf Coast. Monster storms often have very defined eyes.

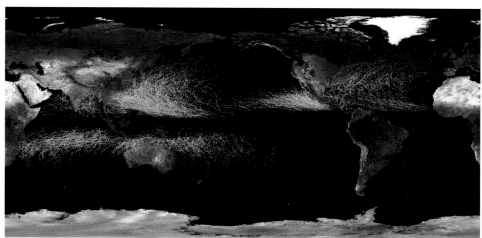

This map shows the tracks of all tropical cyclones which formed worldwide from 1985 to 2005. The points show the locations of the storms at six-hourly intervals. It is easy to pinpoint the world's tropical cyclone breeding grounds and areas that could potentially be affected.

A storm by any other name

Tropical cyclones are the broad class of all low-pressure systems that form in the tropics and have a closed wind circulation with sustained winds of at least 39 mph (63 km/h). When sustained winds reach 74 mph (119 km/h), the storm is classified according to its geographical location:

- **"hurricane"** (the North Atlantic Ocean, the Northeast Pacific Ocean east of the dateline or the South Pacific Ocean east of 160E);
- **"typhoon"** (the Northwest Pacific Ocean west of the dateline);
- **"severe tropical cyclone"** (the Southwest Pacific Ocean west of 160E or Southeast Indian Ocean east of 90E);
- **"severe cyclonic storm"** (the North Indian Ocean); or
- **"tropical cyclone"** (the Southwest Indian Ocean).

Above: A satellite image of Hurricane Ivan taken on September 10, 2004.

Left: Hurricane Rita, at category 5 status, in an image from MODIS instrument on the Aqua satellite. Image from September 21, 2005.

For ease in describing these more intense storms, we'll simply refer to them as hurricanes, unless there's a specific geographical focus.

Once winds in a closed tropical low-pressure system reach 39 mph, the system is named either a tropical storm or tropical cyclone depending upon its location. Then, a name is assigned according to international naming conventions. These names can be drawn from the appropriate six-year naming lists for some ocean basins and/or agreed to multicultural or regional names.

Names are used to help focus attention to particular storms, especially when several storms are occurring at the same time. The names also provide easy recognition for past storminess. Storms that are especially deadly or destructive have their names retired (see Resources for details).

Evolving from a tropical wave, through tropical depression, to a tropical storm and finally a hurricane, tropical cyclones often capture our attention not only because of their power but also because of their wide-ranging societal, economic and physical impacts.

Hurricane season

What distinguishes hurricanes from other types of low-pressure or storm systems is that hurricanes are "warm-core". This means that the entire storm system is composed of warm air. Middle-latitude low-pressure systems often have cold and warm sectors separated by weather fronts.

Most hurricanes form over warm tropical oceans the months of summer and fall. In the Northern Hemisphere, this means June through November. In the Southern Hemisphere, such storms form mainly during the November to May period. Peak hurricane season is usually about two months after the summer solstice or the official astronomical start of summer.

Hurricane season extends far beyond the warmest months of the year because ocean waters warm more slowly and retain their heat longer than either air or land. This is due to their high value of specific heat. Specific heat describes the amount of energy that has to be added to raise the temperature by 33 °F (1 °C). The specific heat of water is higher than any other common substance – you know this from boiling water for cooking. It takes a lot of heat to even warm the water. Specific heat is not defined when describing phase changes.

RARE SOUTH ATLANTIC TROPICAL CYCLONE

Late on March 27, 2004, Hurricane Catarina made landfall on the Brazilian coast near the town of Torres in the southern Brazilian state of Santa Catarina, about 500 miles (804 km) south of Rio de Janeiro. This was apparently the first hurricane to ever strike Brazil and the first hurricane observed in the south Atlantic since satellite technology entered the meteorological scene in the 1960s. There have been at least two documented tropical storms in the south Atlantic.

Lacking any type of ground truth data, with no anemometers in the area where Catarina struck and no hurricane hunter aircraft, satellite data provided the only reliable way to estimate storm strength.

Meteorologists at the National Hurricane Center in Miami estimated that the storm was a full-fledged, Category I hurricane with highest sustained winds of 75-80 mph (120-129 km/h). Viewing the photograph taken by scientists aboard the International Space Station and a polar-orbiting satellite image, winds appear to have been perhaps even a bit stronger because the circulation was so rounded and the eye so well-defined. Again, as with some other hurricanes, there seems to be a multiple vortex structure inside the eye (see direction E2).

Brazilian scientists, however, have disagreed, saying the storm had top winds in the 50 to 60 mph range, far below the 74 mph threshold of a hurricane. Two people were killed by Catarina, the storm destroyed 500 homes and damaged 20,000 others. Some 1,500 people were made homeless. There may also have been casualties at sea as two boats sank in 13-foot (4-meter) seas off the coast.

Catarina evolved over a week-long period after a mid-latitude low-pressure system became separated from the jet stream and sat stationary off the Brazilian east coast over warm ocean waters. Over this time, the low transformed into a warm core system that contained tropical type clouds and thunderstorms. This transformation process has been observed many times with lows off the east coast of the US.

The MODIS instrument onboard NASA's Aqua satellite captured this true-color image of Hurricane Catarina on March 27, 2004.

The state of California has been struck by a hurricane only once in recent history. That's because ocean water temperatures in the region are usually too cold and chill the hurricane's energy before landfall. That didn't happen on October 2, 1858, according to researchers at NOAA's Atlantic and Oceanographic and Meteorological Laboratory. Based on recently found information, the scientists noted that, "...unprecedented damage was done in the city (of San Diego) and (the storm) was described as the severest gale ever felt to that date nor has it been matched or exceeded in severity since."

Recognizing that climate records are incomplete, scientists note that 1858 may have been an El Niño year, which would have better allowed the hurricane to maintain intensity as it moved north along warmer than usual waters.

Scientists estimate that if such a Category I storm were to affect either San Diego or Los Angeles today, damage would likely be in the order of up to several hundred million dollars.

The size of various ocean basins also has to be considered. Due to its large geographical range, the typhoon season in the western Pacific Ocean can sometimes last almost all year. Because of its small size and slightly cooler water temperatures, the tropical South Atlantic Ocean is almost hurricane-free. The first documented south Atlantic hurricane (Catarina) occurred in 2004.

The making of a hurricane

Scientists now realize that hurricanes are most likely to form and intensify when ocean water temperatures are at least 80 °F (27 °C). Warm ocean waters warm the air above them through contact (conduction). When air is warmed from below, the warmer air typically rises and this favors rising air currents known as convection. The ocean is also a source for atmospheric water vapor through evaporation.

However, the heat and moisture, often referred to as the "fuel" of hurricanes, are insufficient alone to spawn them. Other favorable conditions (and an absence of unfavorable conditions) are also required.

In many tropical regions warm, humid air near the ocean's surface or above land areas is "capped" by a layer of warmer and dryer air. This cap prevents convection from developing very high into the atmosphere. Under these stable conditions, the cumulus clouds that form often build to heights of up to 10,000 feet (3,048 meters) or so before being forced to spread out horizontally. It's much like cooking on a barbeque grill with the cover closed. The heat is kept trapped inside the grill. If you open the grill, the heat (and possibly the smoke) bubbles upward (deeper convection).

At times, tropical waves or upper-level low-pressure systems disrupt this stable weather pattern. Then, areas of showers and possibly thunderstorms can develop. In the absence of any mechanism to cause a closed circulation to form, these weather systems simply migrate through the tropics. Upper-level lows can migrate from east to west, but may also move from west to east (even at tropical latitudes) when upper-level winds change direction. Tropical waves (mostly low-altitude features) always move from east to west, embedded in the broad "trade wind" easterlies. The interaction between these upper and lower level systems can sometimes help create a tropical storm, or wipe it out.

Over time, an organized area of showers and thunderstorms in the tropics (typically referred to as an area of "disturbed weather" by hurricane specialists) may develop and then persist for several days. When it does, thunderstorms continually add heat to the mid- and higher-levels of the atmosphere due to the condensation process. While heat is needed to evaporate water, heat is released upon condensation. This heat is known as "latent heat" because it is hidden during the evaporation process and reappears during the condensation phase. When convection occurs, heat and moisture is transported to higher levels of the atmosphere.

Ups and downs

With a large area of thunderstorms and associated rising air, there needs to be a balancing region of sinking air nearby and/or a developing air exhaust atop the thunderstorm area. Lacking one or both of these, too much air will accumulate at higher altitudes, stopping the convective process and destroying the thunderstorm area. This is one reason that most "disturbed weather" areas fail to grow to full potential. However, if either, or both, of these balancing mechanism develop, then the thunderstorm area can fully mature.

The balance between rising air and nearby sinking air goes on in your home each day. If you use a ceiling fan, for example, the fan may force air down in the middle of the room, allowing cooling breezes to reach you. However, there is only so much air that can be forced down until the air atop the room is exhausted. Where does the extra air come from? As air is forced downward, it reaches the floor and spreads out toward the sides of the room. Now there is a deficit of air atop the room (low pressure) and too much near the sides of the room near the floor (high pressure). With this resulting pressure gradient, air is pushed upward along the edges of the room to fill in the deficit. The result is a complete circulation as shown below.

If you have an adjustable fan, you can reverse the flow of air for winter warming. Here the fan lifts cooler air in the center of the room and pushes warmer air from near the ceiling down the sides of the room. This situation (with rising air surrounded by sinking air) is what often occurs in disturbed tropical weather areas.

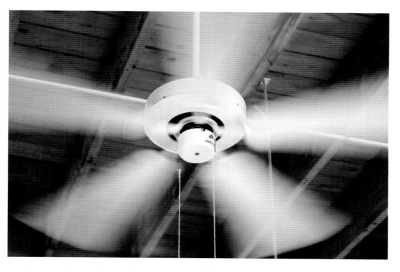

The wind flow created by a fan in its normal summer cooling mode typifies the air flow inside the eye of a hurricane. Air near the center of the room sinks and air at the edges rises.

There are two ways that air can be exhausted from atop a disturbed weather area. One is favorable for tropical storm formation; the other is not.

If upper-level winds blow strongly atop a disturbed weather area, they will blow air away. In this scenario, the winds will also blow the tops of the thunderstorms away or cause the thunderstorms to "lean". This tends to cut off the thunderstorm development process.

A second way to exhaust air atop an area of thunderstorms is to have lighter winds blowing outward in a more gently spiraling pattern. This diverging exhaust process allows the thunderstorms to remain intact and vertical. Such patterns are typically found with upper-level high-pressure systems. Thus, having an area of high pressure above an area of low pressure near the ocean surface in the tropics is a plus for hurricane formation!

It turns out that the release of latent heat can actually help to build an upper-level high-pressure system. This is linked to how temperature, moisture and pressure are linked vertically in the atmosphere, and to the release of latent heat in the condensation process.

A developing hurricane
Disturbed weather or tropical thunderstorm areas can develop in several ways. Perhaps the most common is for thunderstorm clusters to develop over land (e.g. Africa or Central America). Embedded in the trade wind easterlies, these then move westward over ocean waters. Typically ocean waters adjacent to the west coasts of large tropical landmasses are relatively cool (due to ocean current circulations and upwelling). Many times, this chilling effect destroys thunderstorm areas before they reach warmer waters further to the west.

However, some of these land-based thunderstorm clusters may have reached mesoscale convective system (MCS) status before heading out into the ocean. An MCS (see Chapters 6 and 10) is a descriptor for a large thunderstorm cluster that covers at least 100,000 square miles (259,000 square km) – based on cloud top or anvil coverage – has a rounded character and a middle-level low-pressure system beneath an upper-level

A mesoscale convective system (large system of thunderstorms) exits the African coast near Sierra Leone. Weather systems like these are one of the formative mechanisms for Atlantic Ocean hurricanes.

high pressure system. The MCS already has some of the characteristics needed for a tropical low-pressure system to form. What is missing is the low-level cyclonic circulation. That's why hurricane forecasters are always looking at "waves" or MCSs coming off the west coast of Africa as potential incipient tropical storms.

Although MCS development is less likely over Central America, thunderstorm cluster creation is much more common.

In all tropical regions, thunderstorm clusters often develop along the Inter-Tropical Convergence Zone (ITCZ). The ITCZ is a region known for its low-level convergence and upper-level divergence, factors that mesh perfectly with the ceiling fan example described earlier. It is the region in which low-level, Northern Hemisphere, northeast trade winds meet the low-level southeasterly winds from the tropical Southern Hemisphere. The ITCZ normally lies within about 10 degrees latitude of the Equator, mostly in the Northern Hemisphere, due to variations in land and water coverage between the two hemispheres.

As clusters of storms develop, they can migrate poleward from the ITCZ. This is most common in the Northern Hemisphere.

Since African systems have so far to travel, and can be impacted most easily by colder coastal waters, the "Cape Verde" hurricane season (mainly from August 1 to September 30) occurs later in the summer.

Chris Landsea, a past hurricane researcher at NOAA noted that about 65 percent of Atlantic hurricanes develop from easterly "waves" or active regions along the ITCZ. Most of the remaining 35 percent develop from upper-level and/or surface lows that lose their middle latitude characteristics and become "warm-core". Only a handful develop from MCSs alone.

Sometimes tropical storms and hurricanes interact with middle latitude systems. Although the resulting storm may not appear to be tropical, the tropical warmth and moisture can fuel an even more intense middle-latitude storm (the "Perfect Storm" of October 1991) or provide the fuel for incredibly heavy rainfall (Agnes, northeast US, 1972).

When hurricanes move over cold waters or over land (especially mountainous terrain), they usually weaken. However, there are exceptions, for example, Hurricane Wilma, moving across Florida in 2005, was actually stronger after it passed across the Florida Peninsula than it was at landfall. Strong winds at the jet stream level can shear off part of the storm's upper level circulation, also weakening the storm.

Moving along

Storms generally move toward the west or northwest in their formative stages. Then, as upper-level weather systems interact with it, the low-level storm may "recurve" or move toward the northeast, or it may cause the system to take apparently unusual jogs or loops. Lacking any such influences, the storm may move westward at relatively low latitude.

Storms often move westward at speeds below 20 mph (32 km/h). However, some have stalled, while others race forward at speeds of more than 50 mph (80.5km/h). The latter motion is most common as storms enter middle latitudes and either turn northward or recurve northeastward. The Great Hurricane of 1938 (dubbed the "Long Island Express") was moving northward at 60-70 mph (96-112 km/h) when it made landfall on Long Island. This deadly and destructive storm is still the fastest moving hurricane known to have hit the United States.

Close behind is another New England hurricane, known as the "Great Colonial Hurricane", in August 1635. Passing over eastern Long Island, the storm also struck southern New England with such ferocity that journals from the time documented the event extremely graphically. According to Nicholas K Coch, a professor of geology at Queens College (New York City) and a noted hurricane researcher, "The documentation was better than that of any hurricane until the mid-1800s." This is testimony to the impact of the storm on new settlers from England. John Winthrop,

The fast-moving Great Hurricane of 1938 made a virtual clean sweep of Fire Island, NY. The hurricane also brought incredible destruction to large parts of New England.

Atlantic Ocean hurricanes often move westward at lower latitudes and recurve to the northeast when they reach about 25-30 degrees North latitude (top image - 1999). Four years later (bottom image - 2003), the pattern was much more chaotic with some storms making loops.

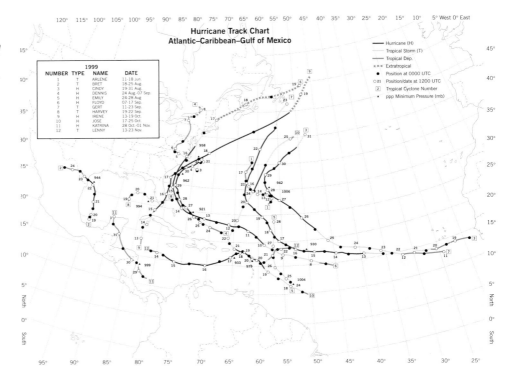

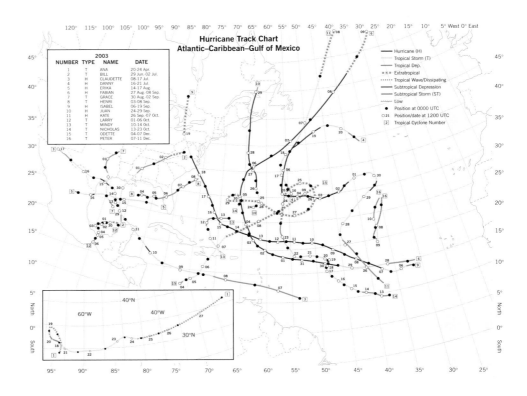

head of the Massachusetts Bay Group, recalled that, "…winds blew with such violence, (and) with abundance of rain, that it blew down many hundreds of trees, overthrew some houses, and drove the ships from their anchors."

In 1999, storms in the Atlantic followed the more usual track. In 2003, the pattern was quite a bit different. Storms in the western Pacific follow a similar pattern. Storms in the eastern Pacific generally move more slowly and often move westward over colder waters and weaken. However, some do move northward and recurve. Some of these have affected Mexico; others have even moved ashore and brought heavy rainfall to Arizona. A few storms (at least their cloud masses) make it through Mexico, interact with other weather systems, and bring heavy rainfall to the southern Plains states. Storm patterns in the Indian Ocean area and near Australia are less well-defined and are less likely to recurve as dramatically as US east coast tropical cyclones.

Hurricanes also exhibit other patterns, in both time and space.

A house destroyed by Hurricane Katrina in Slidell, Louisiana, outside New Orleans.

Time and location

According to maps produced by the University of South Florida, the US Central Gulf Coast was more prone to hurricanes between 1900 and 1919, while Florida was more at risk from 1920-49. Then, during the 1950s, the landfall area shifted to the mid-Atlantic states and the northeast. In fact, my interest in meteorology was kindled when two back-to-back hurricanes struck New York City in 1954. Similar geographical shifts were observed again in the latter part of the 20th Century (Gulf Coast 1960-79; Florida 2000-05).

Regardless, peninsulas and parts of the US that otherwise extend into ocean areas, are most at risk of being struck by a hurricane. Using the NOAA Coastal Services Center's interactive Historical Hurricane Tracks tool it is easy to search and display Atlantic Basin and East-Central Pacific Basin tropical cyclone data for various periods.

A boat sits on a road after being blown ashore by Hurricane Ivan. Ivan made landfall as a strong Category 3 storm in September 2004.

Hurricane dangers and impacts

Hurricanes are not just hurricanes. They bring with them a wide-ranging collection of hazardous weather including high winds, coastal

Residential areas in New Orleans were inundated with flood waters from Hurricane Katrina. This aerial image was taken on August 31, 2005, two days after Hurricane Katrina struck causing a levee to breach.

storm surge flooding and erosion, inland flooding, tornadoes and very high ocean waves. However, in keeping with the framework of balance, without tropical weather systems and storms, many parts of the Caribbean, parts of the western Pacific (including Japan) and even parts of Australia would lose out on a significant part of summertime rainfall.

The negatives, however, dominate. They include loss of life and destruction; affects on health and medical services; limited availability of food, fuel and energy; loss of housing, business and other economic losses; loss of classroom time in school; infrastructure damage; crime; and more. There is also usually an environmental impact (e.g. watershed destruction, nesting habitat loss, loss of animal and plant life, water pollution) that are often lost in the drama of human suffering.

Following Hurricane Katrina in 2005, there was a significant increase in the awareness of how hurricanes affected family pets. Many drowned, but many more were separated from their caregivers. While imbedded identification chips were helpful, some shelters were not equipped with the identification equipment. I learned about "pet driver's licenses" at the time and now have one for my pet. This set of plastic laminated identification tags (one for the dog and one for the owner) ensures easy identification, matching a pet with its rightful owner, and return of the pet to its family.

Hurricane winds

The highest hurricane winds are concentrated in the hurricane's "eye wall," the ring of heavy rainfall that surrounds the eye of the storm. Lesser, albeit still strong, storm force winds can extend outward even a few hundred miles, depending upon storm size and other storm characteristics. Small storms, such as Charley (southwest Florida coast in 2004) and Andrew (southeast Florida coast in 1992), have very localized damage paths. Larger storms, such as Ivan (2004), have broader damage swaths.

By digitally superimposing three satellite images, it is possible to simulate time lapse animation of a hurricane. This image shows Hurricane Andrew in August 1992.

Damage from wind can take many forms, and the damage depends upon building construction and type of building materials, how the winds strike structures, how tall the structures are and so on. Damage can involve roof failure and removal, broken glass, blown in garage doors and downed pool screens (pool screens, common in Florida, are metal structures with screened sections that enclose outdoor pool areas). They are designed mainly to keep insects (like mosquitoes) away from outdoor living areas. As a side benefit, the screens also block some UV (ultraviolet) rays (see Chapter 17). Quality of materials, quality of installation and building code enforcement play major roles in how structures perform during a hurricane. Floridians learned this from Andrew in 1992, when a significant portion of damage was linked to lax building code enforcement and shoddy construction.

Armor Screen™ is one of many types of protective shield products that people can use to protect their homes from wind, driving rain and flying debris. Following Hurricane Andrew, Miami-Dade County, Florida now tests and certifies the products for large missile impact and wind pressure. Because the products work, insurers provide substantial discounts for people who use them to protect their homes.

Engineers are busy trying to find out how they can make buildings safer from hurricanes. One such project is underway at Florida International University's International Hurricane Research Center (IHRC).

Vegetation damage is yet another story. Poorly rooted trees (including new trees that are not supported or braced), trees that are weak or diseased and certain tree varieties are most prone to succumbing to high winds. Banyan trees, for example, have spreading root systems and more easily blow over in high winds. When such trees fall, the exposed root mass is almost as tall as the original height of the tree. Palms and similar trees have little in the way of a canopy and don't catch a lot of wind. While they may lean, few actually fall over due to the wind. Mahogany trees may be strong (at least their wood is), but the tree has no flexibility. Instead of bending, it just breaks. During Wilma, I lost three mahogany trees near my home when the top of the trees simply snapped in Wilma's winds.

The problem with tree failure is that it often takes out power lines and above ground phone and cable lines. This is true, although not as devastating, even when trees are routinely trimmed.

Winds can also knock down traffic signals and signs, seriously disrupting traffic. If strong enough, winds can also down traffic or electric line poles, even the newer concrete variety. Florida Power and Light (FP&L) President Armando Olivera noted that FP&L

Hurricane damage is linked to many factors including wind speed, type and quality of construction and proximity to ocean or lake driven waves. The image on the left shows mainly wind damage while the image on the right shows the combined effects of ocean waves and wind.

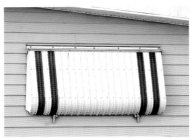

There are many ways to protect windows from flying debris during a hurricane. One such type of system – a removable shutter – is shown here.

customers have gone through two "horrible years" (2004 and 2005) of storms and he said his company understands their frustration and dissatisfaction.

"We can't prevent hurricanes, nor can we prevent all the damage that comes with them," he noted. "But there are additional steps that we can take to prepare for these hurricanes, to reduce the number of storm-related outages and to restore service quickly…"

Following Wilma's massive power outages, FP&L developed a "Storm Secure" plan that included the following:

- Hardening electric transmission and related structures to withstand winds to 150 mph (241 km/h).
- Collaborating more with local governments and contributing toward converting above-ground to below-ground systems. FP&L will also work with developers to install below-ground lines in new subdivisions.
- Carrying out more frequent, scheduled pole inspections.
- Increasing tree pruning activities prior to the onset of hurricane season. Also working with homeowners and others to follow a "Right Tree, Right Place" framework. This includes planting the right kind of trees and planting them away from power lines.

Coastal surge flooding and erosion

Depending upon where the hurricane comes ashore, the angle at which the storm approaches the coast, coastal geography, the underwater bathymetry (shape and depth of water area) offshore, the time period of wind-driven waves along the coast, tide levels (including seasonal and monthly cycles), pressure inside the hurricane, size and strength of high wind speed zones and other factors, water levels along the coast can rise dramatically as a hurricane approaches. Once thought to be a "wall of water", much like a tsunami, scientists now understand that storm surge involves a gradual building of water levels as each successive wave comes ashore but can not drain back to the ocean before the next one comes ashore.

The worst surge situations occur when waves move into coastal bays (submerged river valleys). This is because as the water pushes up the bay, it gets more constrained

HURRICANE AND WIND TESTING

The IHRC is developing an innovative research capability to carry out hurricane wind testing on structures. When completed, its full-scale structural testing system will help to isolate inherent weaknesses of structures when subjected to category 1 to 5 hurricane-force winds and rain. The goal is to determine new technologies, designs and products for addressing building construction and retrofitting practices.

To do this, the IHRC developed the first of its kind – a "Wall of Wind" (WoW) testing facility. Using a two-fan prototype apparatus, the IHRC has already successfully tested roofing shingles and soffits.

Based on this work, the Renaissance Reinsurance Company – the largest re-insurer of hurricane-prone areas in the world including Florida – commissioned the building of a six-fan array which can generate a 140 mph (225 km/h) wind field. This will eventually lead to an 18-fan WoW facility to be constructed on a public parcel of land adjacent to Homestead Air Force Base (HAFB), which is only 20 minutes away from the FIU campus. HAFB was hit hard by Hurricane Andrew in 1992 and has more than a passing interest in this research.

The IHRC is also looking at how to take the WoW results and visually use these to transform the way that people look at building safety. By following the model used to enhance automobile safety, IHRC hopes to do the same for the housing industry.

horizontally due to the bay's configuration. Squeezed on the sides by terrain, the water has no choice but to rise. When Hurricane Camille pushed a surge up Mobile Bay in 1969, this convergence helped create a 24-foot (7-meter) surge at Mobile, Alabama. Katrina, although a weaker hurricane than Camille, was a larger storm. Katrina's surge on the Mississippi coast was believed to be a record breaking 28 feet (8.5 meters). A smaller, but nonetheless significant, surge inundated Baltimore, Maryland as weakening Hurricane Isabel passed to the west of the Chesapeake Bay. The path of the storm paralleled the bay and meshed with the arrival of the high tide. Even 150 miles from the mouth of the bay, the surge created significant flooding in downtown Baltimore, Maryland.

A record-breaking surge was reported along Australia's north coast with the arrival of a cyclone in 1899. In addition to personal observations, fish and other sea creatures were found beached on rocks some 45 feet (14 meters) or more above the normal water levels. Although the report sounds incredulous, scientists carried out an extensive study to ascertain the claims.

Because water is so much heavier than air, the force of moving water overwhelms any wind effects. Crashing waves can smash buildings and destroy any under-the-home supports (e.g. pilings). Television reports almost always include the destruction of at least one coastal home as waves wash it into the ocean.

With homes built too close to the ocean (sometimes even oceanside of protective dunes), even minor storm surges can cause serious property loss. In low-lying coastal areas, overwash of barrier islands can destroy roadways (evacuation routes) even before

Hurricane Wilma paid an unwelcome visit to the resort town of Cancun in northeast Mexico in late October 2005. Some of Mexico's famed Caribbean resorts were knee-deep in water after Hurricane Wilma roared past. Before setting a course for south Florida, Wilma also uprooted trees, smashed homes, damaged hotels and caused the deaths of six people in the area.

Kids attempt to ride their bicycles through flood waters in the Fells Point area of Baltimore, Maryland on September 19, 2003. The flooding was linked to Hurricane Isabel's storm surge that moved up the Chesapeake Bay and into Baltimore Harbor.

the main winds of a hurricane arrive. In places like Bangladesh, with extensive low-lying-flood plain coastal areas, surges can extend much further inland. Hurricanes are known for rewriting the coastal geographical maps by wiping out barrier islands, redepositing sand in new locations and creating inlets.

Inland flooding

Hurricanes, especially if they move inland and slow their forward speed of motion, can deposit large amounts of rainfall. More often, however, they get picked up by the higher speed middle-latitude winds and accelerate northeastward. What is more common is for a weaker tropical system (either a depression or a storm) to move ashore and then "hang around" for a while. Too many storms have done this!

In addition to being slow movers, these storms are more likely to create flooding if one or more of the following conditions are met:
• Terrain is hilly and valleys are narrow.
• There is an interaction with another weather feature (front, upper level low pressure system).
• Antecedent conditions (e.g. wet ground, high river levels) are favorable.
• Rain falls in an area with a large percentage of impervious surfaces (e.g. urban or suburban location).

On many occasions, the heaviest rain will tend to fall overnight. This allows a smaller number of intense thunderstorms to stay concentrated inside the storm circulation rather than have many more weaker thunderstorms develop over a wider area.

In the United States, for example, tropical storm Agnes in 1972 caused devastating flooding in New York, Pennsylvania, and south into the middle-Atlantic region. Although far outside the region of heaviest rainfall, downtown Richmond flooded, while Pittsburgh reported a flood crest more than 11 feet (3.5 meters) above flood stage. Flood waters also rolled westward, causing flooding as far west as Ohio. Some of the flooding was described as a 500-year flood return event. Damage totaled $2 billion (in 1972) and 122 people drowned.

Other significant tropical-cyclone related flood events include Allison 2001 (see page 76), Claudette 1979, where 45 inches (114cm) of rain fell in Alvin, Texas, and Amelia 1978, in which 48 inches (122 cm) of rain fell at Medina in 52 hours with 27 fatalities in the Hill Country.

Once in an idyllic coastal setting, this Jamaican home provides testimony to the forces associated with hurricane-driven ocean waves and storm surge.

In 2006, tropical storm, John, stalled over Baja California, flooding many communities and stranding 10,000 people. In 2005, Hurricane Wilma stalled over the Yucutan and deposited more than 62 inches (157 cm) of rain on the region. A small island off the Yucutan (Isla Mejeres) reported 64 inches (163 cm) of rain during the storm. Japan, because of its mountainous character has frequent flooding events. Typhoon Nabi/Jolina (September 2005) brought record-breaking precipitation (52.01 inches (132 cm) in three days) to western Japan. Typhoon Fran (1976) holds the Japanese national 24-hour precipitation record of 46.22 inches (117 cm).

Damaged homes and downed trees line the shore of Maria la Gorda Beach, Cuba, following the passage of Hurricane Ivan.

Tornadoes

Most landfalling tropical cyclones do not produce large numbers of tornadoes or waterspouts. However, according to scientists studying such phenomena, almost 60 percent of all landfalling US hurricanes spawn at least one tornado; similar research in Japan shows that about 40 per cent of landfalling typhoons produce tornadoes. Tornadoes are most likely to occur in the leading front right quadrant of a moving hurricane. These numbers may be underestimates because so many people are in protective locations and can't see the tornadoes. Afterwards, tornado damage is masked within overall hurricane damage.

Overall, in the immediate area of landfall, tornado damage is minor compared to the damage due to the hurricane itself. However, if the tropical cyclone maintains its identity, even in a weakened state, as it travels far from its landfall location, tornado incidence can cause significant destruction and deaths and/or injuries. Even though most hurricane-spawned tornadoes are weak, some can reach Enhanced Fujita (EF) Scale 2 ratings or higher.

Only three US hurricanes have broken the "100 reported tornadoes" level. Hurricane Beulah (September 20, 1967) produced 115 tornadoes as it affected Texas and northern Mexico. Hurricane Frances captured the record during the September 5-9, 2004 period with 123 twisters (extending from Florida northward into the mid-Atlantic region). Within days (September 15-17, 2004), Hurricane Ivan produced 117 tornadoes as it trekked from northwest Florida to the northeast US.

The costliest hurricane-spawned twister remains the one that struck the Austin, TX, vicinity on August 9, 1980 (Hurricane Allen) inflicting $100 million in damages. Hurricane Hilda did not produce many twisters (only 10), but she has the distinction of having the deadliest tropical cyclone-spawned tornado. That tornado struck Larose, LA on October 3, 1964, killing 22 people.

Ocean waves

Waves on the open ocean inside a hurricane can easily exceed 25 feet (7.5 meters). If the "Perfect Storm" of 1991 is a model, the interaction of a hurricane (or a hurricane-like system) with a middle-latitude storm can generate waves much larger than that.

Flooded 18-wheeler trucks sit stranded on Interstate-10 near downtown Houston after Tropical storm Allison dumped 28 inches of rain in 24 hours. Texas Governor Rick Perry declared a state of emergency in Harris and 28 other southeast Texas counties, June 9, 2001. The truck at the right caught fire and burned despite all the water.

Not surprisingly, mariners were among the first to fear and respect these ocean storms, learning to steer clear of them or to at least be able to find "hurricane havens". Although numerous sunken ships lying along the US east coast may be the result of wars and piracy, many others were sunk by hurricanes and their wintertime mid-latitude cyclone counterparts.

Virginia history archives (prior to 1800) show many shipwrecks and sinkings due to hurricanes. For example, only two ships in Hampton Roads, Virginia, survived George Washington's Hurricane (July 23-24, 1788).

The first report of a hurricane by European explorers to the New World may have occurred in June 1494 near Hispaniola. However, in 1495, Christopher Columbus's exploration suffered three ships sinking. Columbus noted, "Nothing but the service of God and the extension of monarchy would expose me to such danger."

Large waves spreading far from Hurricane Erin buffet the research vessel Atlantis during the first leg of the Deep East Expedition.

Human and other impacts

Even in developed countries, social, economic and health impacts resulting from these storms can be overpowering. The US saw this during, and following, Katrina. At the writing of this book, several years after Katrina's landfall, the US is still dealing with massive urban recovery, rebuilding and infrastructure repair in that city. The economic impact to the city of New Orleans is almost incalculable as half of its population has left. Political and societal fallout may never be fully known.

In less developed countries, disruptions are more frequent and harder to address. Often international aid is required just to take care of the immediate needs of the displaced,

An aerial view of the submerged town of Ibahay, following the passage of Tyhoon Utor. Utor ravaged the eastern and central Philippines, leaving thousands homeless and without electricity. The powerful typhoon also killed at least seven people. Utor also caused the postponement of the Association of South-East Asian Nations (ASEAN) leader's summit in Cebu City.

Storms along coasts can seriously affect fragile ecosystem communities. It is easy to see how intense wave action has damaged dunes at the Archie Carr National Wildlife Refuge (NWR) near Melbourne Beach, Florida. Among its other values, this NWR is the most important nesting area for loggerhead sea turtles in the Western Hemisphere.

let alone address recovery and repair. Medical issues, including disease, injury, and burying or cremating the dead, become consuming.

Once the immediate disaster is past, there is often not enough time, people or resources to make substantive changes before a new disaster strikes.

Environmental losses can be daunting. Flooding, coastal and/or inland, can wipe out watershed communities; waterways can be flooded with chemical, human or other pollutants; nesting habitats may be destroyed and/or the animals and plants themselves may be killed. Too often, such environmental impacts do not receive the same media and political attention that human impacts do.

This makes the overall model followed by the US, the United Nations and other developed counties one that should be applied more fully around the world. Governments (nationally and locally) are quick to review emergency procedures and make changes; engineers look to structural improvements; utility companies try to find ways to lessen customer impacts; emergency managers reassess their plans; television stations try to find better ways to keep viewers informed; and insurers look for ways to reassess risks.

While environmental groups often address impacts to the natural world, it seems that this part of the hurricane picture needs to be better integrated with other parts. As many before have said, it is only by learning from the past that we can better face the future!

Global perspective

Since tropical cyclones occur globally across most tropical regions, let's take a look at these storms in different parts of the world.

Indian Ocean

Tropical Cyclones in the Indian Ocean and Bay of Bengal have a long history of causing incredible loss of life and destruction. The east coast of India was struck by a major cyclone (winds estimated to be above 155 mph (249 km/h)) in October 1999. Ten thousand people were killed and 10 million lost their homes. Tens of thousands were affected by water-borne illnesses such as gastro-ecenaeritis and cholera.

In 1971, and again in 1977, an estimated 10,000 people were killed in the same part of India by storms of similar intensity. In 1991, another big storm hit Bangladesh (formerly East Pakistan), killing an estimated 100,000 people. However, it was the cyclone of 1970 that struck Bangladesh that sets the disaster standard for the region, killing more than a million people!

Reports noted the cyclone was the most devastating to "hit South Asia in the past 50 years." In addition to loss of life, the cyclone wiped out the south coast's entire infrastructure. This included physical structures and its agricultural base. With the effects of both the cyclone and the ensuing war of independence, Bangladesh has earned its label as being one of the poorest countries in the world.

There are many reasons for such high death tolls in the region. The population density is very high; the quality of home construction is poor; there is a lack of warning and preparedness systems; and the people have no place to evacuate to due to the low-lying character of the flood plain region. Sometimes, people stay to protect what little they have.

Changes, albeit slow, are coming. Indian fishermen now carry radios with them so they can receive storm information. Evacuation plans have been improved and shelters have been built. Weather observation systems have been improved (including use of satellite and radar systems).

Cyclones also affect the south Indian Ocean occasionally striking Madagascar near the east coast of Africa. Most, however, remain over open waters of the Indian Ocean between 10 and 25 degrees south latitude.

People flee the aftermath of a cyclone that struck the Indian state of Orissa in 1999.

Atlantic Ocean

Storm losses in other parts of the world are much less daunting. Still, the US suffered seriously at the hands of a hurricane in 1900. Even with unofficial warnings at the time, people did not believe that a storm was coming. The historical marker that commemorates the loss of some 8,000 lives is testimony to how much hurricane warning programs have improved across many parts of the world in just about 100 years.

During the American Revolution, an even worse hurricane struck the Lesser Antilles. It slowly passed through the islands and eventually moved out into north Atlantic after leaving 22,000 dead. This makes it the deadliest Atlantic hurricane in history. Eight British warships were sunk and structures on the island of Barbados were completely flattened.

Australia

Australia has a notable cyclone history. The map right shows the entire north coast is at risk. However, the region is sparsely populated, except for the city of Darwin, in the Northern Territory.

In an average season, 10 tropical cyclones develop over Australian waters, of which six make landfall. Much like the Atlantic Ocean

The 1900 Storm Memorial at Galveston – scene of the worst US natural disaster – created to honor 6,000-plus victims of that hurricane.

15 worst Atlantic Ocean basin hurricanes (based on fatalities)

22,000	Great Hurricane of 1780	1780
11,000-18,000	Hurricane Mitch	1998
8,000-12,000	Galveston Hurricane	1900
8,000-10,000	Hurricane Fifi	1974
2,000-8,000	1930 Dominican Republic Hurricane	1930
7,186-8,000	Hurricane Flora	1963
6,000+	Pointe-a-Pitre Bay Hurricane	1776
4,000-4,163+	Newfoundland Hurricane	1775
4,075+	1928 Okeechobee Hurricane	1928
3,433+	Hurricane San Ciriaco	1899
2,500-3,107	Cuba Hurricane	1932
3,037	Hurricane Jeanne	2004
3,000+	Central Atlantic Hurricane	1782
3,000+	Martinique Hurricane	1813
2,000-3,000	Yucatan Hurricane	1934

climatology, there is considerable year-to-year variability in cyclone numbers. In some years, there are no storms. In other years, numbers can reach well into the teens (the record is 16, set in 1963).

Australia's worst cyclone disaster occurred in March 1912 (12 years after Australia's independence from the British Empire). It is also the worst maritime disaster in Australia's 106-year history – more than 150 lives were lost as the cyclone affected the country's Northwest Territory. In addition to sinking or grounding several large, iron sailing ships, the storm also sunk or wrecked several lighter vessels and pearling luggers. However, it was the loss of the Koombana, a three-year-old coastal steamer, with 140 people aboard, that made this cyclone historical. Although the captain had noted the low barometer, he still sailed straight into the full fury of the cyclone within hours of leaving port.

The second deadliest cyclone also affected the Northwest Territory. On March 26-27, 1935, a cyclone struck the region claiming 141 lives. The worst ever cyclone-related disaster in Australia's history occurred in March 1899, just prior to independence. Known by several names, Cyclone Mahina claimed more than 300 lives. The storm had a very large storm surge and a sudden rise in sea-level near the eye of the cyclone which claimed many lives. Others died as some 100 pearling vessels were destroyed.

Cyclone Vance produced the strongest-ever recorded gust on the Australian mainland – 166 mph (267 km/h) at Learmonth, WA, in March 1999. Cyclone Monica, which struck Australia's north coast, may now hold the record low pressure of any tropical cyclone in the world, 868.5 mb or 25.65 inches of mercury. This pressure can be related to estimated winds of 180 mph (290 km/h) with gusts to 220 mph (354 km/h). However, when she struck the unpopulated and non-instrumented area on April 24, 2006, there was no official record of what her winds were. Australian weather forecasters

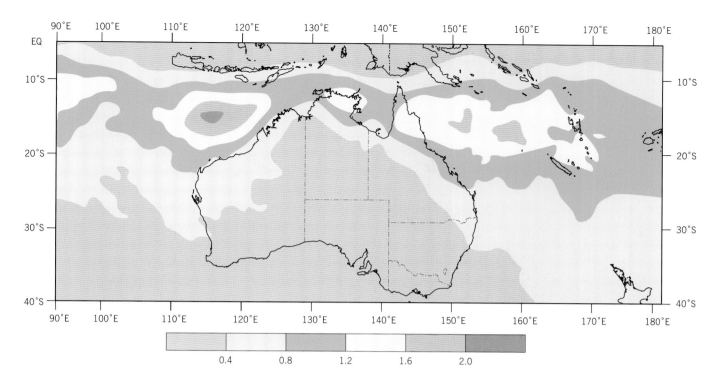

0.4	0.8	1.2	1.6	2.0

Map showing the average annual frequency of tropical cyclones in the Australian region.

were uniformly in awe of how strong she was in Australian cyclone history. Monica (category 5) was one of three storms to strike Australia in 2006. Glenda, category 3 in March, and Larry, category 3-4 on March 20 were the other two. Larry was the worst, striking Queensland with a fury unmatched in that region for perhaps a century. Damage from that storm alone was estimated at over $1 billion US dollars.

Cyclones also affect the island chains to the east of Australia. New Caledonia and Vanuatu are among the places that have seen cyclones in recent years.

Western Pacific

Typhoon history is also noteworthy. Perhaps the deadliest event occurred in 1881 when a typhoon with a large storm surge submerged the city of Haiphong, Vietnam. There were around 300,000 fatalities. In 1975, super-typhoon Niña, category 5, struck China. Heavy rainfall from the storm contributed to the failure of the Banqiao Dam. Casualties totaled around 229,000.

In 2006, the Philippines were struck by six typhoons. Five of these were category 3 or higher – an unprecedented number. Super-typhoon Durian, in late November 2006, was the third deadliest typhoon in Philippine history, and the most damaging. Durian brought torrential rains to the country. Almost 1,500 lost their lives in flooding and mudslides near the Mayon volcano. The storm destroyed more than 200,000 homes, and damaged another 308,000.

Overall, 2,049 people were listed as direct fatalities due to the six typhoons. Property losses reached $1.3 billion (US), making 2006 was the most damaging typhoon season in Philippine history. For comparison, the combined damage from 19 major Philippine typhoons for the period 1980-2002 was more than $400 million (US).

Looking at just one season, 2006 was very active in the western Pacific Ocean but fairly calm in the Atlantic. Although a weak El Niño was at work, this type of balancing of

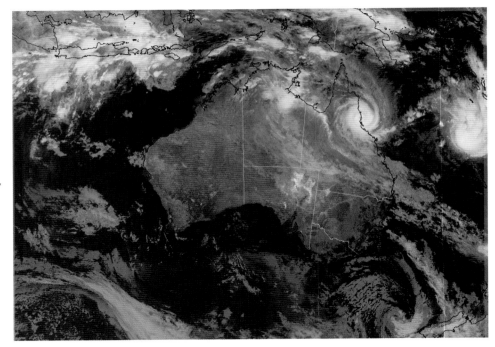

This false colorized satellite image shows Cyclone Larry as it crossed the coast of the Australian state of Queensland late on March 19, 2006. Larry came ashore as a Category five cyclone, destroying homes, leaving hundreds homeless, thousands without power and causing hundreds of millions of dollars damage to agriculture. The satellite image was processed by the Bureau of Meteorology from data received from the geostationary meteorological satellite MTSAT-1R operated by the Japan Meteorological Agency.

tropical cyclones around the world would be expected. In fact, the statistics suggest that, when activity is high in one ocean basin, it is lower in another.

Looking specifically to Australia, scientists note that there is a tendency for tropical cyclones in Australian waters to be less common in El Niño summers, and more common in La Niña summers. This is due to changes in broadscale wind patterns and water temperatures during such events (see Chapter 19). However there is also evidence that the frequency of severe cyclone impacts in northwestern Australia may be similar regardless of El Niño and La Niña situations.

Forecasting hurricanes

Hurricanes are forecast by various governmental weather services around the world. Although NOAA National Hurricane Center is the most recognized, Japan's Meteorological Agency, Australia's Bureau of Meteorology and so on provide similar services for their regions.

Meteorologists today use sophisticated computer models that incorporate physical processes and/or statistical data to improve forecasts. New and improved data types e.g. data buoys, satellites, focused aircraft data probes, coastal reporting sites, and radar, all play roles. While forecast errors in the US have improved dramatically (right), there are still large errors – an average of about 240 nautical miles (275 miles; 445 km) – for forecasting where the center of the storm will be in 72 hours. This is often the key time in which evacuation, and other decisions, need to be made. If the wrong decision is made it can cost millions in unnecessary evacuations or it can place untold numbers of people at risk.

Statistically speaking...

An NOAA study examining the decadal distribution of hurricane strength at US landfall shows several different types of patterns. Even updating the study with the most recent two years of data, hurricane frequency (all categories) was still much higher in the latter half of the 19th Century and again during the 1931-60 period than at any

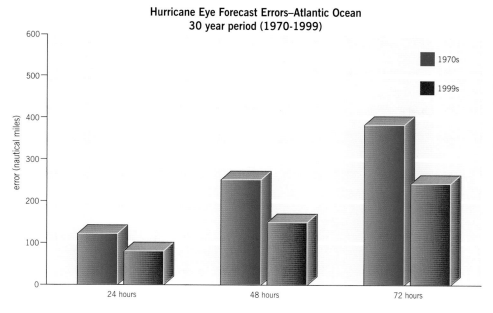

Hurricane Eye Forecast Errors–Atlantic Ocean
30 year period (1970-1999)

Legend: ■ 1970s ■ 1999s

y-axis: error (nautical miles) — 0, 100, 200, 300, 400, 500, 600

x-axis: 24 hours, 48 hours, 72 hours

Due to improved technology and greater scientific understanding, forecasters at NOAA'S National Hurricane Center continue to issue more accurate forecasts of hurricane eye location. Forecast error scores shown here have been further improved through the 2000s. Skill in forecasting hurricane intensity has remained flat through the same time period.

other time in the past 150 years. In fact, the period 1931-60 was the most active in terms of major (category 3 and higher) storms. The period 1961-2000 was relatively tame in terms of hurricanes and number of major hurricanes making landfall.

Even including the records set during the 2004 and 2005 hurricane seasons when "four intense hurricanes made landfall in the USA, breaking the previous record of three set in 2004", it is possible to place a totally different spin on the statistics. Although there were four "major" landfalling storms in 2005, none was a category 4 or 5.

US Hurricane Statistics for the period 2001-2006

| | **Category** | | | | | | |
	1	**2**	**3**	**4**	**5**	**All**	**Major**
2001-2003	4	1	0	0	0	5	0
2004	0	1	2	1	0	4	3
2005	1	0	4	0	0	5	4
2006	0	0	0	0	0	0	0
	5	2	6	1	0	14	7

While 2004 and 2005 were highly active years (in terms of landfalling hurricanes), the other four years were quite tranquil. Thus, when viewing the first decade of the 21st Century (not just a couple of years), it is easy to say that this was not a record period of hurricane activity. If the averages persist, the 2001-10 decade will be comparable to the 1941-50 period. This flies in the face of claims of growing hurricane danger due to ongoing climate change.

NOAA's AOML notes that Atlantic hurricane seasons since 1995 have been significantly more active, with more and more intense hurricanes than the previous two decades.

WHICH IS THE MOST INTENSE TROPICAL STORM ON RECORD?

It is difficult to determine the lowest pressure of tropical cyclones because these storms exist in their most intense phases over ocean waters, often in remote locations. In recent years, however, hurricane hunter aircraft and satellite data have provided better estimates of storm strength.

Until Cyclone Monica captured the record low estimated pressure, Typhoon Tip in the Northwest Pacific Ocean held the honors. Tip's central pressure was measured on October 12, 1979 at 870 mb (25.69 inches of mercury); Tip's estimated surface sustained winds were 190 mph (305 km/h). Typhoon Nancy on September 12, 1961 had estimated maximum sustained winds of 213 mph (343 km/h) with a central pressure of 888 mb (26.22 inches of mercury). However, scientists now recognize that estimated maximum sustained winds for typhoons during the 1940s to 1960s were also strong and that Nancy's record might be inaccurate.

In the Atlantic Ocean basin, Hurricane Gilbert's 888 mb (26.22 inches of mercury) had retained the record for the lowest estimated pressure (mid-September 1988) for the Atlantic basin. However, that value was eclipsed by Wilma's 882 mb (26.04 inches of mercury) reading on October 19, 2005.

Earlier periods, such as from 1945-70 (and perhaps earlier), were as active as the most recent decade. More hurricanes have made US landfall in the past decade, but periods of even higher landfalls occurred early in the 20th Century.

Although the most recent period seems to mesh with global warming, the earlier peaks in storminess and landfalling storms fly in the face of such claims. In fact, the period of high incidence of storminess in the early part of the 20th Century is during a time when ocean sea surface temperatures were significantly below average.

It should be noted that anomalous sea surface temperatures in the tropical Atlantic were significantly warmer than the global average from about 1930-70 and after 2000. This warming is attributed to the Atlantic Multi-decadal Oscillation, a slow cycle of natural fluctuation in atmospheric conditions and water temperatures.

Other factors, such as changes in data quality, density, sources and methodologies for estimating hurricane strengths, lie at the heart of arguments whether or not a climate change-related trend in hurricane intensities can be detected.

Hurricanes respond to a variety of factors besides local ocean temperatures. In particular, the vertical wind structure is of crucial importance; favorable wind conditions in conjunction with warmer ocean temperatures contribute to active periods. The Atlantic Multi-decadal Oscillation and the El Niño/La Niña cycle are important factors in determining the environmental conditions for seasonal to multi-decadal extremes in hurricane activity.

Property losses

Property losses are, however, breaking records and this has nothing to do with storminess but everything to do with demographics. In the US and other hurricane-prone areas, people are flocking to coastal zones (see also Chapter 3), and this means an associated growth in business and personal property at risk to wind and flood damage.

Losses from each US hurricane flood disaster further strain the US Federal Flood Insurance Program, while escalating wind damage losses push more private insurance companies to reduce their exposure in such high-risk areas. What this means is fewer choices to the homeowner and overall burgeoning insurance costs. Some states have had to create state-run or state-supported insurance programs. Other states have negotiated with insurers to allow homeowners to assume greater risk (higher deductibles) in an effort to lower costs.

Meanwhile, homeowners are increasingly suing their insurance company when claims are not settled in their favor. One recently adjudicated case involved an insurance company that split a home's damages between flooding and wind during the damage adjustment process. There are now several pending cases, at least one before the US Supreme Court, that claim global warming as the cause of hurricane property damage or increased coastal erosion. The basis of these lawsuits is that the warming is caused by uncontrolled greenhouse gas emissions.

In less developed nations, insurance is not an option. International relief efforts often have to come into play after the disaster is recognized. The concern is that the number of disasters is growing, as are the resources needed to respond to it.

The eyewall of Hurricane Katrina taken on August 28, 2005, as seen from a NOAA P-3 hurricane hunter aircraft before the storm made landfall on the US Gulf Coast.

"Raleigh, NC, June 22, 2005: Scattered thunderstorms moved across the metro area this afternoon bringing reports of large hail, torrential downpours and high winds. Trees were knocked down in southeast Durham and portions of Interstate 40 near Cary were closed due to high water. Two-inch diameter hail fell near Leesville in Northwest Raleigh."

H Michael Mogil

STS-64, astronauts aboard the Discovery captured this image showing thunderstorms near Hawaii in the central Pacific Ocean.

If the hurricane is the "greatest storm on Earth," then the thunderstorm must be the "greatest, little storm on Earth." One reason for this is that thunderstorms drive hurricanes with energy release linked to the evaporation-condensation process. But at any given time, there are an estimated 2,000 thunderstorms in progress on the Earth. Each of these small engines transports heat and moisture from the ground to the higher levels of the troposphere. They also transport cooler air from high altitudes down to Earth.

In addition, these thunderstorms bring needed precipitation to many places. In some locales, more than 50 percent of the annual rainfall is linked to thunderstorms.

A thunderstorm along the Italian coast. In the foreground, the curved cloud marks where cold winds from within the thunderstorm are blowing outward toward the photographer. In the distance, behind this cloud, a rain curtain shrouds the horizon. In the far left, lightning flashes outside the rain area.

Thunderstorms also help to keep the Earth's overall electrical field in balance. In the process, they can bring some artistic skies and/or some unique cloud formations. If the story stopped here, we wouldn't need this chapter. But there is a deadly and destructive side to these storms.

Deadly weather

Sometimes thunderstorms, either individually or collectively, bring too much rain and cause flooding (see Chapter 10). Many thunderstorms can become "severe", meaning that they bring winds above 58 mph (93 km/h), hail larger than 1/4 inch (2 cm) in diameter (see also Chapter 9), cause wind damage and/or produce tornadoes (see Chapter 7) at least during part of their lifetime. By definition, any storm that produces thunder must also have lightning. And even though not considered a formal severe weather warning event, lightning can create its own destructive impacts and kill and injure (see Chapter 8 for more information). Annual thunderstorm fatality statistics for the US alone are quite impressive.

Aerial view of a thunderstorm amid a field of cumulus and towering cumulus clouds. These clouds are embedded in a thick pollution layer, masking them from a ground view. Once above this layer, the clouds are easily seen against a bright blue sky.

A mile-wide tornado beneath a picturesque striated supercell thunderstorm threatens north-central Nebraska on May 29, 2004. This photograph is a composite of two images shot only a few seconds apart. Except for overlapping the two images, no manipulation has occurred. Distance, size, shape and color are authentic.

International hazard statistics are harder to come by, but more and more nations and severe weather-focused organizations are starting to compile them. Estimates for lightning casualties in Australia, for example, indicate about five to 10 deaths and 100 injuries per year.

It is important to recognize that thunderstorms hazards do nut occur in isolation. If the thunderstorm is strong or organized enough to produce one type of hazard, it is often capable of producing others. Only the tornado requires a very special thunderstorm structure; but once tornadoes are in the offing, just about every possible severe weather type is also likely.

The thunderstorm

Each year, in the United States alone, some 100,000 thunderstorms occur. These have lifetimes that range from a few minutes to up to few hours (super-cell variety). About 10 percent of these storms become "severe". Thunderstorms can occur individually (weak or strong "super-cell") or they can congregate in clusters, lines, or systems. Each of these brings its unique type of thundery weather.

Individual thunderstorms can "pop up" due to slight variations in surface heating from one place to another. In short, where it gets hottest is usually where air rises most easily and becomes the breeding ground for thunderstorms. This can be due to variations in the character of the surface (e.g. urban/suburban locations heat up more quickly than nearby forest areas; land heats up more quickly than nearby water areas) and due to variations in elevation (thunderstorms will tend to form over mountain peaks sooner than nearby valley areas because warmer air and winds push air up the mountain). Because forest fires and volcanoes are also heat sources, they also spawn thunderstorms (see Chapters 13 and 18 respectively)

Thunderstorm-related flooding along the River Rye in North Yorkshire, UK. A Royal Air Force rescue helicopter flies above the area searching for people trapped by the flooding.

Thunderstorm lines

With a well-defined surface wind pattern, low-level convergence zones (winds blowing together) can often line up along and ahead of cold fronts. When this occurs, individual thunderstorms may develop, but frequently they reform into one or more lines. If the line is more or less continuous, it is often referred to as a squall line (due to the

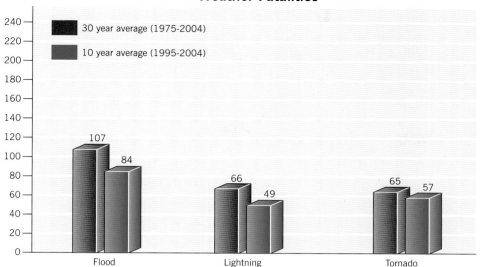

Weather Fatalities

■ 30 year average (1975-2004)
■ 10 year average (1995-2004)

Flood: 107, 84
Lightning: 66, 49
Tornado: 65, 57

Fatalities and injuries caused by thunderstorms. These and other statistics are compiled by the National Weather Service and the National Climatic Data Center from information contained in Storm Data, a report comprising data from NWS forecast offices in the 50 states, Puerto Rico, Guam and the Virgin Islands. Flood statistics have been corrected to remove most large-scale flood events that may not be thunderstorm-induced.

squalls or wind gusts that often accompany it). But the line can also have significant gaps in it. In this case the line may be composed of many isolated "super-cell" thunderstorms.

Super-cell thunderstorms frequently occur outside of a line structure. In fact, that is their preferred habitat. Not competing for heat, moisture, inflow winds and other storm ingredients, super-cell storms typically grow quickly and are long-lived. The easternmost and southwesternmost squall lines in in the map below have a super-cell spacing. The line closest to St Louis is more-or-less continuous.

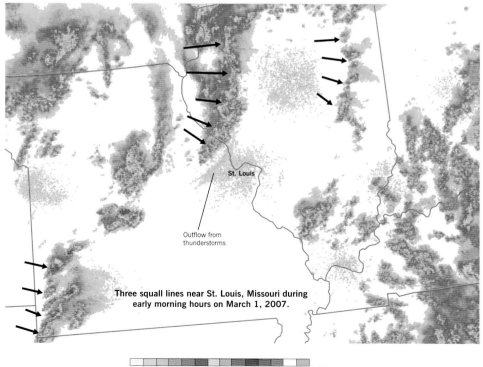

St. Louis

Outflow from thunderstorms

Three squall lines near St. Louis, Missouri during early morning hours on March 1, 2007.

0 5 10 15 20 25 30 35 40 45 50 55 60 65 70 dBZ

Three squall lines march across the central Mississippi River Valley (March 1, 2007). The line of thunderstorms, near St Louis, Missouri, has an outflow boundary (where winds race ahead of rain area). Radar reflectivity (the amount of radar wave energy reflected back to radar by precipitation) is shown by a colored scale. Red indicates the greatest reflectivity and, hence, the heaviest rain and/or the largest hail.

RADAR

The most effective tool to detect precipitation is radar (RAdio Detection And Ranging). Radar has been utilized to detect precipitation, and especially thunderstorms, since the 1940s. Recent improvements in radar technology now enable meteorologists to see three dimensional winds and detailed storm structure. This "Doppler" technology has helped to extend severe weather warning times and provide accurate warnings for severe weather events.

The US operates more than 160 Doppler radars, including the US Territory of Guam and the Commonwealth of Puerto Rico. It includes radars operated by the military, the NWS and the Federal Aviation Administration. These aviation-focused radars are known as Terminal Doppler Weather Radars (TDWR). Similar radar technology is used throughout the world.

How does a radar work?
The radar houses a complex array of equipment including transmitters, receivers, computers, and associated image interpretation algorithms. These algorithms evaluate radar image information and derived data sets to highlight high-risk severe weather areas. The algorithms distinguish between straight winds and tornadoes and even between hail and rain.

As the radar antenna turns, it transmits extremely short bursts of radio waves, called pulses. Each pulse lasts about 0.0000016 seconds, with a 0.00019-second "listening period" in between. The transmitted radio waves move through the atmosphere at speeds similar to the speed of light.

By recording the direction in which the antenna was pointed, the direction of the target is also known. Generally, the better the target is at reflecting radio waves (for example, a large number of raindrops, big hailstones, etc.), the stronger the reflected radio waves will be. The reflected waves are called "echoes", much like a human voice may be bounced back to the sender in a cave. The process is repeated up to 1,300 times per second. By keeping track of the time it takes the radio waves to leave the antenna, hit the target and return to the antenna, the radar can calculate the distance to the target.

Doppler radar pulses have an average transmitted power of about 450,000 watts. By comparison, a typical home microwave oven will generate about 1,000 watts of energy. However, because of the very short period the radar is actually transmitting, the time the radar is "on" is only about seven seconds each hour. The remaining 59 minutes and 53 seconds are spent listening for any returned signals.

Doppler radar brings big benefits
By comparing where echoes are over time, Doppler radar systems can provide information regarding the movement of targets as well as their position.

By measuring the shift in phase between a transmitted pulse and a received echo, the target's radial velocity (the movement of the target directly toward or away from the radar) can be calculated. A positive phase shift implies motion toward the radar and a negative shift suggests motion away from the radar (motion in a radial direction only).

The larger the phase shift, the greater the target's "radial" velocity (the velocity of the object in the direction of the line of sight). The phase shift effect is often referred to as the "Doppler shift". An object emitting sound waves will transmit those waves in a higher frequency when it is approaching your location (inbound velocity = positive shift) as the sound waves are compressed. As the object moves away from a location, the sound waves will be stretched and have a lower frequency (outbound velocity =

This structure (which looks like a golf ball on a tee) houses the latest in radar technology. Doppler radar allows meteorologists to see inside storm clouds and detect hail, heavy rainfall, high winds, and even tornadoes. This is a National Weather Service radar; many TV stations also have Doppler radars.

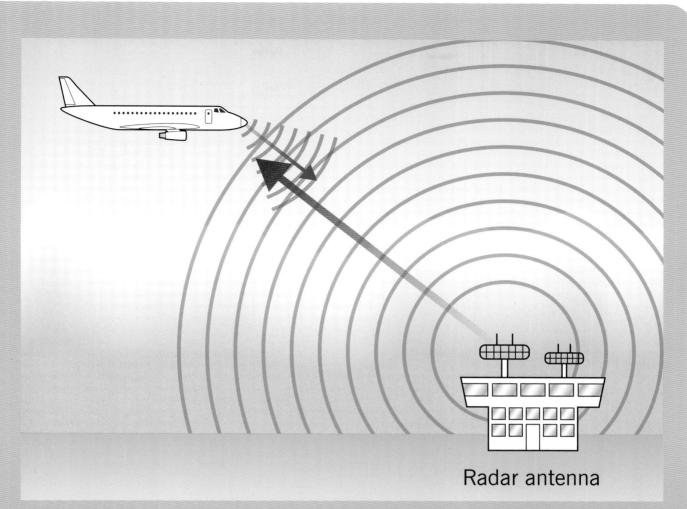

Radar antenna

negative shift). You have probably heard this effect when an emergency vehicle drives past you with its siren blaring – after the vehicle has passed your location, the frequency (pitch) of the siren sound is lower than it is as the vehicle approaches you.

Looking up, too!
Doppler radars employ scanning strategies in which the antenna automatically raises to higher and higher preset angles, or elevation slices, as it rotates. These elevation slices comprise a volume coverage pattern (VCP). Once the radar sweeps through all elevation slices then a volume scan is complete. In precipitation mode, Doppler radar completes a volume scan every four to six minutes depending upon which VCP is in effect, providing an updated three dimensional view of the atmosphere around the radar site.

Weather radar works much like this aviation radar. The radar sends out a signal (radio wave) that bounces back from a target. The return signal is called an "echo" because the energy from the initial signal bounces off a target and is returned to the source.

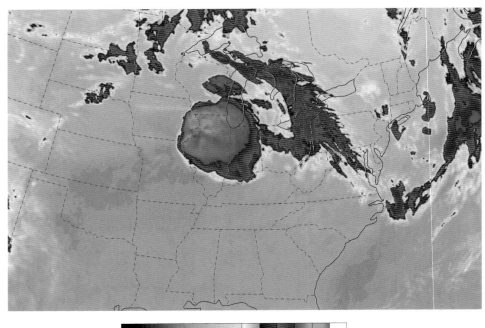

Mesoscale Convective System (MCS) located over Midwestern United States. MCS's tend to develop at night and quickly assume a rounded cloud top shape.

Sometimes thunderstorm lines don't move very far or very fast. As individual thunderstorms move along this "quasi-stationary" line they bring a series of heavy rainfalls to small areas. Meteorologists often refer to it as "train echoing", because the individual thunderstorms move along the line like train cars move past a railroad station. Many "flash floods" (floods that, literally, happen in a flash) develop in this setting (see also Chapter 10).

Mesoscale convective systems

Finally, thunderstorms may form a cluster (often appearing as a round cloud mass when viewed from a satellite in space see page 88). If the cluster becomes organized, large in size and lasts for a long enough time, it gets labeled as a Mesoscale Convective System (MCS). MCSs often develop overnight (not the usual time of day for thunderstorm formation) and then persist well into the next day. Sometimes, MCSs will develop in the same general area for several nights in a row. Much as the "train echo" situation, this scenario can lead to serious flash flooding.

Cumulus clouds with classic flat bottoms and puffy tops. Cumulus form as air rises often due to sunlight heating the ground. Clear areas between clouds are where compensating sinking air is taking place.

MCSs are not unique to the US. In fact, many MCSs develop in the Inter-tropical Convergence Zone (ITCZ) across Africa. Some of these eventually become the formative cloud masses that spawn Atlantic Ocean hurricanes.

Thunderstorm formation

Thunderstorms can form under many varied conditions, but the most common involves the presence of a warm, humid air mass and some mechanism (e.g. fronts, sea breezes or mountains) to help lift the air. Sometimes, the air mass is so unstable (i.e. air can rise or sink more freely) that just solar heating of the ground can get the rising air process (convection) going. If the air is cold at high

altitude and other conditions are favorable, temperatures at the ground do not have to be like summer for thunderstorms to form. In fact, some thunderstorms have even been reported while snow is occurring. These "thundersnows", while unusual, are not uncommon.

The first sign of convection is the formation of cumulus clouds. These are the flat-bottomed, puffy clouds that often dot the sky by (local) midday. Cloud bases are often between 3,000 and 5,000 feet (915 and 1524 meters) above the ground, but can be as high as 8,000 feet (2438 meters) over dry regions. If the atmosphere is more stable, the cumulus clouds will not build very high before they start to either spread out horizontally or break apart. These flat cumulus clouds often evaporate by evening.

If the atmosphere is unstable, the cumulus can quickly develop vertically. As updrafts (rising air currents) power these towering cumulus, cloud tops can reach altitudes of five miles (8 km) above the ground, or more. Since temperatures at high altitude are so much colder than those near the ground, the top of the storm literally becomes frozen. Instead of just being filled with raindrops, the tops of these developing storm clouds can have frozen rain (hail or even "soft" hail known as graupel), ice crystals, snowflakes or super-cooled raindrops (raindrops that are still liquid even though they are below freezing). Super-cooled raindrops can exist even at temperatures as low as -40 °F (-40 °C).

Thunderstorm demise

As the storm continues to mature, downdrafts (sinking air currents) develop. These form as the precipitation loading of the cloud overcomes the rising air currents or the rising air currents themselves weaken. Sometimes dry air from just outside the thunderstorm becomes entrained into the upper parts of the storm, causing evaporation (cooling) to occur. Cold air, being denser (heavier) than warm air, wants to sink.

Most of the time, downdrafts are relatively weak. When they reach the ground, they have no choice but to spread out horizontally (much like the winds in a high-pressure system). These "outflow" winds often

Thunderstorm/cumulonimbus cloud. The cloud top is starting to assume the classic anvil shape (see pointed cloud pattern at right). A well-defined precipitation curtain, known as a rainshaft, is evident from right hand portion of the cloud.

Life Cycle of a Thunderstorm

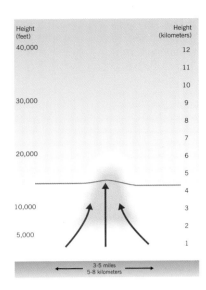

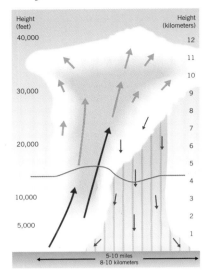

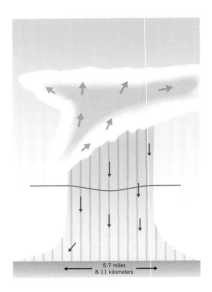

Developing Stage

Towering cumulus cloud indicates rising air.

Usually little if any rain during this stage.

Lasts about 10 minutes.

Occasional lightning.

Mature Stage

Most likely time for hail, heavy rain, frequent lightning, strong winds, and tornadoes.

Storm occasionally has a black or dark green appearance.

Lasts an average of 10 to 20 minutes but may last much longer in some storms.

Dissipating Stage

Rainfall decreases in intensity.

Can still produce a burst of strong winds.

Lightning remains a danger.

Life cycle of a non-severe thunderstorm in middle-latitudes. In the mature stage the updraft is on the western or southwestern side of the storm and the rainfall falls to the east or northeast of the storm. Storms in lower latitudes (or other places where winds are from the east) have to view the storm in reverse. The updraft lies to the east and rain falls to the west of the advancing thunderstorm. Even if thunder and lightning are not observed, the convective process is similar for heavier rain showers.

bring cooling breezes, as well as welcome rainfall. For this reason, the outflow winds can be likened to the arrival of a cold front, except on a more localized or miniaturized scale. Also, much like cold fronts, these outflow boundaries also provide a low-level wind convergence zone and can help to create new thunderstorms. In short, as the parent storm dies, offspring can grow.

Sometimes the outflow races far ahead of the parent thunderstorm. When this occurs, the thunderstorm's warm air source is cut off and the thunderstorm weakens. New thunderstorms can develop on the outflow boundary. If you are in this transition region, the approaching storm can fail to reach you, while a new storm forms past you. In such cases, the storm literally "jumps" forward.

Finally, the storm rains itself out, leaving middle- and high-level clouds behind as evidence of its movement of moisture to higher altitudes.

During this process, the thunderstorm has brought warm moist air to high altitudes and colder, rain-laden air to the ground, stabilizing the atmosphere. The sinking air has also balanced the rising air, keeping it from collecting too much or too little in one place.

Downbursts and microbursts

Sometimes, however, the sinking air takes on stronger characteristics. When this happens, sinking air currents can race toward the Earth at 50 mph (80.5 km/h) or more.

Dying stages of a thunderstorm. Mid and high level clouds, in the shape of an anvil cloud, remain, but the lower portion of cloud has rained itself out.

Upon striking the Earth, they transform into damaging horizontal winds (perhaps winds that meet official severe storm criteria of 58 mph (93 km/h) or more). Depending upon how and where they strike, the winds can be enhanced by local topography or buildings, creating even stronger localized winds.

If the downburst or microburst (see below) is strong enough, it can destroy or severely damage homes and knock down or snap tops off large stands of trees (criteria that make the event "severe"). These localized sinking air currents can also knock planes from the sky (especially those landing or taking off). For ease in discussion, we'll use the term "downburst" to refer to both downbursts and microbursts.

Downbursts often strike the ground and spread out in a star-burst pattern. The pattern is often skewed in the direction of thunderstorm movement since there is a combined motion effect – downburst spreading plus storm motion. This pattern of mostly straight-line winds helps meteorologists distinguish downburst damage from that caused by its rotating tornadic cousin.

Downbursts are areas of sinking air in thunderstorms. They form when air inside the thunderstorm cloud becomes colder than air outside the storm and/or the storm becomes "overloaded with precipitation." In either case, air starts to sink. When the sinking air reaches the ground it spreads out. It's much like dropping something breakable on the floor. The image on the left is a schematic emphasizing aviation implications; the image on the right shows precipitation falling from a thunderstorm but not reaching the ground (virga) with a microburst in progress at the ground (blowing dust).

Downburst: an area of strong, often damaging winds produced by sinking air inside a thunderstorm. At any instant, it affects an area less than about 1/3 square miles (1 sq km) to slightly more than 35 square miles (100 sq km).

Microburst: a downburst that covers an area less than about six square miles (16 sq km) with peak winds that last two to five minutes.

Derecho: a widespread (240-mile long/386-km) convectively induced straight-line windstorm (i.e. one in which the surface winds do not have any significant curvature).

Tornado: a violently rotating column of air, in contact with the ground, either pendant from a cumulus-type cloud or underneath a cumulus-type cloud, and often (but not always) visible as a funnel cloud.

The "derecho"

Downbursts can occur individually or in a series. A downburst from a localized thunderstorm is more likely to begin and end within a few minutes. A downburst, or series of downbursts, from a super-cell thunderstorm, a multi-cell thunderstorm, or a strong line of thunderstorms may last for half an hour or more. A "derecho" – a Spanish origin word meaning "direct" or "straight ahead" – is now defined as a downburst event with a track of least 240 miles (386 km) and lasting for several hours.

The word "derecho" was coined by Dr Gustavus Hinrichs, a physics professor at the University of Iowa, in a paper published in the American Meteorological Journal in 1888. He used it to describe a significant derecho that moved across Iowa on July 31, 1877.

The longer-lived downbursts – derechoes – are often more deadly and more damaging. However, this should not detract from isolated downbursts which can be noteworthy.

Clouds at the base of a thunderstorm cloud provide information about the potential for downbursts. On the left, a curved rain shaft suggests that a downburst is in progress. On the right, a layered set of clouds (known as a shelf cloud) foretells of outward blowing thunderstorm winds. In the distance (lower left), a rain curtain within the thunderstorm hugs the horizon.

Visual downburst clues

Visually, there are clues that the thunderstorm may be producing strong outflow winds. For so-called "wet microburst" events, the localized, well-defined rainshaft tells where air (and precipitation) are leaving the cloud and heading toward the Earth. A rainshaft that is curled or bowed away from the rain area near the ground suggests an outflow in progress. The curved shelf cloud – which looks like shelves in a closet or pantry – at the leading edge of the thunderstorm shows outflow beneath the shelf cloud and rising air above it. The shelf cloud marks the interface between the two. An arcus cloud – a horizontal roll cloud detached from the parent thunderstorm – also signifies thunderstorm outflow.

Remember, however, there can also be "dry" microbursts in which precipitation is limited or even non-existent. Sometimes blowing dust or smoke can signify the downburst. I recall seeing thunderstorm winds blowing smoke from a fire near Tallahassee, Florida. The downburst winds took a vertically rising smoke plume and transformed it into a horizontal plume. When the downburst weakened, the smoke plume again rose vertically. Moments later, a second downburst blew the plume horizontally again.

Radar and satellite downburst clues

There are downburst signatures in radar and satellite data. These all hinge on the fact that the downburst thunderstorm operates in the reverse from the regular thunderstorm life cycle shown on page 96. Here, the strongest rising air occurs at the leading edge of the storm and once the main storm passes by then there is an extended period of moderate or heavy rain.

Bow echoes (fast-moving, radar signatures characterized by a backwards letter "C" in the highest reflectivity values) mark a well-defined and stronger downburst. This is often matched with high wind speed values in Doppler radar velocity images. Reflectivity data shows a more intense thunderstorm at the leading edge of the rainfall area. Doppler velocity data often shows a wind shift accompanying the intense rainfall area. Weaker downbursts and outflow winds can often be seen on longer lines of thunderstorms in radar data.

Satellite data can also help isolate downburst events or situations in which thunderstorms exhibit broad outflow wind patterns. As with radar, when the most intense part of the thunderstorm is at the leading edge of the storm, that signifies that the strongest upward motion is ahead of the main rain area. On satellite images the leading edge gradient (LEG) signature is also shaped like a backwards letter "C".

And some can't be seen at all….
Still, due to size and lack of easy-to-see signatures, radar and satellite may not detect smaller, shorter-lived downbursts until after the event has occurred.

In the western United States and in other locations at certain times, downburst thunderstorms may be very hard to detect. These so-called "dry-microbursts" have little precipitation (and what they do have often evaporates before reaching the ground, further fueling the downburst through evaporative cooling). Such downbursts (and the lightning caused by the "dry thunderstorm") are a significant factor in the creation and spread of forest fires in dry climates, (see also chapter 13)

Aviation + downbursts = recipe for disaster…
In the aviation arena, larger thunderstorm systems, squall lines and MCSs are easy to spot and act against. It is not uncommon for an airport to close or carefully monitor aircraft on landing and takeoff when thunderstorms are nearby (see below). However, some airports in the US and in other countries operate more or less normally even with thunderstorms at the airport. In other situations, it is often difficult to know when a microburst or even a downburst threatens.

Another part of the downburst equation centers on aircraft operations (including pilot training). Aircraft performance hinges on landing and taking off into the wind. This provides additional lift. When there is a sudden wind shift, lift can be compromised. With little altitude or time available to adjust, the pilot may not be able to keep his/her plane airborne.

The table below highlights some significant commercial aviation downburst-related events from 1970-2005. Notice that all of these (and most other) downburst events occur in summer – prime thunderstorm season.

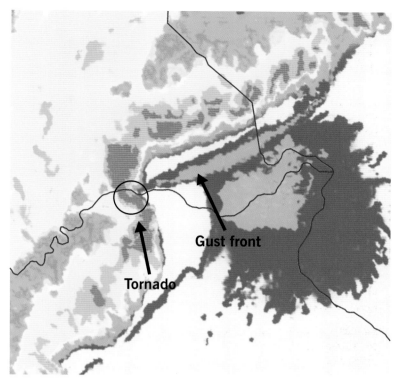

Tornado

Gust front

In this radar image, thunderstorm outflow winds (gust front) have moved far ahead of the actual thunderstorm rain area. That rain area shows the heaviest rain (in red) located at the leading edge of the advancing line of thunderstorms. To the southwest, a bow echo (backwards letter C) is bowing toward the east. The bulge is caused by high winds blowing out of the thunderstorm. Where the two highly active lines of thunderstorms meet, there is a tornado signature.

The Toronto incident occurred exactly 20 years after the similar incident at Dallas-Fort Worth Airport (Texas). No one was killed in the Toronto incident; 137 people died in the Dallas crash. The thunderstorm at Toronto was fairly large but they may have had radar signatures that suggested wind shear potential. The thunderstorm that brought down the Delta plane in Dallas was a small thunderstorm that was deemed insignificant based on larger storms nearby.

This series of commercial aviation downburst/microburst events set the stage for the US to embark on an aggressive program to combat these deadly winds. Terminal Doppler Radars (designed to focus on the airspace in the immediate vicinity of the airport), a wind shear detection system (a series of sensors to monitor small scale wind shifts in the runway areas), pilot observations and improved weather forecasting are all important. Currently 44 US airports have TDWRs and more than 100 have wind shear detection systems. In the US, the Federal Aviation Administration has mandated that commercial aircraft have onboard wind shear/downburst detection systems.

Internationally, the focus on wind shear is tempered. However, Canada, Denmark, Italy, Hong Kong, China and Sweden are among other countries that have airport wind shear detection systems in place.

COMMERCIAL AVIATION DOWNBURST-RELATED EVENTS 1970-2005

The wreckage of Air France flight 358 lies in a gully off the end of the runway in Toronto, August 3, 2005. Canadian investigators recovered the black black box flight recorders from Toronto's "miracle" plane that crashed and burned in heavy thunderstorms. Investigators rely on the information in these flight recorders to help them reconstruct the weather events leading up to an accident. All 309 passengers and crew survived after the Airbus A340 overshot the runway and burst into flames as it landed in extreme weather conditions.

June 24 1975 Eastern Air Lines Flight 66 Boeing 727 encountered wind shear (another commonly used name for a downburst) on final approach and struck approach lights at New York Kennedy.

July 9 1982 Pan American flight crashed on takeoff from New Orleans Airport. More than were 150 killed.

July 2 1985 Windshear was blamed for the crash of a Delta Airlines during landing at Dallas Fort Worth International Airport, killing 137 people.

July 2 1994 US Air jet crashes at Charlotte, North Carolina. The aircraft encountered heavy rain and wind shear during approach several miles from the airport. The crew attempted to abort landing and go around for another landing attempt, but the aircraft could not overcome the wind shear. All five crew members survived, but 37 of the 52 passengers were killed.

August 2 2005 An Air France aircraft landing at Toronto Airport was effected when a localized microburst transformed a light headwind into a 40 mph (64 km/h) tailwind. Apparently compensating for the windshift, the pilot may have increased airspeed; this, in turn resulted in the plane landing further down the runway, running off, catching fire, and not being able to stop before rolling off the runway. Miraculously, no one was killed and most injuries were minor.

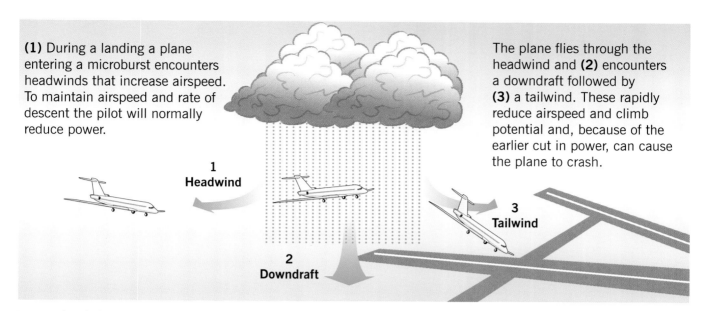

(1) During a landing a plane entering a microburst encounters headwinds that increase airspeed. To maintain airspeed and rate of descent the pilot will normally reduce power.

The plane flies through the headwind and **(2)** encounters a downdraft followed by **(3)** a tailwind. These rapidly reduce airspeed and climb potential and, because of the earlier cut in power, can cause the plane to crash.

1
Headwind

2
Downdraft

3
Tailwind

We've already looked at some derecho and aviation-focused downburst events. Now, let's take a look at some other significant downburst/microburst events to get a better idea of what these events look like and the other types of weather that accompany them.

Squall line high wind event

During the late afternoon and early evening of February 13, 2000, a strong squall line moved across the southern United States. The weather system included widespread severe thunderstorm activity. There were at least 72 reports of wind damage or high winds.

During landing and takeoff, changes in airspeed and lift due to microbursts can lead to plane crashes.

When tornadoes strike, the resulting damage pattern is chaotic; in downburst situations, nearly all the damage is in straight lines.

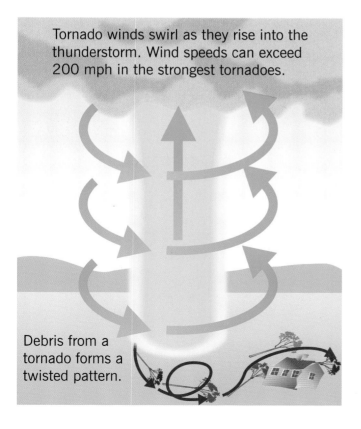

Tornado winds swirl as they rise into the thunderstorm. Wind speeds can exceed 200 mph in the strongest tornadoes.

Debris from a tornado forms a twisted pattern.

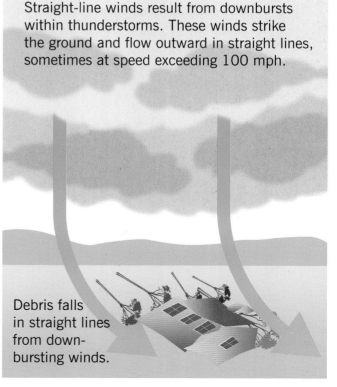

Straight-line winds result from downbursts within thunderstorms. These winds strike the ground and flow outward in straight lines, sometimes at speed exceeding 100 mph.

Debris falls in straight lines from down-bursting winds.

Straight-line wind damage has brought down one tree (atop a car) while leaving homes and other trees untouched.

Several wind damage events occurred in a populated area near Huntsville, in northern Alabama. As is customary in the US NWS staff conducted a damage survey to determine if the event was tornadic or involved only straight-line (downburst) winds.

Hurricane high wind event

A downburst event (with a well-defined bow echo) struck the Raleigh-Durham, North Carolina, (RDU) area on September 17, 2004. The storm event broke the record for the highest recorded wind at the RDU Airport. The trailing end of the thunderstorm line, however, moved much less during the period, setting the stage for "train-echo" rainfall.

Metal roofing is more easily damaged or removed by high winds than conventional composition roof shingles.

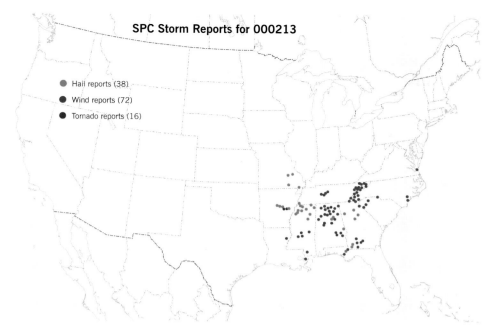

SPC Storm Reports for 000213

- Hail reports (38)
- Wind reports (72)
- Tornado reports (16)

Storm reports for February 13, 2000 (shown as 000213 on map) across the United States. NOAA's Storm Prediction Center (Norman, Oklahoma) posts thunderstorm-related severe weather reports in near-real time. All of the "action" on this day was in the southeast part of the nation.

Personal safety comes first

When strong straight-line thunderstorm winds strike, there's little a homeowner can do to protect their property. The winds may knock down trees, rip off parts of roofs, and send debris flying into the side of buildings, onto cars and into signs.

For your protection, the best advice is the same as that given for tornadoes; you want to be inside a sturdy structure and as far away from the winds as possible. Cars, mobile homes and similar structures are not safe places to be.

If you are boating, fishing, playing golf or otherwise enjoying the outdoors, reading the sky is your best defense against these storms. If a storm approaches, and you are boating, get to shore and then to a safe structure. For any other outdoor setting, get indoors.

We'll look at specific safety aspects for other thunderstorm related hazards when we examine these phenomena in later chapters.

Climatic implications

While damage from severe thunderstorms and reports of severe weather continue to increase, there is no indication that storms are any more severe than they have been in the past. Much of the increase is due to a greater population at risk and better reporting of the phenomena. If anything can be said, it would be that better education and understanding of all types of severe weather phenomena have lead to a decrease in casualties from almost all types of thunderstorm hazards.

> *"...We can't stop tornadoes,*
> *but with early warnings, we can save lives..."*
> *Al Moller, a Warning Co-ordination Meteorologist*
> *for the National Weather Service*

Tornado in southwestern Oklahoma during May 1999.

In drier climates, tornadoes often pick up lots of dust. This tornado on August 20, 2006, between Bennett and Watkin, Colorado, is no exception. The rope-like character of the tornado indicates that it is in its weakening stage.

In Oklahoma City, May 3, 1999, a massive tornado outbreak – with at least 70 twisters – devastated parts of Oklahoma, Texas, and Kansas during the afternoon and evening. The twisters claimed more than 45 lives and injured almost 800 people. Preliminary damage estimates exceeded one billion dollars. It was the most active tornado day in recorded Oklahoma history.

Tornadoes are officially defined in the American Meteorological Society's Glossary of Meteorology as, "a violently rotating column of air, in contact with the ground, either pendant from a cumuliform cloud or underneath a cumuliform cloud, and often (but not always) visible as a funnel cloud." While most tornadoes develop from cumulonimbus (also known as thunderstorm clouds), some can come from cumulus (flat-based, puffy clouds) or towering cumulus clouds.

Setting the stage

Tornadoes are perhaps the most visible, and typically the most frequently reported, extreme weather event. They know no time of day, no season and no geographical location – twisters don't even seem to mind striking cities or mountain areas. Because they are so small on a meteorological scale, people can more easily see and photograph them. For larger-scale events like snowstorms, it is easy for our eyes to see the impacts, but not see the overall storm itself. Similarly, the damage that tornadoes cause, considering their size and lifetime, makes their destructive power incredible. And sometimes their unusual damage patterns can raise interesting questions about how tornadoes actually work.

Just about every tornado brings stories about survival, quick thinking, and harrowing experiences – human interest stories that live far beyond the damage tornadoes cause. Unsurprisingly, interest in and reporting of tornadoes continues to increase and, while some may suggest that the growing numbers of reported tornadoes from around the world may be linked to changes in weather and climate patterns, it is possible to look to other causes. The higher numbers are possibly tied to greater detection (thanks to media reporting and technological advances), greater availability of photographic devices (including cell phones and digital cameras), increased public awareness and a growing population that is spreading into tornado-prone areas that before had few people.

Looking at tornadoes

Tornadoes and other family members, such as waterspouts, funnel clouds and dust devils often have a classic funnel shape that makes them very easy to recognize. Yet, every tornado has characteristics that make it unique. In fact, meteorologists have an uncanny ability to catalog and recall tornado images. Looking at many images of tornadoes, variations in size, shape, color, background lighting and airborne debris patterns enables observers to see different aspects of storms. Some tornadoes even have multiple tornadoes spinning around a main tornado (known as "multi-vortex tornadoes").

Tornadoes look the same around the world. This twister forming from very low clouds near McLeans Ridges, New South Wales, Australia, indicates that the air mass is extremely moist.

With a funnel cloud this well formed and not visibly reaching the ground (left), the tornado circulation (and hence the actual tornado) has likely reached the ground. To the right, a well defined dust cloud marks the presence of a tornado even if the condensation funnel remains aloft.

TORNADO TOURISTS

If you want to see tornadoes, but aren't quite confident enough to chase them yourself, you can take a tornado chase vacation. While driving with an expert meteorologist (much like some scientists do in their research jobs), you can see firsthand how stormy weather evolves. Even if you don't see a tornado, you will likely experience other severe thunderstorm weather (high winds, hail and intense lightning) and return highly knowledgeable in severe weather science. Tours are currently available in both the US and Australia.

A large tornado, rated at F4, passes in front of Tempest Tours near Seward, Nebraska.

Tornadoes are often classified by their shape. The cone-shaped tornado and the elephant trunk tornado are usually seen earlier in the tornado's lifetime and are often the strongest. The rope-shaped tornado typically signifies a dying tornado. A tornado may exhibit many shapes throughout its lifetime. In fact, if you think about it, there's really nothing different here than what we do everyday recognizing people – we look to their attributes.

Tornado shapes appear elsewhere in our lives. Check the water draining from a bathtub or flush an older style toilet and you'll see similar patterns. For all of these reasons, and more, tornadoes are possibly the ultimate extreme weather fascination. Not surprisingly, several companies now offer tornado tours.

Tornado geography

Tornadoes can occur throughout the world. However, they are most likely to form in middle latitudes where different air masses clash. An "air mass" is defined as a large volume of air with similar temperature and moisture characteristics. The battleground typically occurs along boundaries known as "frontal zones". In the United States, another type of boundary, known as a "dry line" – separating warm, dry air from warm, humid air – also plays a significant role in "tornadogenesis".

Since the United States has the most favorable geography for having these three disparate air masses clash, it is not surprising that it's the Central Plains region that has been dubbed "Tornado Alley". The shape of the affected region can be likened to a bowling alley, a long, thin area extending from the southern Plains into the Midwest. There is a second tornado alley known as the "Southern Tornado Alley" that extends along the Gulf Coast from east Texas into the Carolinas. It is here that many late fall to early spring tornadoes occur because the frontal boundary battle zone has seasonally shifted so far to the south.

Although thunderstorm and tornado frequency is linked to solar heating, the relationship cannot be extended to the seasons – most tornadoes do not occur in summer (the time of maximum annual surface temperatures) – but rather the change in seasons. That's

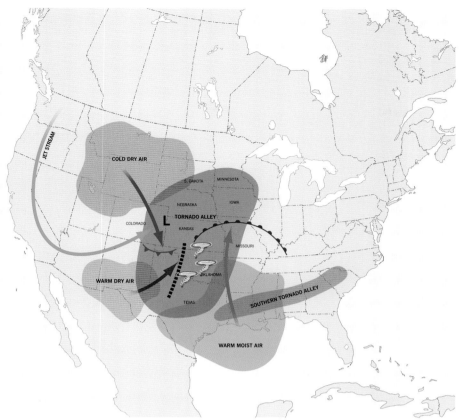

Tornado Alley's name comes naturally. The elongated region in the Central Plains area encompasses a high percentage of all US twisters. The Alley is where widely differing air masses often clash. A second "alley" exists along the Gulf Coast states.

because summertime often comes with warmer air at high altitudes (less instability) and the lack of air mass clashes. In summer, colder air from northern latitudes doesn't advance towards the Equator as often or as far as in colder months. Jet stream winds are also not as strong in the summer. Thus, spring (April to June) and fall (September to November) in the Northern Hemisphere are the two peak tornado seasons, with spring dominating the statistics. Fall tornadoes can also be linked to land-falling hurricanes and tropical storms (see Chapter 5, Hurricanes).

Tornado incidence migrates with the sun. In North America, during late fall to early spring, the Gulf Coast region has most favorable conditions for tornado formation. As the sun's position shifts towards the North Pole, the region of greatest expected incidence follows it. Not surprisingly, the highest tornado incidence in summer is along and either side of the US-Canadian border. Being "down under", Australia's seasons are exactly opposite those in North America. Hence Australia's peak tornado season is November to January.

Most US tornadoes move from the southwest to northeast, tied to a jet stream (a jet stream is a relatively narrow zone of high-speed winds) steering winds. However, some weather situations, especially over the northeast part of the country, favor tracks moving from the northwest.

Where do they occur?

Because of its high number of reported events, the US has become known as the "tornado capital

Tornado Frequency: UK versus US

	Area (sq mi)	Approx # of tornadoes per year	Tornado density per 10,000 sq mi
United Kingdom	94,526	35	5.42
United States	3,537,441	1,300	3.67

of the world". However, despite its smaller size (only 2.7 percent of the area of the US) and lack of a "classical" tornado air mass clash setting, the United Kingdom actually has a higher incidence of twisters per square mile. Considering recent statistics from the UK (some 60 reported tornadoes per year between 2004 and 2006), their lead in tornado density continues to grow. The US, however, has many more stronger, more deadly tornadoes.

Similar statistics are now being developed for other parts of the world as scientists realize that tornadoes are not unique to the US. Much of this effort has begun during the past 30-40 years. For example, on average, Canada reports about 80 twisters a year; the UK logs about 35; Australia and Japan about 20 each; and Bangladesh records about three.

A selection of worldwide tornado data

Country	Location	Date	Description
Argentina	San Justo, Buenos Aires Province	January 10, 1973	Large, strong twister killed 54 and injured 350.
Australia	Bucca, Queensland (300 km north of Brisbane)	November 29, 1992	The strongest tornado ever officially recorded in Australia.
Australia	Brisbane, Queensland (northern and western suburbs)	November 4, 1973	Australia's most damaging tornado. On its 32-mile (51-km) long track, it destroyed or badly damaged 1,400 buildings.
Australia	Kin Kin, Queensland (approx.130 km north of Brisbane)	August 14, 1971	Possibly Australia's deadliest tornado with three fatalities.
Bangladesh	Daultipur-Salturia region	April 26, 1989	With more than 1,300 fatalities, the deadliest tornado in recorded world history.
Czechoslovakia	Prague, Czechia	April 8, 1255	This twister struck the Prague Castle.
Germany	Black Forest Region	July 10, 1968	Three people reported dead.
India	Calcutta (east side)	April 8, 1838	First documented tornado in the country. There were 215 fatalities.
India	New Delhi	March 17, 1978	28 fatalities reported.
Italy	Parma	July 4, 1965	25 fatalities and 160 injuries reported.
Japan	Saroma, Hokkaido	November 6, 2006	Tornado killed nine and injured 26.
New Zealand	Frankton-Hamilton	August 25, 1948	F2 tornado killed three and injured 80. Twister caused significant property damage.
Philippines	Southern part of country	June 14, 1990	30 people reported killed.
Russia	Ivanova-Yaroslav	June 9, 1984	Tornado outbreak situation with more than 400 fatalities reported.
South Africa	Heidelberg (28 miles (45 km) south of Johannesburg)	October 21, 1999	An F3 (part of tornado outbreak) struck a rural area. Still 40 injured (10 seriously) and 300 homes destroyed.
United Kingdom	London	December 7, 2006	Six injuries reported.

Tornadoes have even been reported in Egypt, Scandinavia and Japan. Antarctica, with its chilly climate, is the only continent that has no documented tornadoes. These statistics do not, however, address waterspout frequency.

Some countries, such as Germany, have reconstructed a rudimentary tornado database dating back to the Ninth Century; other countries have only just now started to document these events. In the process, various groups (some based at universities or government agencies, others not) have started to compile and post tornado data online.

According to TorDACH (a European-based group studying German, Austrian and Swiss tornadoes), there have been 863 recorded tornado events in Germany since the year 855. Some of these even involved damage to castles! A few events in Switzerland involved waterspouts coming ashore from large lakes.

Jonathan Finch – a weather forecaster at the National Weather Service (NWS) office in Dodge City, Kansas (in collaboration with a colleague in Bangladesh) – has been conducting research into tornado events that have occurred in the eastern part of the Indian sub-continent. Based on a review of newspaper clippings, literature and Internet searches, the team has documented about 85 tornadoes from 1955-99. Most of the tornadoes occurred in a relatively small area of central, south central and southeast Bangladesh, or an area about the size of the state of New Hampshire (slightly less than 10,000 square miles). The average tornado density for the whole country is about 0.54 per 10,000 square miles (or about one-tenth that of the UK).

According to Finch, tornadoes are much less common in northwest India and Pakistan. This differentiation is linked to geographical factors such as the Himalayan Plateau, formation of a dry line across northern India and the availability of warm, moist air from the Bay of Bengal (conditions similar to those found in the central US although more localized). While many central US tornadoes race from the southwest toward the northeast, tornadoes in this region tend to move from the northwest to southeast.

The US storm spotter network, its extensive government-university linked research programs, and high interest level in tornadoes ensures that a greater percentage of storm events are documented. Even with the low population density in portions of the Central Plains, hundreds of storm chasers fill in the gap. That cannot be said to the same degree in many other countries. Unsurprisingly, the number of tornadoes in the US has grown steadily over the years, although there have been significant year-to-year fluctuations.

Humble beginnings

Although clashes among large-scale air masses favor most tornado developments, other factors have to be considered. These include the location and orientation of jet streams, atmospheric stability, wind shear patterns (how winds change with altitude), the amount and type of cloud cover, the strength of nearby frontal zones (including the dry line), the presence of thunderstorm outflow boundaries and even topography. In short, the recipe for tornado formation is quite complex.

Before most tornadoes can develop, a parent thunderstorm – often the super-cell variety – must form. Super-cells are large, long-lived thunderstorms that often occur miles away and isolated from the influence other thunderstorms. This isolation let's them "feed" on favorable atmospheric conditions without having to compete with neighbors. (See also, Chapter 6, Thunderstorms.)

The starting point for any tornado discussion, therefore, involves the formation of cumulus clouds. Cumulus are the puffy topped, flat-bottomed clouds that often develop

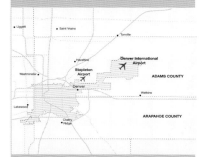

Classic supercell thunderstorm over rural Kansas. The striated shelf cloud marks the boundary between colder air from within the storm and warm humid air outside. In the distance at right, heavy rain is falling from the storm.

during the morning or early afternoon. The process starts when sunlight heats the ground, which in turn heats the air. Warmer air, being less dense than colder, wants to rise. Cumulus clouds indicate the presence of the resulting rising air currents. The flat base marks the altitude at which condensation occurs and the rising turrets pinpoint rising air currents known as updrafts.

If the atmosphere is stable, then the cumulus will be more flat than tall. The more unstable the atmosphere, the more the clouds tower. If the upward motions are strong enough, the cloud can reach altitudes of 5-10 miles (8-16 km) or more, above the ground, and at which the top of the cloud becomes frozen. The resulting fuzzy cloud top character and a flattened cloud top that spreads across the sky – also known as an anvil because it looks like a blacksmith's anvil – make it easy to tell whether the cloud has become a cumulonimbus or thunderhead.

Thunderstorms reach peak intensity during the afternoon – just after maximum solar heating – and can continue well into evening hours. Tornado occurrence closely follows this timetable. Fortunately, most thunderstorms are not severe weather or tornado producers.

Creating a tornadic thunderstorm

Tornadoes are most likely to occur when the upper level jet stream, found at 5-10 miles (8-16 km) above the ground, is linked to a U-shaped circulation pattern (see tornado alley figure on page 105). This particular wind flow pattern is called an upper level trough because it resembles a watering trough from which horses drink. It favors bringing warmer air poleward in advance of the low pressure system and colder air towards the Equator behind it.

However, in the US, a low level jet, similar to the upper level variety, only at around one mile (1.6 km) above the ground, often develops from Texas northward into the Central Plains. This low level jet can also play a role in aiding severe storm formation. When the two jet streams criss-cross in a certain way, they create a vertical wind profile that is very favorable for tornado development.

The low level jet often creates a horizontal roll in the atmosphere. This resembles a tornado lying on its side. As localized areas of upward and downward motion develop (due to instability and other factors), parts of this rolling circulation can be tipped into a more vertical position. In this orientation, the circulation becomes more tornado-like.

Tipping the tornado

Atmospheric instability provides the tipping mechanism noted above. Typically, air near the ground is warm compared to the air at high altitudes. If the temperature difference is large enough and the atmosphere is "capped" with a layer of warm air (usually around two miles (3.2 km) above the ground), upward and downward air motion is inhibited. The "cap" is much like putting a lid on a pot of boiling water – if the lid is removed the condensation cloud will rise quickly. In the atmosphere, when the cap is broken or the upward motion overcomes it, rising air currents can be just as dramatic as will happen when lifting the lid of the pot. The sky can then transform within an hour or two from being cloud-free to the formation of a super-cell thunderstorm with a tornado on the ground.

Sometimes, cloud coverage variations help to focus the upward motions. If a cloudy region lies adjacent to a cloud-free region, everything else being equal, solar heating and upward motions will be greater in the cloud-free region. This helps to focus the region of upward motions (also making them stronger) rather than having them exist over a larger area and be less intense.

Outflow boundaries

Sometimes, thunderstorms produce localized cold "air masses" that can move 10-50 miles (16-80 km) or more away from the parent storm. These are caused by rain-cooled air and its associated strong, sinking air currents. As these sinking air currents meet

Idealized view of a "classic" supercell, looking west

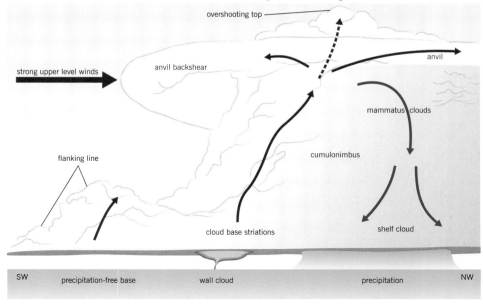

Shortly after storm chasers began observing and photographing severe thunderstorms, they realized that many of the storms had similar attributes. Following the police sketch artist approach, they arrived at a composite picture (shown here). The lowered and rain-free cloud base (wall cloud) and overshooting top both link to the updraft (core of rising air inside the thunderstorm; the precipitation area to the northeast marks the downdraft (or sometimes, downburst area). The flanking line shows where moisture and winds are feeding into the storm. And the anvil shows exhaust at the top of the storm (where strong winds are present). The key to the supercell is that the updraft and downdraft remain separate. This prevents the rain from falling back into the updraft and destroying it. As a result, a supercell thunderstorm lasts much longer than an average thunderstorm. If you want to see how the atmosphere can separate an updraft and a downdraft, just look to a large spouting lake or display fountain on a windy day.

the Earth, they spread out from their impact region on the ground with the leading edge of the outflow acting as a miniature cold front (boundary). For example, pour water quickly onto a sidewalk and you can see what an outflow wind pattern looks like. As the boundary passes a point on the ground, the temperature can drop dramatically and the wind can shift and gust strongly. Some outflow boundaries can bring winds as high as 70 mph (112 km/h) or more, although most produce much weaker winds.

These outflows are often short-lived and may disappear as solar heating, or other factors, come into play. Sometimes, the boundaries act to focus super-cell thunderstorm formation because they force low level winds to converge (converging winds near the ground act to force air upwards). Sometimes outflow boundaries act to destroy a storm because they bring in cold air at low levels, making the atmosphere more stable.

Some weather situations favor the development of long thunderstorm lines – 100 miles (161 km) or more in length – known as squall lines. These lines got their names because they tended to produce strong straight-line winds. Other settings allow for the development of individual, long-lived and large thunderstorms known as super-cells that can last for hours rather than the usual 20-30 minute thunderstorm lifetime. Super-cells are the most likely type of tornado-producing thunderstorm and typically produce the longest-lived and strongest tornadoes.

Under and over the thunderstorm

As a super-cell thunderstorm develops, things are happening that foretell if it will become a tornado producer.

Because condensation of atmospheric water vapor is linked to several factors (including changes in air pressure), the condensation level beneath the thunderstorm will lower as a low-pressure center starts to form inside the storm. Lowered cloud bases in just one part of the thunderstorm mark this low-pressure center. Once formed, the low-pressure center allows winds to start to spin gently into it. The resulting lowered cloud base is known as a "wall cloud". If there is rotation in the wall cloud, tornado evolution is more likely.

Two tornadoes strike simultaneously on the farmland and spin across Turner County in South Dakota on June 24, 2003.

Inside this wall cloud, a smaller region of even lower pressure may develop. When this happens, the cloud base locally may continue to lower and a tornado funnel may appear. The funnel at this stage is nothing more than a cloud (condensed water vapor). If the funnel cloud continues to reach toward the ground, it may become a tornado.

Because the radius of the spin is now even less, rotational speed is greater. Conservation of angular momentum (in which the velocity of the spin and the radius, or size of the spin, are inversely related) is responsible for the increase in rotational speed. Just think about what happens when a spinning ice skater brings her arms tightly in towards her body. (Note, however, that tornadoes can also occur even without the classic formation of a wall cloud.)

A region of lower pressure forms in the thunderstorm because there is a net loss of air from within the atmospheric column above the thunderstorm's base. This often occurs when more air leaves the top of a thunderstorm than comes in from below to replace it. This can be likened to a fireplace chimney. If there is plenty of exhaust at the top, the heated fireplace will draw air from other parts of the house.

In tornadic thunderstorms, there is more outflow on top of the thunderstorm than inflow beneath. This can create a very low atmospheric pressure inside the tornado. Although hard to measure due to instrument survivability and placement, storm chasers obtained a reliable tornado measurement of 850 millibars (mb) or 25.17 inches of mercury during the F4 Manchester, South Dakota tornado on June 24, 2003. Average sea level pressure is slightly above 1000 mb (29.92 inches of mercury). For comparison, Hurricane Katrina's lowest minimum central pressure was 902 mb (26.64 inches of mercury) on August 28, 2005, while a Category 5 hurricane.

The difference between the central pressure of a storm and its surroundings defines the wind flow back into the storm. The lower the pressure, the more the atmosphere wants to fill the low pressure area back with air and the higher the wind speeds. You can

WHICH WAY DO THEY SPIN?

Since the low pressure center inside the thunderstorm spins in the same sense as other low pressure systems for its location on Earth (counterclockwise in the Northern Hemisphere and clockwise in the Southern Hemisphere), the direction of tornado rotation follows. Still, a relatively small number of tornadoes spin the "wrong" way. This can happen when tornadoes occur in pairs, one spinning in each direction. If an updraft and a downdraft within the thunderstorm tilt the tornado's roll, it is possible to create two vortices spinning in opposite directions.

TORNADIC THUNDERSTORMS

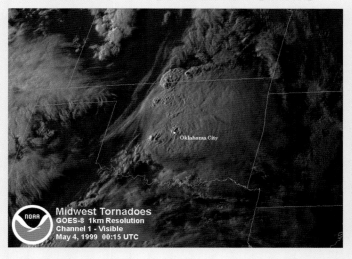

Midwest Tornadoes
GOES-8 1km Resolution
Channel 1 - Visible
May 4, 1999 00:15 UTC

At the time of this visible geostationary satellite image, several large tornadoes were on the ground. Geostationary satellites orbit the earth in such a way that they always look down on the same point on the Equator. Visible imagery matches what you would see if you were onboard the satellite.

Here, numerous severe thunderstorms cover the state of Oklahoma. Their anvils have merged to create what looks like one huge storm cloud. Look carefully and you can see the tops of strongly rising cloud towers that are overshooting the anvil canopy.

Waterspouts can be very weak or they assume the character of a real tornado over water. This waterspout over the Caribbean would be classified as tornadic. Notice the water spray being kicked up the spout.

simulate this effect by spinning up a vortex in a tall glass using a spoon or other thin stirrer. Stop spinning and watch how fast the depression in the water surface returns to a level or average height.

Other ways tornadoes form

Some tornadoes form when strong straight-line thunderstorm outflow winds spread out and away from a thunderstorm. As they do, the winds may start to curl and even spin small to moderately sized tornadoes. On August 28, 1990, a series of microbursts (very intense sinking winds that spread out at high speed near the ground) helped to spin up an F5, the strongest tornado ever reported in the Chicago, Illinois, area. The twister struck the Plainfield-Crest Hill corridor around 35 miles (56 km) to the southwest of the "Windy City". The tornado, moving from northwest to southeast, was on the ground for more than 16 miles (25 km). It killed 29 people, injured 350, and destroyed a high school and scores of homes and apartments.

Sometimes, very cold air at high altitude moves over relatively warm water. In this highly unstable situation, waterspouts (tornadoes over water) can develop. The Great Lakes region, coastal Southern and Central California, south Florida, the Hawaiian Islands, parts of the Mediterranean region and Australia are among the places where waterspouts can maintain their intensity even as they come ashore as tornadoes. Hawaii, for example, experiences about 20 waterspouts/funnels annually, but only about one tornado, according to University of Hawaii Professor Thomas Schroeder. For comparison, Japan averages about five waterspouts per year. Waterspouts can sometimes even develop from small cumulus clouds and are generally not as strong as tornadoes on land.

It's more than just a cloud

Even if you can't see the tornado's circulation extending from the thunderstorm cloud base to the ground, there can still be a tornado occurring. This is most common in drier locales (e.g. the western High Plains from western Oklahoma to the western Dakotas and then westward to the Rocky Mountains). Here, cloud bases are higher, dew points (a measure of the absolute moisture content of the air) are lower and the central pressure inside the tornado is often not low enough to support a large condensation funnel. Sometimes, however, you can see the connection between a funnel cloud (tornadic circulation aloft) and the tornado's dust swirl on the ground.

It may look like a tornado and spin like one, but lacking any clouds, this is just a "dust devil". Dust devils form because the ground becomes very hot during the day and small scale rising air currents develop.

More typically, however, the tornado is visible as a cloud (condensation funnel) or as a cloud of dust and/or debris. A waterspout can be composed of lake, river, or ocean spray.

Moving along

Once the tornado forms, it will follow the movement of the parent thunderstorm. However, it will also rotate around the parent wall cloud. If there are some smaller tornadoes rotating around the main tornado, they too will move along and undergo additional rotational and translational movements.

One way to visualize this is to consider the rotation of a reflector on the wheel of a bicycle as the cyclist moves across your field of view.

If the reflector were at the center of the wheel (spoke), the path of the reflector would trace out a straight line. If the reflector were in any other position, it would trace out a cycloidal pattern that includes the combined effect of both straight line and rotational motions. Now, imagine this motion placed into a horizontal frame of reference as a tornado moves across the landscape, but the tornado isn't exactly in the center of its wall cloud. This motion could result in the overall tornado path encompassing an array of damaged homes; but careful examination of the actual damage pattern would show many homes with minimal or perhaps no damage. Meteorologists often find cycloidal patterning when tornadoes move across wheat and corn fields. This is because the tornado flattens plants and scours out soil without depositing structural debris to camouflage the pattern.

If one considers the effect of multiple vortices moving within an overall tornado path, it becomes easy to see that highly unusual tornado damage patterns can result.

The tornado and its parent thunderstorm cloud act in consort. However, greater friction at the ground slows the movement of the bottom part of a tornado. With the parent cloud moving faster, it is only a matter of time before the tornado starts to stretch out across the sky as a classical "rope-like" twister. When stretched too much, the tornado dissipates.

If the wall cloud circulation remains intense, a new tornado can develop vertically from the wall cloud, even as the old tornado is dying. For some very long-tracked and intense thunderstorms, there can be several life cycle tornadoes (also known as a tornado family) comprising what appears to be a single long-tracked tornado. There may be small, hard-to-distinguish gaps in the path of this series of tornadoes.

The Tri-State Tornado

The longest tornado ever documented in the United States occurred on March 18, 1925. Known as the "Tri-State Tornado", it raced at some 60 mph (96.5 km/h) or more across Missouri, Illinois and Indiana; the record-breaking twister killed almost 700 people and literally wiped some communities from the map. The tornado was an average one mile (1.6 km) wide (a record) during its three-hour lifetime. Many reported that the tornado looked like a giant cloud on the ground.

Tornadoes can be spectacular but they are also extremely destructive. This double tornado demolished the Midway Trailer Park in Dunlap, Indiana, on April 11, 1965.

Residents comb through the wreckage of their homes in the town of Griffin, Indiana, in the wake of the Tri-State Tornado in March 1925. The tornado began in Missouri on March 18 and tore through Illinois and Indiana, killing 689 people.

The Six Greatest US Tornado Outbreaks (based on significant tornadoes)

Dates	U.S.States	Total	Significant	Violent (F2 and F3)	Killer (f4 and F5)	Total Deaths
April 3–4 1974	IL IN KY OH AL TN GA NC Ontario	148	95	30	48	315
Nov 21–23 1992	MS TN GA KY IN SC NC	95	42	5	9	26
April 11–12 1965	IA IL IN MI OH	48	38	19	21	256
Sept 19–23 1967	TX	111	15	0	2	5
May 18–19 1995	TN AL KY IL MO	80	17	2	2	4
May 3 1999	OK KS	76	18	6	2+	45

A "once in several lifetimes" event, the Superoutbreak Tornadoes of April 3-4, 1974 remains the largest such event in recorded US history.

However, this tornado occurred before scientists more fully understood the tornado family life cycle and before technology (including aerial photography) was available. Thus, it is possible that this record-breaking event and others (see table above) were actually a series of long-tracked, large tornadoes.

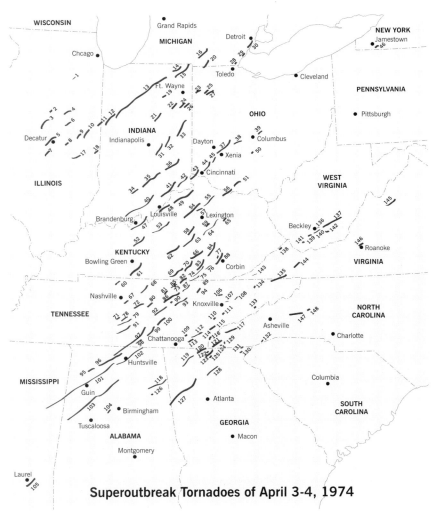

Superoutbreak Tornadoes of April 3-4, 1974

Outbreaks

Tornado outbreaks (defined in various ways but always with at least "many" significant or stronger tornadoes) are more common than isolated long-track tornadoes. Often, an outbreak contains several long-track twisters. These two types of tornado events account for the majority of tornado-related deaths, injuries, and major property damage.

Recently, Greg Forbes, a severe weather expert at The Weather Channel™, developed an index for comparing the impact of tornado outbreaks. His index uses 11 variables (including the number of tornadoes, deaths, and injuries; number of long-tracked tornadoes; number of significant (F2 or greater) tornadoes and number of violent (F4 or greater) tornadoes; and an inflation-adjusted value for actual property damage.

Major outbreaks happen, on average, every few years in the United States. Between 1880 and 1995 there were 40 outbreaks that contained more than 20 significant tornadoes. Seventeen of these occurred from 1916-49. Thirty outbreaks, during the same period each resulted in more than 50 deaths. The April 3-4, 1974 outbreak remains the largest ever reported with 148 documented twisters in about a 24-hour period.

Forty-seven outbreaks contained four or more violent tornadoes; three had 10 or more violent tornadoes associated with them. Although rare, hurricane-related outbreaks can also occur.

The 1967 Texas outbreak (see entry in the table below) was associated with Hurricane Beulah; other hurricane outbreaks with more than 100 twisters have included Frances and Ivan in September 2004. Ivan's tornadoes were not just concentrated around the time of landfall but actually spanned a three-day period as the storm moved northeastward from Florida to New England.

Although not shown here, each state has its own records for most significant tornado outbreaks and most significant tornadoes. To maintain the statistical focus, recognize that there are tornado incidence statistics for many metropolitan areas (large and small), too.

Other deadly statistics

Tornadoes in the United States are probably best known for how deadly and long-lived they are. Though many of the tornadoes on these lists lie in "Tornado Alley", some do not. The April 5-6, 1936 killer pair in Mississippi and Georgia occurred in the "Southern Tornado Alley". Many, like this pair, occur after dark or early in the morning (uncommon times); many are shrouded in heavy tropical-like downpours and are hard to see; and large spans of trees often block the view to the horizon even when heavy rain is not present.

While people in "Tornado Alley" can take shelter in mobile home park shelters, basements, or inside substantial structures, such safety areas are lacking in many parts of the rural South due to high water table levels and economic factors. A disproportionate number of tornado fatalities occur in mobile homes and cars.

While outbreaks are impressive, so too are individual intense and/or long-tracked twisters. Often these storms are part of an outbreak, but they don't have to be. Notice that in the deadliest tornado listing, all but the undefined Natchez tornado (1840) are either F4 or F5 intensity. Also, three of the top four longest, tracked tornadoes are also among the deadliest.

10 Deadliest US Tornadoes

Rank	State(s)	Date	Time	Dead	Injured	F-Scale	Town(s)
1	MO-IL-IN	March 18, 1925	1:01 PM	695	2027	F5	Murphysboro, Gorham, DeSoto
2	LA-MS	May 7, 1840	1:45 PM	317	109	F?	Natchez
3	MO-IL	May 27, 1896	6:30 PM	255	1000	F4	St. Louis, East St Louis
4	MS	April 5, 1936	8:55 PM	216	700	F5	Tupelo
5	GA	April 6, 1936	8:27 AM	203	1600	F4	Gainesville
6	TX-OK-KS	April 9, 1947	6:05 PM	181	970	F5	Glazier, Higgins, Woodward
7	LA-MS	April 24, 1908	11:45 AM	143	770	F4	Amite, Pine, Purvis
8	WI	June 12, 1899	5:40 PM	117	200	F5	New Richmond
9	MI	June 8, 1953	8:30 PM	115	844	F5	Flint
10	TX	May 11, 1953	4:10 PM	114	597	F5	Waco

The 10 Longest US Tornado Tracks

Rank	Path Length	Date	State(s)
1	219 miles (352 km)	March 18 , 1925	MO/IL/IN
2	170 (274)*	April 9, 1947	TX/OK/KS
3	160 (257)*	February 21, 1971	MS
4	155 (250)*	April 24, 1908	LA/MS
5	155 (250)*	May 26, 1917	IL
6	135 (217)*	May 27, 1973	AL
7	130 (209)*	April 20, 1920	MS/AL
8	125 (201)*	April 29, 1909	MS/TN
9	121 (194)*	April 3, 1974	IN
10	115 (185)*	March 30, 1938	IL

Indicates that these tracks may have been due to a family of tornadoes instead of one single tornado.
Some researchers argue that the Tri-State Tornado may have also been a family of tornadoes.

Canada has had its share of deadly tornadoes and tornado outbreaks too. This listing is not based on the Forbes Impact Index, but is keyed to the number of fatalities. Note that the April 3-4, 1974, US Super Outbreak even reached into Canada.

- **Regina, Saskatchewan,** June 30, 1912, 28 dead (The Regina Cyclone)
- **Edmonton, Alberta**, July 31, 1987, 27 dead (Edmonton Tornado)
- **Windsor, Ontario**,
  June 17, 1946, 17 dead (Windsor-Tecumseh, Ontario Tornado of 1946)
- **Pine Lake, Alberta**, July 14, 2000, 12 dead (Pine Lake, Alberta Tornado)
- **Salaberry-de-Valleyfield**, Quebec, August 16, 1888, nine dead
- **Windsor, Ontario**, April 3, 1974, nine dead (Super Outbreak)
- **Barrie, Ontario**, May 31, 1985, eight dead (US-Canadian Outbreak)
- **Sudbury, Ontario** - August 20, 1970, six dead (see Sudbury, Ontario Tornado)
- **Sainte-Rose, Quebec**, June 14, 1892, six dead
- **Portage la Prairie, Manitoba**, June 23, 1922, five dead

Although these statistics portray the worst tornadoes, most are very weak and short-lived. In fact, of all the US tornadoes reported annually, only about two percent would fall into the violent (F4 or F5) classification. Most tornadoes are weak (F0 or F1) and most only stay on the ground for a few miles and affect an area 50-100 yards (45-90 meters) wide. Yet, the few violent tornadoes account for the majority of US tornado fatalities.

Tornadoes and climate change?

A look at the preceding listings of significant events, and also record-breaking and other significant tornado statistics from around the world, raise interesting questions. For example, if the climate is warming and storminess is increasing, then shouldn't tornado incidence and impact be greater now than in previous years?

Many of the continental records (e.g. longest track, deadliest etc.) all occurred in the early 1970s (although Europe has a much longer tornado history than even the US). Although tornado incidence continues to climb, frequency of F3-F5 tornadoes is actually lower since 2000 than at almost any time since 1950. While improvements in warning systems and public awareness can be used to explain the reduced fatality count, greater population density would temper that assessment. Still, the lack of recent tornado-only attributes (like the longest track) cannot be easily explained. Even just in the greater Chicago, Illinois area, the peak incidence period (1950s-70s) has been replaced by a relative lull since.

Classifying twisters

Currently, tornadoes can't be classified in real-time, but they can be classified after the fact. It was Professor T Theodore Fujita, a meteorologist from Japan (see page 125), who revolutionized the science surrounding tornadoes. Fujita developed his Tornado Classification Scale that linked estimated winds to damage patterns based on intensive examinations of storm damage. This included ground and aerial surveys and examination of building construction. Fujita's maps of storm damage patterns helped meteorologists understand how unusual damage patterns occurred. Fujita did not see a tornado first-hand until late in his research career.

In recent years questions have surfaced about the Fujita Scale. These are related to the link between estimated wind speeds and damage (engineers have found that lesser winds can cause the damage that was earlier attributed to stronger winds), variations in construction quality (some homes not built to code) and even situations in which only natural things (not buildings) have been destroyed.

And while Doppler radar can give good estimates of wind speed within the tornadic circulation aloft, it cannot tell what the wind speed was at ground level. Many wind instruments are destroyed even before the twister reaches them.

FUJITA TORNADO DAMAGE SCALE
Developed in 1971 by T Theodore Fujita of the University of Chicago

SCALE	WIND ESTIMATE (MPH)	TYPICAL DAMAGE
F0	< 73	Light damage. Some damage to chimneys; branches broken off trees; shallow-rooted trees pushed over; signboards damaged.
F1	73–112	Moderate damage. Peels surface off roofs; mobile homes pushed off foundations or overturned; moving autos blown off roads.
F2	113–157	Considerable damage. Roofs torn off frame houses; mobile homes demolished; boxcars overturned; large trees snapped or uprooted; light-object missiles generated; cars lifted off ground.
F3	158–206	Severe damage. Roofs and some walls torn off well-constructed houses; trains overturned; most trees in forest uprooted; heavy cars lifted off the ground and thrown.
F4	207–260	Devastating damage. Well-constructed houses leveled; structures with weak foundations blown away some distance; cars thrown and large missiles generated.
F5	261–318	Incredible damage. Strong frame houses leveled off foundations and swept away; automobile-sized missiles fly through the air in excess of 100 meters (109 yds); trees debarked; incredible phenomena will occur.

Please note that wind speeds are approximate.

Enhanced Fujita (EF) Tornado Damage
Based on the original scale developed in 1971 by T Fujita (page 119) and updated in 2007 by NOAA

FUJITA SCALE			DERIVED EF SCALE		OPERATIONAL EF SCALE	
F Number	Fastest ¼-mile (mph)	3 Second Gust (mph)	EF Number	3 Second Gust (mph)	EF Number	3 Second Gust (mph)
0	40-72	45-78	0	65-85	0	65-85
1	73-112	79-117	1	86-109	1	86-110
2	113-157	118-161	2	110-137	2	111-135
3	158-206	162-209	3	138-167	3	136-165
4	207-260	210-261	4	168-199	4	166-200
5	261-318	262-317	5	200-234	5	Over 200

The wind values here are based on three-second wind gusts not the classical wind speeds used in weather reports. Damage is now derived based on damage estimated using 28 damage indicators.

As a result, the US National Weather Service has implemented a new tornado classification system (February 2007) and the TORRO Group in England has developed an independent tornado classification system (see pages 120-121).

Forecasting and warning

While reconstructing the past is important, so too is safeguarding people. Although the use of "tornadoes" in forecasts was once officially banned by the US Weather Bureau, that practice ended after meteorologists started to demonstrate skill in forecasting these violent storms.

One of the earliest tornado forecasting programs was initiated in March 1884 by Sergeant John Finley of the US Army Signal Corps. He issued twice-daily tornado forecasts for 18 regions in the United States, east of the Rocky Mountains for a three-month period. Although Finley claimed a high skill level, current forecast verification procedures would have given Finley much lower marks.

In early 1948, two Air Force officers – Capt (later Col) Robert Miller and Major Ernest Fawbush – noticed striking similarities in the developing weather pattern to other situations which produced tornadoes (including the Tinker AFB, OK, tornado several days before). The two forecasters advised their superior officer of a tornado threat in central Oklahoma that evening. Compelled to issue a yes/no decision on a tornado forecast after thunderstorms developed in western Oklahoma, they put out the word of possible tornadoes and the base carried out safety precautions. A few hours later, despite the tiny odds of a repeat direct hit in five days, a second tornado struck. Following this first successful documented tornado forecast, the US Weather Bureau moved forward and established a formal tornado forecast program in March 1952.

Tornado forecasting in the United States actually includes three stages of alerts. When a major outbreak is expected, the National Weather Service will issue a public statement or outlook about the risk. At this stage, nothing may be happening (or even expected to happen for hours). It is designed to ensure that storm spotters, emergency managers, television stations and others are aware of the potential.

As the day continues, tornado watch areas (usually defined as rectangles or parallelograms for easier mapping and visualization) are issued. Watches may cover areas as large as 25,000 square miles and include several states. Watches mean to "watch out" and keep an eye to the sky and an ear and eye to the media and they mean that tornadoes are possible in (and also near) the watch area. When a tornado has been

DOPPLER RADAR SEES TORNADOES

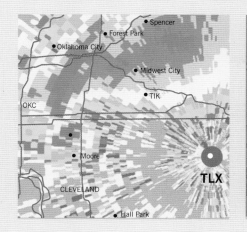
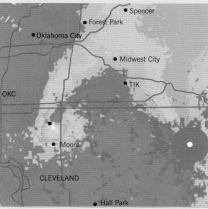

Shortly after the satellite image showing tornadic storms over Oklahoma (page 111), Doppler radar captured these images of an F4-F5 tornado just to the south of Oklahoma City (time 7.27pm CDT). The radar, operated by the National Severe Storms Laboratory (NSSL) was located at Twin Lakes (TLX), Oklahoma. The tornado, shown near the center of a large red-yellow coil (upper image), was located about nine miles (14.5 km) to the west-northwest of TLX. The reds and yellows indicate the most intense radar reflectivity (or energy reflected back to the radar by precipitation). The lower images show wind speed measurements. Greens indicate wind speeds blowing toward the radar; reds indicate motion away from the radar. The color couplet (with whitish tints near each colors center) shows very high wind speed and directional changes across a very small distance and is known as a tornado vortex signature (TVS).

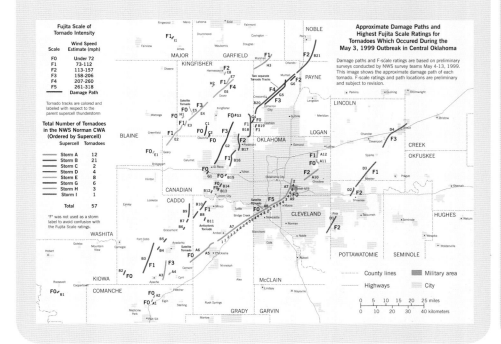

TORNADOES DO STRANGE THINGS

Tornadoes have been known to do "strange" things. For example, amid the destruction of one of the May 3, 1999 tornadoes in Oklahoma, a baby – Aleah Crago – was carried away by the tornado. She was found after the tornado in a field, some distance away from the family's home, with only a small scratch and some minor bruises. Following the Shinnston, WV, tornado of June, 1944, Mrs Cena Mason was found sitting, unhurt on a chair in the middle of the wreckage from her home.

Tornadoes can pick up and carry smaller objects far from their source. As the tornado weakens, these objects then rain to Earth. One tornado rained money (most likely from a destroyed bank vault) while another dropped frogs from the sky (probably picked up from a lake). On July 4, 1995, soft drink cans picked up in a soda plant 150 miles (241 km) to the south of Keokuk, Iowa, were deposited on people in the town. Following some tornadoes, farmers have even found naked chickens stumbling about. (It is now believed that the loss of their feathers was not due to wind plucking, but rather a defensive reaction due to fear, known as "flight molt!").

sighted, or one is indicated by Doppler radar, the NWS issues a warning. The warning (for imminent danger) is site specific, often specifying cities or communities in the storms' path. The overall warning may encompass one or two counties. It means get to safety immediately.

It is important to note that warning systems in other parts of the world can be quite different than those in the US. These are based on local risk, availability of improved detection technologies, storm spotters etc.

THE INTERNATIONAL TORNADO INTENSITY SCALE
(Developed by TORRO)

Tornado Intensity	Description Of Tornado and Windspeeds	Description Of Damage (for guidance only)
T0	Light Tornado 17-24 m s-1 (39-54 mi h-1)	Loose light litter raised from ground level in spirals. Tents, marquees, awnings seriously disturbed. Some exposed tiles, slates on roofs dislodged. Twigs snapped; trail visible through crops. Wheelie bins tipped and rolled. Garden furniture and pots disturbed.
T1	Mild Tornado 25-32 m s-1 (55-72 mi h-1)	Deck chairs, small plants, heavy litter becomes airborne. Minor damage to sheds. More serious dislodging of tiles, slates. Chimney pots dislodged. Wooden fences flattened. Slight damage to hedges and trees. Some windows already ajar blown open breaking latches.
T2	Moderate Tornado 33-41 m s-1 (73-92 mi h-1)	Heavy mobile homes displaced. Light caravans blown over. Garden sheds destroyed. Garage roofs torn away and doors imploded. Much damage to tiled roofs and chimneys. Ridge tiles missing. General damage to trees, some big branches twisted or snapped off, small trees uprooted. Bonnets blown open on cars. Weak or old brick walls toppled. Windows blown open or glazing sucked out of frames.
T3	Strong Tornado 42-51 m s-1 (93-114 mi h-1)	Mobile homes overturned / badly damaged. Light caravans destroyed. Garages and weak outbuildings destroyed. House roof timbers considerably exposed. Some of the bigger trees snapped or uprooted. Some heavier debris becomes airborne causing secondary damage breaking windows and impaling softer objects. Debris carried considerable distances. Garden walls blown over. Eyewitness reports of buildings physically shaking.
T4	Severe Tornado 52-61 m s-1 (115-136 mi h-1)	Motorcars levitated. Mobile homes airborne / destroyed. Sheds airborne for considerable distances. Entire roofs removed from some houses. Roof timbers of stronger brick or stone houses completely exposed. Gable ends torn away. Numerous trees uprooted or snapped. Traffic Signs folded or twisted. Some large trees uprooted and carried several yards. Debris carried up to 2km leaving an obvious trail.

Getting the word out

The media, especially television, plays a major role in keeping people informed. In addition to broadcasting official forecasts and warnings, television can interrupt broadcasting for immediate storm reports, displays of radar information and other life-saving information. Often key safety rules (such as those shown on the next page) are broadcast frequently. Sometimes, there can even be live broadcasts of the tornado in progress. On July 18, 1986, a helicopter news crew provided live coverage as a tornado moved across the Minneapolis, MN, area. That live footage became the evening newscast!

Tornado Intensity	Description Of Tornado and Windspeeds	Description Of Damage (for guidance only)
T5	Intense Tornado 62 - 72 m s-1 (137 - 160 mi h-1)	Heavier motor vehicles (4x4, 4 Tonne Trucks) levitated. Wall plates, entire roofs and several rows of bricks on top floors removed. Items sucked out from inside house including partition walls and furniture. Older, weaker buildings collapse completely. Utility poles snapped.
T6	Moderately -Devastating Tornado 73 - 83 m s-1 (161 - 186 mi h-1)	Strongly built houses suffer major damage or are demolished completely. Bricks and blocks etc. become dangerous airborne debris. National grid pylons are damaged or twisted. Exceptional or unusual damage found, e.g. objects embedded in walls or small structures elevated and landed with no obvious damage.
T7	Strongly-Devastating Tornado 84 - 95 m s-1 (187 - 212 mi h-1)	Brick and Wooden-frame houses wholly demolished. Steel-framed warehouse-type constructions destroyed or seriously damaged. Locomotives thrown over. Noticeable de-barking of trees by flying debris.
T8	Severely-Devastating Tornado 96 - 107 m s-1 (213 - 240 mi h-1)	Motorcars carried great distances. Some steel framed factory units severely damaged or destroyed. Steel and other heavy debris strewn over a great distances. A high level of damage within the periphery of the damage path.
T9	Intensely-Devastating Tornado 108 - 120 m s-1 (241 - 269 mi h-1)	Many steel-framed buildings demolished. Locomotives or trains hurled some distances. Complete debarking of any standing tree-trunks. Inhabitants survival reliant on shelter below ground level.
T10	Super Tornado 121 - 134 m s-1 (270 - 299 mi h-1)	Entire frame houses and similar buildings lifted bodily from foundations and carried some distances. Destruction of a severe nature, rendering a broad linear track largely devoid of vegetation, trees and man made structures.

Tornadoes of strength T0, T1, T2, T3 are termed weak tornadoes.

Those reaching T4, T5, T6, T7 are strong tornadoes.

T8, T9, T10, T11 are violent tornadoes.

The NOAA Weather radio system, a network of special frequency radio receivers, brings information in real-time, from the NWS into homes, schools and even cars. Its tone-activated feature allows for the radio's alarm to be turned on even when people are asleep. There have been numerous events in which the Weather Radio has been credited with saving lives.

Some communities, especially those in the more tornado-prone areas, have community warning systems in place, including sirens mounted on tall poles. Sometimes police will actually drive through communities with loudspeakers to alert residents to take cover.

During a tornado outbreak in north central Kansas, in late September 1973, a police officer was caught in the tornado as he drove around trying to warn residents. Taken to a nearby hospital, he was again attacked by a twister a few hours later. He survived.

Wrapping up tornado statistics

Although statistically rare for a tornado to strike the same place several times, there have been instances (other than the one noted above). Recall the Tinker Air Force Base situation that helped create a national US tornado forecast program. Back in 1916, 1917, and 1918, Cordell, Kansas, was struck by tornadoes on exactly the same calendar date, May 20th!

How statements, watches, and warnings differ in timing, size and specificity

	Time	Area	Specificity
Statement/Outlook	Up to 24 hours	Multi-state, tens of thousands of square miles	Broadly defined risk area
Watch	Up to 6 hours	Up to about 25,000 square miles	Focused area of potential tornadoes
Warning	Up to an hour	Typically hundreds of square miles	Specific counties or communities are in immediate danger

While some settings clearly favor repeat tornado visits, there is considerable folklore that suggests certain areas are "safe" from these storms. Some legends (dating back to Native American times) suggest that tornadoes will avoid where rivers fork. Other tales note that tornadoes won't occur over mountainous terrain. There is even lore that suggests that large cities are immune. Time continues to dispel these and other myths.

Tornadoes continue to move across rivers and mountains. On May 31, 1985, a long-tracked twister went up and down many mountains in its trek across Pennsylvania. During the period May 12, 1997-March 28, 2000, tornadoes struck five major US cities:

• May 12, 1997 Miami, FL;
• April 17, 1998, Nashville, TN;
• May 3, 1999, Oklahoma City, OK;
• August 11, 1999, Salt Lake City, UT; and
• March 28, 2000, Fort Worth, TX;

Most recently, on December 7, 2006, a twister hit northwest London, England, injuring six people.

Within a 20-mile (32-km) radius around the center of Oklahoma City, OK, (population at least 100,000), the city has been struck by tornadoes more than 115 times since 1888. So much for the myth about tornadoes not striking cities large or small!

If mountains and even cities can't disrupt a tornado's circulation, then what can? Some have proposed firing missiles at tornadoes. The hypothesis is that the explosion will disrupt the tornado's circulation and cause it to weaken. In reality, by the time a tornado is detected and a missile fired (most likely from a distant location), the tornado may have already dissipated. Further, if the missile goes off-track, it might actually strike the community it was designed to protect.

Scientists and emergency planners agree that the best defense against tornadoes is understanding these storms and the damage they can cause and knowing the weather around you. Toward this end, many states hold tornado or severe weather awareness weeks to spread the word about weather safety.

Keeping safe

The same Kansas tornado situation in which the police officer was struck twice, demonstrates the value of community storm shelters. In 1972, 18 months prior to that tornado event, the mobile home park in the same area was devastated by a killer tornado. The park did not have a community storm shelter. Following that event they built one. About eighteen months later, a tornado tore through the park again. All residents were in the shelter and no one was killed. After the second tornado struck, the park closed.

There are many tornado safety rules (and unfortunately many old and incorrect ones), but perhaps the easiest rule to remember is to "get inside, get as many walls between you and the outside, and get as low down (even into a basement) as you can." Get out of easily transportable objects such as mobile homes and cars.

You should also be able to recognize, and avoid, following antiquated rules about storm safety. One such rule was that you should open windows when a tornado threatened.

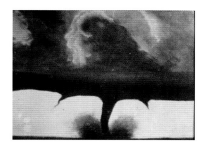

This is the oldest known photograph of a tornado. It was taken on August 28, 1884 near Howard, South Dakota. The name of the photographer is not known. Notice that the main tornado appears to be surrounded by two smaller tornadoes or funnel clouds.

A tornado moves through parts of the Miami, Florida, metropolitan area on May 12, 1997. Five people were injured and approximately 20,000 residents lost power when the storm struck.

This was supposed to allow air pressure to be equalized and to prevent the house from exploding. This myth is founded on films that showed homes disintegrating when a tornado struck them. However, the destruction is actually caused by flying debris, winds weakening the structure and other factors. The flying debris aerated the house, breaking windows and puncturing the side walls. That rule has now been replaced by "get away from windows and put as many walls between you and the outside of your home as possible."

In schools and other places you should avoid large-spanned rooms since support for the roof is limited.

If caught out in an open area, get as low down to the ground as you can (or get into a drainage ditch). Obviously be aware that the ditch can fill with floodwaters.

Keeping atop the day's weather (your own eyes and ears, the media, and the Internet) will ensure you will know when to act. Then, when a tornado does strike your area or clouds suggest one may be coming, you won't be at as much risk and hopefully you won't become a deadly statistic.

This aerial view shows destroyed houses in a suburban neighborhood south of Oklahoma City. The tornado struck on May 3, 1999 and was part of a major tornado outbreak. At least 45 people were killed and hundreds injured when dozens of twisters smashed through parts of Oklahoma and Kansas.

Although the number of reported tornadoes continues to increase, tornado deaths for the most part are decreasing. In fact, when normalized for population, the incidence of fatalities per million US citizens has dropped dramatically since the tornado warning system was implemented. Fortunately, tornado deaths from individual tornadoes have dropped markedly, too. Of the top 10 killer tornadoes, none have occurred since (one year) after the advent of the Nation's tornado warning system in 1952.

A tornado heads for the downtown area of Fort Worth in March 2000.

TRIBUTE TO TORNADO RESEARCHER TETSUYA THEODORE "TED" FUJITA

Theodore Fujita, Professor of Meteorology at the University of Chicago, works in his lab with a special tornado simulator.

Born in Japan in 1920, Ted Fujita became "hooked" on tornadoes when he surveyed the damage from a tornado event on Kyushu Island in 1948. Although fascinated by weather and related sciences before this, Fujita's writings on weather events earned him an invitation to affiliate with the University of Chicago in the early 1950s. From that point, until his death in 1998, Fujita advanced the knowledge base of severe thunderstorms and tornadoes as no one else before or since. His efforts earned him the nickname "Mr Tornado".

Fujita was able to assess tornado damage patterns and from these determine storm strength, wind type (tornadic or straight-line) and other storm attributes. His damage classification – the storm strength classification system known as the F-scale – set the standard for classifying storms. He introduced the concepts of tornado families and multi-vortex tornadoes, used photographs to define thunderstorm structure and coined terms like wall cloud and microburst. His work at small-scale (mesoscale) weather map analysis set the stage for how tornadoes were forecast in the United States. To say the least, Fujita revolutionized the way meteorologists and the public both look at tornadoes and severe thunderstorms.

Perhaps Tom Grazulis, Head of The Tornado Project, summed up Fujita's accomplishments best when he wrote, "One of Ted's fortés has been to make it 'look easy'. To take the most bewildering assortment of damage and wind, and make such simple sense out of it. That ability is the mark of a true genius..."

8 LIGHTNING

> **"Thunder is good, thunder is impressive;
> but it is lightning that does all the work."**
> *Mark Twain*

What is lightning?

Most of us are in awe of lightning. We often sit by the window and watch, enthralled, as lightning dances across the heavens. But it isn't until the thunder and lightning happen simultaneously that we begin to appreciate the power of our world's electrical forces. Even in this setting, however, the true impact of lightning escapes us. That's because lightning rarely grabs the headlines in the same way as hurricanes, winter storms and tornadoes. Lightning does its work in small numbers and in fractions of a second. However, when it strikes, it really strikes!

Wherever you are on Earth, lightning looks the same. Here are two images showing cloud-to-ground lightning. The image on the left is from Nova Scotia, Canada and the image on the right from Switzerland.

Lightning is basically a gigantic spark that is an electrostatic discharge. It is much like the spark that occurs when you scuff your sock-footed feet across a carpet, in very dry air, and you touch a doorknob and receive a small shock. The only difference is that the lightning is much, much more powerful.

The Earth is jolted by lightning a lot! Each day, about 44,000 thunderstorms occur across the planet, with about 2,000 in progress at any moment. Each second, there are about 100 lightning discharges. But the distribution of lightning across the planet favors tropical and middle latitudes and the land masses that lie in these bands.

Thanks to NASA's space probes, lightning has been detected on other planets in our solar system: the Galileo spacecraft photographed lightning on Jupiter; in 2006, lightning was detected on Saturn thanks to the Cassini spacecraft's radio instruments. The radio equipment "crackled", much like an AM channel has "static" when thunderstorms are nearby.

Atmospheric electricity

Our atmosphere is full of positive and negative ions (freely moving electric charges), and thunderstorms add greatly to this supply. Violent up-and-down motions that occur close

126 Lightning

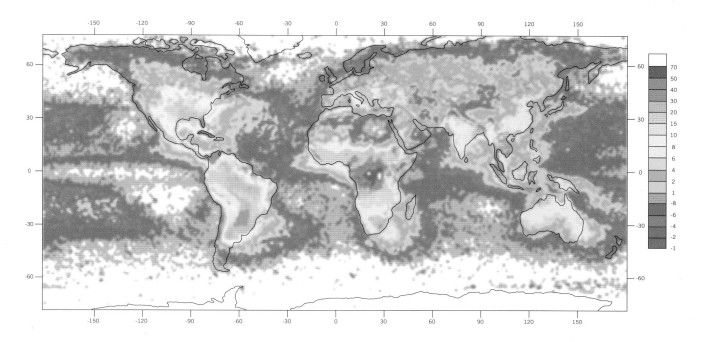

to each other inside a thunderstorm (see Chapter 6), allow hail, soft hail (graupel), raindrops and snowflakes to rub past each other. In the process, electrons are stripped off some precipitation particles. Wind currents then carry precipitation to different parts of the storm. Positive charges tend to concentrate in the upper portions of the cloud and negative charges gather nearer the storm's base. This creates a much larger charge differential than is normally found elsewhere in the atmosphere.

Because like-charged particles tend to repel each other, the positive charge atop the thunderstorm induces a negatively charged region in the atmosphere above the storm. Below the storm, positive charges gather in response to the negatively charged storm bottom. Meanwhile, nearby thunderstorm clouds are creating their own charged regions.

This map reveals the uneven distribution of worldwide lightning. Color variations indicate the average annual number of lightning flashes per square kilometer (with yellows and reds indicating higher frequencies. Note that most of the lightning occurs over land (where solar heating is stronger and hence rising air currents are often stronger). The map was produced by NASA Marshall Space Flight Center's Lightning Imaging Sensor Science Team, and includes data taken over an 11-year period from from two space-based observing platforms – NASA's Optical Transient Detector and TRMM's Lightning Imaging Sensor.

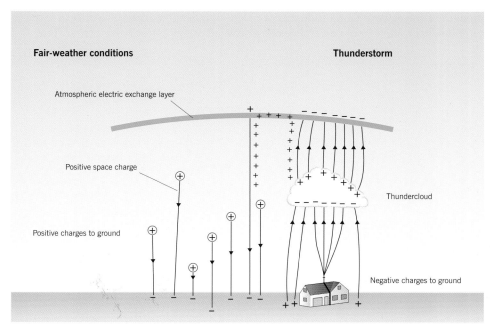

Fair-weather conditions

Thunderstorm

Atmospheric electric exchange layer

Positive space charge

Positive charges to ground

Thundercloud

Negative charges to ground

The combined effects of the Earth's fair weather electric field and that created by thunderstorms.

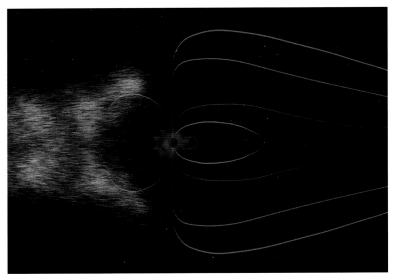

Thunderstorms transfer large amounts of negative charge to Earth, making the Earth negatively charged and the ionosphere positively charged. The ionosphere is the highly charged upper part of the upper atmosphere (altitudes of 50 miles (80 km)) and higher that responds to the solar wind – a streaming plasma composed of hydrogen and helium that has escaped from the Sun. A plasma is a superheated and highly ionized gas.

As the "wind" flows past the Earth (at an average speed of 250 miles (402 km) per second), it affects the Earth's magnetic field (magnetosphere) and the ionosphere. In turn, the magnetosphere and ionosphere help to divert the solar wind around Earth, preventing the wind from eroding our atmosphere.

View from NASA's Imager for Magnetopause to Aurora Global Exploration (IMAGE) spacecraft. The yellow mass (left) is the solar wind; the curved lines represent the Earth's magnetic field. Notice how the solar wind is forced to curve around the Earth.

A typical vertical atmospheric charge differential near the ground is about 200 volts/ six feet (1.8 meters) (the height of a "short" basketball player). The charge differential becomes less the higher up one goes in the atmosphere. Overall, this translates into a differential of 200,000-500,000 volts between the Earth's surface and the ionosphere. The voltage tells how much push there can be when electricity starts to flow. For comparison, the typical household voltage is between 100-240 volts.

Outside the thunderstorm, a fair weather electric "circuit" bleeds electric charges from the Earth up to the ionosphere in an attempt to balance the work of thunderstorms. Without thunderstorms to recharge the electric field, the atmospheric electric field would disappear in about an hour.

Lightning
While the thunderstorm process creates electric charges that then try to group themselves into charge centers, the atmosphere acts as a giant insulator, preventing the transfer of electricity and allowing huge amounts of charge to build up. When the strength of the electric field inside and/or near a thunderstorm reaches a critical level, current starts to flow from one charge center toward an opposite charge center. The

Multiple cloud-to-ground lightning flashes in Luton, United Kingdom. The time exposure image was taken from a hilltop outside the town. These strikes occurred between the shelf cloud and rain shaft (see Chapter 7 for storm structure details).

outcome, lightning, occurs when charge centers connect. In the process, the temporary charge centers in the atmospheric electric field come into balance and the ionospheric-Earth voltage differential is restored.

Lightning discharges can occur between a cloud and the ground (or anything on the ground), cloud and air, and inside and between clouds. When lightning occurs inside a cloud or even between clouds, you may not be able to actually see the flash, just the brightening of the cloud. Lightning between a cloud and ground is typically referred to as "cloud-to-ground" lightning.

Today, we refer to the lightning discharge as a flash. Years ago lightning was referred to as "bolts", probably a throwback to historical Greek, Roman and Norse times when gods were thought to literally throw "bolts" of lightning (see box on page 131).

Types of lightning

There are two categories of ground flashes; natural (those that occur because of normal electrification in the environment), and artificially initiated or "triggered". Triggered lightning includes strikes to very tall structures, airplanes, rockets and towers on mountains. Natural lightning is cloud-to-ground while triggered lightning goes from ground-to-cloud. Natural lightning looks like an inverted tree, while triggered lightning, emanating from a point at the ground and extending upward, looks like a tree with its branches extending toward the sky. According to Ron Holle, a consultant to Vaisala Group, "triggered flashes are an extremely small category – on the order of less than one 10th of a percent of all flashes."

When a ground flash starts, a channel of negative charge, called a step leader, will zigzag downward from the thunderstorm in roughly 50-yard (45-meter) forked segments. This step leader is invisible to the human eye, and reaches the ground almost instantaneously. As it nears the ground, a channel of positive charge, called a streamer reaches up to meet it. Often, this streamer will start from a taller object like a tree, house, or telephone pole. That's because the taller object is already closer to the step leader. Unfortunately, the streamer can also start from you, if you are among the taller objects in the area.

A daytime lightning flash over Hill City, Kansas. Aside from a small amount of rain falling behind the flash, this dark cloud base is rain-free.

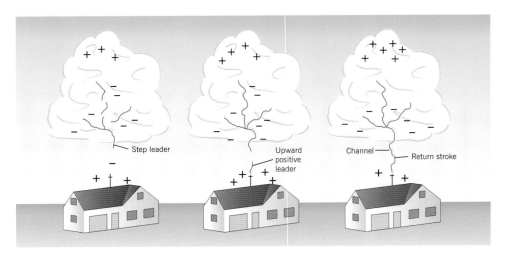

Formation of a cloud-to-ground lightning flash. Note that the flash involves paths from both the cloud and a ground based object.

Step leader

Upward positive leader

Channel — Return stroke

When the two connect, making an electric circuit, a powerful electrical current begins flowing. A return stroke of bright luminosity travels at speeds of about 60,000 miles (96,560 km) per second back toward the cloud. A lightning flash consists of one or as many as 20 return strokes. You may see lightning flicker when the process rapidly repeats itself several times along the same path. Although it appears to be huge, the actual diameter of a lightning channel is only one to two inches.

As the lightning channel briefly becomes a plasma, it can reach temperatures of up to 50,000 °F (27,760 °C) – almost five times the temperature of the Sun's surface). It can also discharge 30,000 amperes of electricity (current or electric flow), at up to 100 million volts (push) with electron densities exceeding 1022/ft³. In the process, the lightning emits light, radio waves, x-rays, and even gamma rays.

Not all lightning forms in the negatively charged area low in the thunderstorm cloud. Some originates at or near the top of the thunderstorm or in the cirrus anvil. This area carries a large positive charge and thus lightning produced here is called positive lightning.

This "bolt from the blue" image (Santa Fe, New Mexico) shows that lightning can really strike the ground far away from the main thunderstorm. Although "flashes" have replaced "bolts", the expression is an important lightning safety reminder.

Positive lightning is particularly dangerous for several reasons. It frequently strikes away from the rain core, either ahead or behind the thunderstorm, sometimes by as much as 10 miles (16 km). Sometimes this type of lightning can even appear with blue skies overhead (hence, the term "bolt from the blue"). Positive lightning typically has a longer duration of continuing current, so its heat and electric charge are better able to ignite fires. Positive lightning also usually carries a high peak electrical current, which increases the risk of death or injury to an individual.

Sheet lightning occurs when strong winds blow the lightning channel. Vaisala's lightning network near the Dallas-Fort Worth Airport (Texas) commonly detects in-cloud flashes more than 60 miles (96 km) long. Small fingers called sprites, jets, and elves can extend upward from the tops of the thunderstorm.

Ball lightning is a glowing, floating ball-like mass that moves slowly near the ground. Ball lightning has been known to enter homes and slowly move around inside before dissipating.

LIGHTNING: WEAPON OF THE GODS

Hopi Indian Kachina lightning figure.

To explain natural events, ancient cultures often "invented" gods or heavenly forces that explained the unexplainable.

Scandinavian mythology alludes to Thor, the thunderer, who was the foe of all demons. Thor "tossed" lightning bolts at his enemies. Early Greeks believed that lightning was a weapon of Zeus, the god of thunder during the day. Summanus was the Roman god of nightly thunder. Jupiter was Thor's counterpart, known for "throwing lightning". Donar was a Teutonic (Old German) god of thunder (comparable to Thor)and had war-like strength.

In the Hindu religion, Indra was the god of heaven, lightning, rain, storms and thunder. The Maruts (minor deities) used thunder bolts as weapons.

The Japanese had a pair of deities to explain lightning and thunder. Futsu-Nushi was the god of fire and lightning; Kaminari was the "Thunder Woman", who brought "Heaven's Noise".

Namarrkun was an Australian lightning god-figure. He came out of the sky and made thunder and lightning by striking clouds with stone axes attached to his elbows and knees. When men and women disobeyed the law, he would hiss and crackle or strike wrongdoers with fiery spears. Australian aborigines had Mamaragan, a man of lightning who rode on a thundercloud and threw bolts of lightning to the ground; thunder was his voice.

Umpundulo is the lightning bird-god of the Bantu tribesmen in Africa. Even today their medicine men go out in storms and bid the lightning to strike far away.

Some of these cultures revered the ground where lightning struck. Greek and Roman temples often were erected at these sacred sites, where the gods were worshipped in an attempt to appease them. In other cultures, people who were struck by lightning were shunned or thought to have angered the gods. In some less developed parts of the world, even today, a person struck by lightning is to be avoided.

However, almost every culture that had a lightning deity recognized that lightning and thunder brought much-needed rainfall.

Biblical references to lightning also abound. In Judeo-Christian bibles, thunder and lightning are viewed as tokens of God's wrath, God's glorious and awful majesty, or some judgment of God on the world. Similarly, the Koran notes, "…that Allah…sends mountains (of clouds) down from the sky, filled with hail, and with this he hits whoever he wants… The brightness of his lightning almost blinds the eye."

(See list of resources at end of this book for further information)

Lightning produces the thunder we hear. But the sounds can come from many sources, even including different parts of the same lightning flash. When this happens, we hear a rumbling sound that can last for many seconds. If lightning strikes close by, the sound can resemble an almost instantaneous "crack".

Thunder

As lightning occurs, its high temperature causes the air adjacent to the channel to expand rapidly, generating a sonic boom that we call thunder. Since sound and light travel at different speeds in the atmosphere, it is easy to estimate how far away the lightning is. Lightning travels at the speed at light (186,000 miles (300,000 km) per second) and the light from lightning reaches your eyes almost instantaneously. Sound travels much more slowly (around 760 mph (1,223 km/h)). Thus, it takes about five seconds for the sound to travel one mile. (The actual speed of sound is dependent to an extent upon atmospheric temperature; the sound wave can also be affected by wind.)

To get an approximation of how far away the lightning was from you in miles, count the number of seconds between the lightning flash and thunder sound and divide by five. Be careful not to assume that this is a horizontal distance – the thunder may have been produced inside a cloud or in an arc across the sky from one cloud to another.

When the lightning is very close, the sound may appear to be a crack or a boom. Thunder further away, and whose path is at a varying distance from you, will allow sound from many parts of the lightning channel to reach you over several seconds, causing a rumbling or rolling sound. Such variations in sound can also be caused by sound echoing off buildings, mountainsides and other large objects. Sound can diminish as it moves through the atmosphere due to attenuation (the sound wave itself loses strength the further it gets away from its source), the curvature of the Earth and even interaction with other thunders. As a result, thunder is rarely heard more than 10 miles (16 km) from its source.

Distant thunderstorms, especially if they are "over the immediate horizon" can provide a lightning show without visible lightning channels and without thunder. When this occurs, people may refer to it as "heat lightning". It was once thought that the air got so warm that it expanded and sparked.

In some places where there are few obstructions to visibility (e.g. parts of Kansas, Arizona and Florida), it is easy to see night time flashes from thunderstorms more than 100 miles away. This usually happens in the summer when skies overhead are cloud-free and temperature and humidity values are very high.

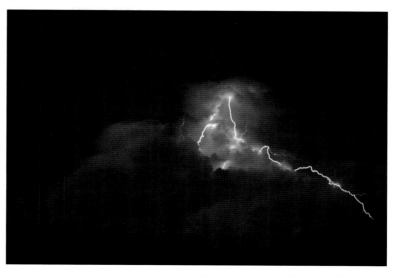

A weak, late evening thunderstorm along the coast near Byron Bay, New South Wales, Australia. The storm shot out several clear-air lightning bolts during a 10-minute period before rapidly decaying.

The human factor

Lightning causes significant damage to property, forest and grass lands, airplanes and electrical systems each year. It is also responsible for scores of deaths and hundreds of injuries around the world. Let's look at the human factors first.

Although casualty figures are much lower than they were 30-50 years ago, almost 50 people are killed and several hundred injured each year by lighting in the US alone. Military statistics indicate about 75 injuries a year from lightning, many at coastal military sites, especially in the southern parts of the United States. Discussions with lightning experts suggest that these group injury statistics are disproportionately high compared to the population at large. This may be based on how the military responds when lightning strikes near a group (i.e. sending everyone for medical evaluation). As has been the case for the past several decades, males tend to be victims about three times more often than females. Victims are also more likely to be below the age of 50 and a majority of lightning deaths involve people being out in the open or under tall trees. Golfers, hikers, farmers and even motorcyclists are often lightning targets.

Not surprisingly, Sunday is the day most likely to record a lightning death. Sunday has nearly 25 percent more deaths than any other day of the week. Sunday, followed by Saturday, are the top two days of the week for lightning injuries.

Across the United States, Florida is the thunderstorm capital (see Chapter 6) and the lightning capital for cloud-to-ground discharges (see page 130 of this chapter). Florida is first when it comes to lightning casualties, averaging nine deaths per year during the period 1990-2003. Florida's Gulf and Atlantic Coast neighbors are not far behind.

When normalized for population (excluding tourism), however, the Rocky Mountain states jump out as the prime danger area (six out of the top 10 states) while Florida drops to fourth. Wyoming soars to first with a risk four times greater than that of Florida. What this means is that anyone in Wyoming is four times more likely to become a lightning casualty than anyone in Florida.

Data for Australia shows a similar pattern with the greatest number of thunder and lightning days concentrated along the north coast. The annual average thunder-day map

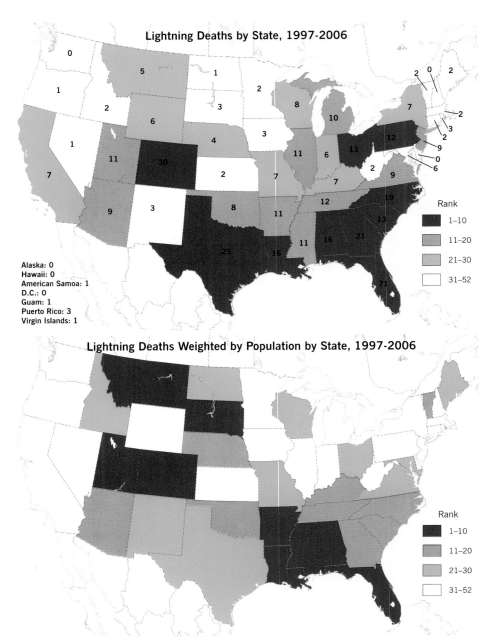

Lightning Deaths by State, 1997-2006

Alaska: 0
Hawaii: 0
American Samoa: 1
D.C.: 0
Guam: 1
Puerto Rico: 3
Virgin Islands: 1

Rank
1–10
11–20
21–30
31–52

Lightning Deaths Weighted by Population by State, 1997-2006

Rank
1–10
11–20
21–30
31–52

is based on observed thunderstorm activity at about 300 weather stations over a 10-year
period (1990-99). The lightning map, which is based on five years of satellite-derived
data, provides estimates of average ground flash (lightning) density across the country.
Both data sets have limitations that could misrepresent actual values. Drought
conditions in parts of Australia may also affect recent thunderstorm and lightning
occurrences.

However, there is a shift inland along the north coast of higher flash density (number of
strikes per unit area) values. This is most likely related to the movement inland of the
local sea breeze and the effect of elevated terrain (e.g. Kimberly Plateau).

Due to the way in which lightning is reported, and how NOAA collects data, lightning
experts believe that there is a significant underreporting of lightning incidents. This

involved at least two studies that showed that lightning deaths in Texas and Colorado were actually 33 percent and 28 percent higher, respectively, than shown in NOAA's data. Personal experience attests to this – I am aware of at least two people who were struck, but not seriously injured, and never reported the event. Since the incident was not reported, there was no newspaper trail and hence no entry into NOAA's database.

There is also the uncertainty of cause of death when someone is just found without any lightning burn marks. Could it have been a natural cardiac arrest or a lightning strike?

Lightning statistics internationally are even harder to come by, but a new web site (see Resources on page 300) is now posting many news stories about lightning in other parts of the world.

Ron Holle has also studied international lightning incidences since 1992. Some of his findings are summarized in the table (direction). By using US casualty-population ratios and considering the strong rural/agricultural base of many lightning-prone parts of the world, he estimates that global lightning casualties could be as high as 24,000 annually (with a corresponding 240,000 injuries).

Incidences since 1992:
• EGYPT: 430 fatalities when lightning struck an Army fuel depot (1994).
• KENYA: nine killed and 12 injured while taking shelter in a church with a corrugated roof (2002).
• CHINA: 22 farmers killed while planting fields (1994).
• INDONESIA: similar farming situation with 10 killed and five injured (2002).
• VIETNAM: 10 killed and four injured while searching for shellfish on a beach (1995).
• YUGOSLAVIA: one soccer player killed and several injured during practice (2000).
• GUATEMALA: two soccer players killed and 10 injured (2001).

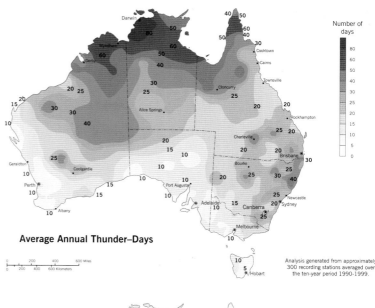

Average Annual Thunder-Days

Analysis generated from approximately 300 recording stations averaged over the ten-year period 1990-1999.

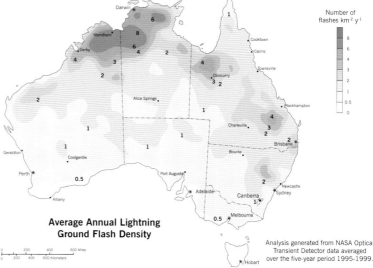

Average Annual Lightning Ground Flash Density

Analysis generated from NASA Optical Transient Detector data averaged over the five-year period 1995-1999.

Australian thunderstorm days and ground flash density.

In February 2007 there were at least eight deaths and 11 injuries resulting from lightning strikes around the world. Many of these occurred in the Southern Hemisphere (where it was summer. Note that there is less land area and a smaller population at risk). These included six people who were injured in a lightning strike on a South African vegetable farm; seven killed when lightning struck thatched homes in India; and a man was injured in Melbourne, Australia. Perhaps the most bizarre lightning event occurred in New South Wales, Australia when a man, a member of the Chinese paragliding team, was struck and killed by lightning while training for the Paragliding World Championships.

Even in the Northern Hemisphere, lightning made the news when it struck, and injured, a landscaper while he was replacing bushes near Tampa, Florida (US).

Average Annual Lightning Deaths by Country

Brazil	100
India	59
Zimbabwe	150
Singapore	3
United States	49

ROY SULLIVAN AND HIS MANY LIVES

If lightning strikes a place, there's no reason it can't strike the same place many times. The Empire State Building in New York City, which is 1,453 feet (423 meters) high, is designed to capture and dissipate lightning strikes – a good idea as the building is struck about 100 times each year.

But what about a moving target? According to the Guinness Book Of World Records, former Park Ranger Roy "Dooms" Sullivan is the most lightning-struck person on record. Between 1942 and his death in 1983, Sullivan was struck by lightning seven times. The first lightning strike shot through his leg. Then, 27 years later, the attacks began in earnest. In 1969, a second strike burned off his eyebrows and knocked him unconscious. Another strike just a year later, seared his shoulder. In 1972 lightning set his hair on fire and Sullivan had to dump a bucket of water over his head to put out the flames. In 1973, another lightning flash hit him on the head, again setting his hair on fire. A sixth strike in 1976 left him with an injured ankle. The last lightning incident in 1977 sent him to the hospital with chest and stomach burns. Sullivan could never offer any explanation for this strange and unwelcome electrical attraction.

While seven strikes is unusual, being struck once or knowing someone who has been struck is not. In fact, given an 80-year life span, a person living in the United States has about a 1 in 10 chance of either becoming a lightning victim or being closely related to one.

In the UK, the average number of lightning fatalities per year is about four, so the chance of being struck by lightning is roughly one in 14 million, which is about the same as the chance of winning the jackpot in the National Lottery! Australia logs about five to 10 deaths per year due to lightning, according to Australia's Bureau of Meteorology.

Not everyone struck by lightning dies. In fact, most don't. Some regain consciousness on their own; others recover when cardio-pulmonary resuscitation (CPR) is administered quickly. Even if someone isn't killed or seriously injured by lightning, people struck by lightning can suffer from a variety of long-term, debilitating symptoms. These include memory loss, attention deficits, sleep disorders, numbness, dizziness, stiffness in joints, irritability, fatigue, weakness, muscle spasms, depression and an inability to sit for long.

Prior to the 1990s, most of the efforts undertaken in cataloguing lightning centered on meteorology. Then, in the early 1990s, Dr. Mary Ann Cooper started efforts to better assess the medical side of lightning injuries. Around the same time, a support group, Lightning Strike and Electric Shock Survivors International (LSESSI), was founded by a lightning victim. LSESSI provides on-line, phone and face-to-face conference support for survivors of both lightning and other electrical shocks. Efforts are underway in other places (South Africa and Southeast Asia) to similarly address medical issues of lightning victims. (See Resources on page 300 for more information.)

Lightning damage

It is hard to estimate damage losses from lightning. Insurance companies often group lightning with other fire incidents. The National Oceanic and Atmospheric Administration (NOAA) compilations from its Storm Data publication have to meet certain monetary values or be reportable in local newspapers in order to be tallied. Further, much damage is "minor" and falls within, or close to, deductible limits. In these situations, many people simply pay for the loss instead of having a claim against them on their policies.

Perhaps the most comprehensive assessment of property losses to date was undertaken by Richard Kithil, President and CEO, National Lightning and Safety Institute (NLSI) in 2004. Although somewhat dated, his Annual USA Lightning Costs and Losses Report (see Resources on page 300) noted that although, "US government official figures describe losses at some $35 million annually…on-going research suggests realistic lightning costs and losses may reach $4-$5 billion per year." A late 1990s report by lightning researcher Ron Holle, noted that the number of Storm Data events was under-reported by 367-to-1 based on a review of insured personal property losses in three western states.

Here is a partial listing of some of Kithil's findings:

Description	Annual estimated losses
Lightning is responsible for more than $5 billion in total insurance industry losses annually, according to Hartford Ins Co. Source: TMCNet Newletter, Sept 14, 2006; this is five times the estimate provided by the Insurance Information Institute in 1989.	$5 billion
Some 30 percent of all power outages are lightning-related on annual average. Source: Ralph Bernstein, EPRI; Diels, et al (1997).	$1 billion
From May to July, 1999 lightning started most of the 2,282 Florida fires. Suppression costs were $160 million. Dollar losses amounted to $394,600,000. Source: NFPA Journal Nov/Dec 1999.	$555 million
Lightning accounted for 101,000 laptop and desktop computer losses in 1997. Source: Computer Security News, www.secure-it.com/newsletters.Statistics98.htm.	$125 million
Half the wildfires (10,000) in the western USA are lightning-caused. Source: Dale Vance, BLM, US Dept. Interior.	$100 million

Based on the updated entry from Hartford Insurance in the table above, Kithil appears to have underestimated losses even with a $4–$5 billion annual figure.

Lightning-caused forest and grass fires are a major concern in the southwestern United States, Australia, and other places that experience prolonged dry periods (see also Chapter 12). The reason is that often, "dry" thunderstorms bring lightning, but not rainfall. Such thunderstorms are also prone to bringing gusty, shifting winds.

A time exposure image showing many cloud-to-ground lightning flashes over Istanbul, Turkey.

Notwithstanding the fire risk, lightning can also damage or kill trees. By superheating the sap and causing it to boil, lightning can cause tree bark to literally "explode".

Although not "physical damage" in the classical sense, lightning claims the lives of many cattle each year. The cattle often congregate under trees during thunderstorms and succumb after lightning strikes the tree.

When lightning strikes homes, it can produce power surges and blow out walls. The power surges can spark electric fires, overload electric equipment, blow out televisions and cause other damage. Even surge protectors may not be enough to prevent damage to sensitive electrical equipment.

Lightning protection systems

Home lightning protection systems may help to channel a lightning strike to the ground where it can dissipate. These evolved from Benjamin Franklin's famous "kite in a thunderstorm" experiment in 1752 and his later efforts at developing a lightning conductor. Franklin's "conductor" was a metal rod or wire through which electrical discharges were led harmlessly to earth. Others who studied lightning around the same time were Thomas Francois D'Alibard of France and GW Richmann, a Swedish physicist. Richmann was the most unlucky of the three as he was struck and killed by lightning during his experiments.

Although some recent studies suggest that such protection systems may not be effective, personal and professional experience suggest otherwise. And Ron Holle noted that, "…properly designed, installed and maintained lightning protection systems are used successfully around the world for power stations, hospitals, police stations, oil refineries and the like."

A friend who owns a home on North Carolina's Outer Banks experienced a lightning strike one summer. She was home at the time. Following the strike, she smelled smoke and called the fire department. The fire was minor and quickly extinguished. The fire department noted that had she not had a lightning system on her home, her home would almost certainly burned down.

I, too, have a home with a lightning system. The same summer, my home was struck by lightning, but not directly. The surge came in from the television cable system. It blew out six televisions and two VCRs but otherwise did not cause any damage. As indicated earlier, much like others, I did not report the loss to my insurance company. Instead, I paid the $1,500 to replace the electrical equipment. Both incidents were never included in the official lightning data base.

If lightning can strike homes, it is an easy matter for it to affect power transmission lines, transformers and other parts of the electric power and electric system infrastructure. In mid-July, 1977, several lightning strikes, coupled with human and system error, helped to create a massive power blackout in New York City. However, given the large number of lightning strikes to electrical power systems, power losses are typically confined to small numbers of homes or businesses and then only for a short period of time.

Tree split by lightning during a summer thunderstorm. The heat of lightning can literally superheat the sap and water in a tree, blowing off bark and/or splitting the tree.

A wind turbine is seen with one of its wings missing after it was struck by lightning during a thunderstorm early on June 9, 2004 near Wulfshagen, Germany, causing a 2 million-euro damage. Fierce storms lashed northern Germany, snarling traffic and sparking hundreds of fires.

Aviation considerations

It is estimated that, on average, each airplane in the US commercial fleet is struck slightly more than once per year. In the UK, in a 12-month period during 2000-01, about 24 planes were struck by lightning out of two million flights in UK airspace. Even without natural lightning, aircraft typically trigger their own flashes when flying through a heavily charged region of a cloud.

When lightning strikes an airplane, it usually attaches itself at an extremity (e.g. nose or wing tip) and detaches itself at another extremity. In between, the plane substitutes for part of the lightning's path through the air, allowing the current to travel through the conductive exterior skin and structures of the aircraft.

Fortunately, nothing much usually happens to the plane or its passengers (other than temporary light flickering and minor instrument interference) because of onboard lightning protection systems and the plane's shell. However, there have been several instances in which electronics have been seriously affected. One involved an Icelandair Boeing 757 en route from Keflavik, Iceland to New York on March 7, 2006. Struck minutes after takeoff, the plane returned to Iceland because of severe damage to the onboard radar and nose. (See resources.)

In February 2006, hundreds of passengers were stranded overnight at Santander Airport in Spain after four Ryanair planes were hit by lightning.

There have been several commercial plane crashes directly attributed to lightning strikes causing catastrophic fuel tank explosions. These include Elkton, Maryland (December 1963), Milan, Italy (1959) and the Amazon Rainforest (1971). Military aircraft and helicopters have also been struck and have suffered some electrical and physical damage. Near Los Angeles (1969) several commercial aircraft were struck, damaging their onboard radar systems.

Smaller, private planes and planes operated in less developed countries remain most vulnerable. In May 2006, Senator Ted Kennedy's eight-seat Cessna Citation plane lost all electrical power, including communications, following a lightning strike. The pilot had to fly the plane manually and make an emergency landing. No one onboard was hurt.

Although not airborne, the metal cage around an automobile affords similar protection as that experienced by airplanes. The alleged "rubber" tires (tires today are made mainly from artificial materials and often contain steel belts) offer no protection from the power of lightning.

Petrified lightning

Lightning has been around for eons and there is fossil evidence that lightning occurred 250 million years ago. "Fulgurites", rocky formations shaped of sand fused or solidified by lightning (Fulgurite is derived from fulgur, the Latin for lightning) provide some of the clues.

REPELLING LIGHTNING

It was once thought that ringing church bells during a thunderstorm would protect the community and the church against lightning. People believed that ringing church bells would ward off evil spirits associated with the lightning, and that the sound from the bells would cause the lightning to disperse. The belief was so strong that the words, Fulgura frango, "I break the lightning", were inscribed on many church bells.

The process didn't work and bell-ringing as a lightning protection system was eventually abolished. However, this did not happen until lightning struck 386 French church towers between the years 1753 and 1786, killing 103 French bell ringers.

Many tall buildings include lightning rods and conductors as part of their design. This is the lightning conductor on the northern tower of Notre-Dame cathedral in Paris.

Churches, during the 18th century, were often used to store large quantities of gunpowder. The combination of a high steeple and an explosive arsenal often proved dangerous. In 1769, lightning struck the tower of St Nazaire in Brescia, where 100 tons of gunpowder were stored. The resulting explosion destroyed one-sixth of the city and killed 3,000 people. As late as 1856, lightning struck the church of St Jean on the island of Rhodes near Turkey – the powder stored in the vaults exploded and 4,000 were killed. (See also Resources on page 300.)

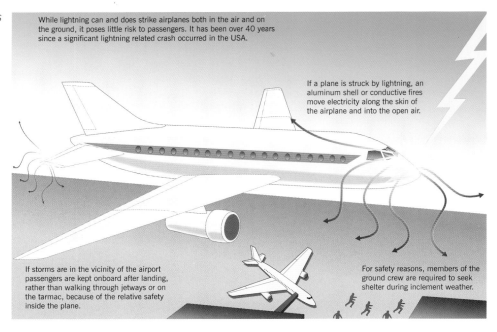

While lightning can and does strike airplanes both in the air and on the ground, it poses little risk to passengers. It has been over 40 years since a significant lightning related crash occurred in the USA.

If a plane is struck by lightning, an aluminum shell or conductive fires move electricity along the skin of the airplane and into the open air.

If storms are in the vicinity of the airport passengers are kept onboard after landing, rather than walking through jetways or on the tarmac, because of the relative safety inside the plane.

For safety reasons, members of the ground crew are required to seek shelter during inclement weather.

This simulation shows what happens when lightning meets an aircraft. The lightning strikes one part of the aircraft's outer metal shell and then leaves from another part of the plane. Basically, the plane just provides an additional conducting channel for the lightning. Most of the time, the plane is not affected by the lightning strike.

Silicon dioxide (the main component for most sands) melts at 2,950 °F (1621 °C), well below the temperature of lightning. When lightning strikes the sand, the sand easily melts. Once fused, the sand mass cools down, leaving a lightning path record in the sand. The fulgurite is actually "petrified lightning".

Although encased in fused sand, the fulgurite tube is fairly narrow and fragile. Since it can easily break apart, most fulgurite sections are fairly small.

Lightning safety

Lightning safety rules abound on the Internet. But the best way to avoid becoming a lightning statistic is to know that "When Thunder Roars, Go Indoors". That's the message of struckbylightning.org, an organization formed by lightning victim Michael Utley. The message, simple by design, is targeted specifically at young children. It works for adults, too! In other words, when you see lightning and/or hear thunder, you should be in a safe place indoors.

Sporting arenas (even high school stadiums), outdoor concert sites and large campgrounds (including Boy Scout Jamborees) pose special risks because so many people are in proximity and evacuation to safe places can be an unattainable and lengthy process. As Joel Gratz, Center for Science and Technology Policy Research at the University of Colorado, Boulder noted in his study of several major venues, people are typically jammed into evacuation routes and unable to find safe shelter during such storms. At some events, like Scout jamborees, substantial structures, that might afford some measure of lightning safety, are lacking.
Early detection, awareness and planning can address many of lightning's dangers. Retrofitting facilities or building structures may prove too costly.

Climate implications

Earth was not always as we know it now. A review of different geologic periods (see Resources page 300 and Chapter 20) shows that there have been periods of extreme cold and other periods of extreme warmth. There have also been dry and wet periods. However, perhaps most telling is that Earth's atmosphere (currently 80 percent nitrogen

Fulgurite in its original setting near Starke, Florida. As part of his lightning research at the University of Florida, Professor Martin Uman, documented this record 16-17 inches (40.5-43cm) long in-ground fulgurite.

THE MANY SIDES OF LIGHTNING

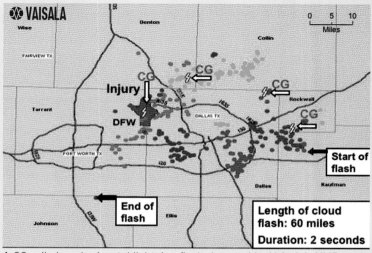

A 60-mile long horizontal lightning flash detected by Vaisala's VHF total lightning detection network LDAR II on August 17, 2001. The flash, lasting two seconds, started east of Dallas and ended southwest of Fort Worth. Colors show time starting with light blue, followed by purple, green, orange, and ending in red. There were four cloud-to-ground flashes associated with this long flash; one caused an injury at Dallas-Fort Worth International Airport.

Cloud-to-ground lightning flashes detected by Vaisala's National Lightning Detection Network (NLDN). These cloud-to-ground flash summaries maps are for two, two-hour time periods during a day in April 2007. Lightning count increased from about 2,000 for the period ending at 7am CDT to almost 35,000 for the period ending at 7pm CDT (image not shown). In the 1pm image, the storm system (as seen in lightning data) assumed a classical wave cyclone pattern with thunderstorms concentrated along both cold and warm fronts. During the time period of most rapid lightning count increase, heavy snow and tornadoes were both reported across eastern Colorado.

Lightning strikes can seriously affect space mission launches. One major incident occurred during the 1969 launch of the Apollo 12 mission when lightning briefly knocked out vital spacecraft electronics. Fortunately, the astronauts regained control of the spacecraft. Lightning also struck near the launchpad during the launch of STS-8, the eighth space shuttle mission, on August 30, 1983. Floodlights give the appearance of a direct lightning strike to the launch pad. (STS stands for Space Transportation System, the formal name for the Space Shuttle program.)

Since, approximately 30 percent of weather delays and scrubs for the STS program are related to lightning avoidance rules, NASA is understandably cautious when it comes to lightning.

Some additional lightning facts
Lightning can travel large horizontal distances.

Radial horizontal arcing has been measured at least 60 feet (18 meters) from the point where lightning hits ground. The distance lightning travels depends on soils characteristics, soil wetness and other factors.

Lightning can cause forest fires. Smoke and carbon micro-particles, when introduced into the upper atmosphere, can become the initiators of static. Sufficient atmospheric static can spark lightning discharges. Reports of massive lightning storms in coastal Brazil, Peru and Hawaii have been linked to the burning of sugar cane fields. In the late 1990s, smoke from Mexican forest fires may have helped to cause unusually high lightning activity in the US High Plains.

Recent studies suggest that the Houston, Texas, petrochemical industry, discharging copious amounts of hydrocarbons into the upper atmosphere, may be responsible for higher-than-normal lightning activity (and rainfall) in that area.

Lightning is an ever-present concern at the Kennedy Space Center in Florida. Launches are scrubbed when lightning is nearby. Notice the contorted shape of the lightning path. This is similar to (but not as long as) the flash described above that occurred near DFW Airport in Dallas, Texas.

This lightning strike near a college football stadium in Virginia sparked increased efforts at ensuring public safety from lightning at major outdoor events.

and 20 percent oxygen) is relatively new, geologically speaking. Some 3.8 billion years ago (early in the Earth's history), according to NASA scientists, our atmosphere was roughly a 50-50 carbon dioxide-nitrogen mix.

Then something happened and carbon dioxide levels dropped dramatically. Since plants need carbon dioxide, water, light and nitrogen to survive, scientists theorized that plant forms of this geologic time had to make significant adaptations.

However, lightning had been fixing (creating) nitrogen compounds from atmospheric nitrogen gas throughout Earth's history. In trying to understand lightning's role in plant evolution, scientists needed to know if the change was something related to the lightning process or to changes in the atmospheric gas composition. Enter a team of scientists from NASA Ames Research Center and the Universidad Nacional Autonoma de Mexico (UNAM) who simulated all possible ancient atmospheric combinations of carbon dioxide and nitrogen in the lab. Using a high-powered laser they "created" lightning and had that lightning interact with

MYTHS

Of all the weather hazards, lightning is perhaps the most misunderstood. Myths go back to times when lightning and its effects were beyond explanation due to rudimentary scientific knowledge. Here are just a few of the myths and scientific corrections to them.

MYTH: If it is not raining, then there is no danger from lightning.

FACT: Lightning often strikes outside of heavy rain and may occur as far as 10 miles (16 km) away from any rainfall. Lightning can even strike from storms that are not producing rain. I once witnessed a lightning flash from a small towering cumulus cloud.

MYTH: The rubber soles of shoes or rubber tires on a car will protect you from being struck by lightning.

FACT: Rubber-soled shoes and rubber tires provide no protection from lightning. In the case of an enclosed car with a hard-topped roof and windows rolled up, it is the steel frame (likened to a Faraday cage) that provides increased protection. Most of the time, the lightning charge stays on the outside of the car, with the charge being safely carried to ground. If you are on the flatbed of a pick-up truck, you are not in a safe place. Some years ago, a truck carrying some farm workers in Texas was struck by lightning. The three workers inside the enclosed cab were okay; their four co-workers who were on the back of the truck were all killed.

MYTH: People struck by lightning carry an electric charge and should not be touched.

FACT: Lightning-strike victims carry no electrical charge and should be attended to immediately. CPR and first aid can and should be applied quickly.

MYTH: "Heat lightning" occurs after very hot summer days.

FACT: What is referred to as "heat lightning" is actually lightning from a thunderstorm too far away for thunder to be heard.

MYTH: Lightning is a random, chaotic and dangerous fact of nature.

FACT: Based on everything described in this chapter, hopefully you'll agree that lightning is much more organized than we may think!

See Resources on page 300 for more information.

potential ancient atmospheres. Their work proved that lightning really did produce nitrates in the early Earth atmosphere. They also proved that, as carbon dioxide levels decreased, nitrate production also went down (one could argue that as carbon dioxide levels in the Earth's atmosphere continue to increase in the 21st century that this will enhance the production of natural nitrogen fertilizer!).

Thus, if lightning wasn't able to fix (create) enough nitrogen compounds from atmospheric nitrogen gas, plants had to find a way to do it on their own. Bacteria became more self-sufficient, more productive, and more adaptable in the process.

Chris McKay, one of the NASA researchers working on this project, noted, "We are used to thinking of the environment and life as steady and unchanging, but the early Earth was quite different. Major changes in the atmosphere occurred and life had to adapt…nitrate fertilizers were once made naturally, but that was depleted, so plants were forced to develop the capacity of developing their own fertilizer."

When a thunderstorm is electrically active, it is possible to capture a large number of lightning flashes in one time exposure image. This thunderstorm occurred over Sydney, Australia.

While there is no direct evidence that plants actually did change, McKay noted, "the system obviously didn't die."

With plants making this adaptation, they were better able to colonize more environments on Earth, eventually raising oxygen levels in the air. This set the stage for making an environment suitable for animals. One key point made by McKay is that, "Once life invented the ability to fix (produce) nitrogen, life's new invention made it stronger. When life responds to changes, that life is stronger than before."

Today, legumes (plants with nodules containing nitrogen fixing bacteria) play a major role in adding nitrogen compounds to soils, and there is growing interest in re-establishing these types of plants as nitrogen providers rather than relying on commercial fertilizers.

Still, lightning discharges provide about 10 percent of the world's natural soil nitrogen production (some 120 million tons). It does this by breaking the strong chemical bonds among nitrogen molecules, causing them to react with oxygen. Through a series of reactions with atmospheric water (which produce nitric and nitrous oxides), nitrate and nitrite ions, basically nitrogen fertilizer, are produced.

Recently, some scientists have concluded that lightning may have played a part in the evolution of living organisms. Nobel prize winning chemist Harold Urey proposed that the earth's early atmosphere consisted of ammonia, hydrogen, methane, and water vapor. One of his students, Stanley Miller, used an electric spark to duplicate lightning and introduced it into the chemical brew. He was careful to exclude any living organisms from the experiment. At the end of a week, he examined the mixture and found it contained newly-formed amino acids, the very building blocks of protein. Perhaps lightning did play a role in creating life itself!

"...Ice balls from the sky when summer winds blow from the east..."
*Native American folklore (referring to weather
in Colorado and the nearby High Plains)*

HAILSTORM DEVASTATES GARDEN CITY, KANSAS

Garden City, Kansas, July 1, 1999; an overnight thunderstorm pummeled this small southwest Kansas community (population around 28,000) with widespread golf ball to softball size hail. Almost every house in the city was damaged and all vehicles "left in the elements" received hail dents and/or broken windows. Some residents even reported that hail punctured roofs and bounced on the floor of their homes. Damage exceeded $50 million.

Hail of various sizes surrounds a small plant.

Small balls of ice

Most instances of hail are far less devastating than the event described (to the left). However, hail is a widespread type of severe weather and its size can sometimes border on impressive. Hail can, occasionally, be larger than a softball.

Hail is frozen precipitation in the form of small balls or lumps usually consisting of concentric layers of clear and cloudy ice.

Hail forms in thunderstorms as super-cooled liquid water (water that remains in liquid form even though it is in a below-freezing environment) first freezes and then collects additional super-cooled water on its surface. The alternating clear and cloudy ice can be linked to how fast the water freezes. Clear ice involves a longer cooling period; cloudy ice freezes faster, trapping air bubbles inside. Super-cooled water can exist in its liquid form at temperatures as low as -40 °F (-40 °C).

The hail formation process was once thought to entail several up and down transits of individual hailstones within a thunderstorm updraft. Now scientists believe that while this can occur, super-cooled water can also be transported past hailstones that remain suspended in updrafts in a cumulonimbus (thunderstorm) cloud. Depending upon the orientation(s) and/or movement of the hailstone in the updraft, interactions with adjacent stones, temperature variations, and so on, hail can assume some unusual shapes (e.g. spikes or flatness).

Geographical considerations

Hail occurs in many of the same geographical regions as tornadoes and severe thunderstorms. This isn't surprising since hail needs the same strong updraft-generating situations, atmospheric instability and moisture in which to form. In fact, hail, strong winds and tornadoes often occur within the same thunderstorm. However, hail (especially small hail) can fall even from less intense storms.

In the US, the region of maximum hail frequency extends across the southern central plains westward to the front range of the Rocky Mountains. This region extends northward into the western Canadian prairie region of Alberta. This encompasses the high plains westward to the front range of Rocky Mountains. These regions are more hail prone because hail is more likely to survive its fall from a cold thunderstorm cloud when either or both the cloud base and ground are closer to the cloud's freezing/melting level (usually around 10,000 feet above sea level). Hail is also more likely to survive its downward journey when it falls through drier air because cooling through the evaporation and sublimation processes keep the air temperature cooler.

Although hail typically occurs from the late spring to early summer, hail can occur even in winter if favorable thunderstorm conditions are present.

Other places around the world with a high frequency of hail include Australia, New Zealand, China, Northern India and Bangladesh (near the Himalayas), South Africa, northern Italy (near the Alps), the United Kingdom (in a northwest-southeast swath from Lancaster southeastward to near London), France (from southwest to northeast across the Centrale Massif region) and Russia. Many of these align closely with plains agricultural regions.

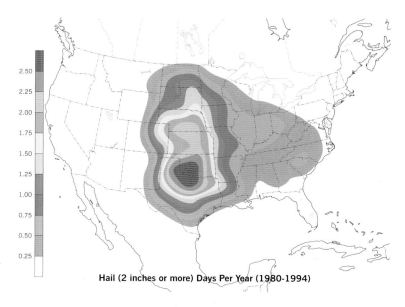

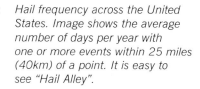
Hail (2 inches or more) Days Per Year (1980-1994)

However, mountain areas can also be prone to frequent hail occurrence. Even though it lies on the Equator, Quito, the capital of Ecuador (at an altitude of over 9,000 feet (2,743 meters), has many hail days each year.

Hail frequency across the United States. Image shows the average number of days per year with one or more events within 25 miles (40km) of a point. It is easy to see "Hail Alley".

Hail size

Although small hail is most frequently reported, it is larger hail that is most destructive and deadly. Hail the size of grapefruits has been observed in Canada. Golf ball size hail falls at least annually in New Zealand. In the United Kingdom, hailstorms with tennis ball-size hailstones have occurred, but the last report was on September 5, 1958.

The largest hailstone recorded in US weather history fell in Aurora, Nebraska on June 22, 2003. Measuring seven inches (17.5 cm) in diameter with a circumference of 18.75 inches (47.63 cm), this hailstone easily eclipsed the Coffeyville, Kansas stone that had held the size record since September 3, 1970. Because an accurate weight could not be determined for the Aurora hailstone, the Coffeyville stone still remains the heaviest hailstone (1.68 pounds). In addition to their size, hailstones from this storm left craters in the ground three inches (7.6 cm) deep and 14 inches (35.5cm) in diameter.

The larger the hailstone, the stronger the updrafts have to be in order to keep it "floating" in the air. When the hail gets too large for the updraft to support it, or the updraft weakens, gravity wins and the hail falls. Very large hail can have terminal fall velocities approaching 100 mph (160 km/hour). If accompanied by strong horizontal

Images of the Aurora, Nebraska hailstone. It measured seven inches (17.5 cm) in diameter with an 18.75-inch (47.63 cm) circumference.

HAIL SIZES

Hailstone size	Measurement		Updraft Speed	
	in	cm	mph	m/s
softball	4 ½	11.4	103	46
grapefruit	4	10.1	98	44
tea cup	3	7.6	84	38
baseball	2 ¾	7.0	81	36
tennis ball	2 ½	6.4	77	34
hen egg	2	5.1	69	31
golf ball	1 ¾	4.4	64	29
walnut	1 ½	3.8	60	27
half dollar	1 ¼	3.2	54	24
quarter	1	2.5	49	22
nickel	⅞	2.2	46	21
penny	¾	1.9	40	18
dime	⁷⁄₁₀	1.8	38	17
marble	½	1.3	35	16
pea	¼	0.64	24	11
bb	<¼	<0.64	<24	<11

Updraft velocities needed to keep hail of various sizes suspended in a thunderstorm. Note that tennis ball and larger hail requires updraft wind speeds that rival horizontal winds observed with hurricanes.

winds of 50 mph or more, it is easy to see how hail can strike objects with speeds high enough to puncture even thick roofing material. The effect is the same as if winds were calm, but an automobile was traveling along at 50 mph (16km/hour) or more when hail strikes it.

Large hail moving at high speed can injure or kill. Although rare in the US and other developed countries, hail fatalities are more likely in places lacking substantial structures or where many people are outside working. Farm animals can also be killed or injured by large hail.

Sometimes the combined effect of strong updrafts and strong horizontal winds near the top of thunderstorm clouds allows hail to be exhausted from the anvil portion of the cloud. When this happens, hail may reach the ground at some distance from the main thunderstorm cloud.

Hail losses

According to Stanley Changnon, the consummate US hail researcher, losses to property due to hail have escalated in recent years and averaged about $1.2 billion annually during the 1990s (adjusted to 1997 prices). As with other hazards, this is linked to a growing population with more belongings and more expensive possessions at risk. Not surprisingly, hail reports and hail damage are often concentrated in more populated areas. Even so, damage from hail can vary significantly across a local area due various weather factors (e.g. location of largest hail or strength of winds) and also structural variations (roof angle or type and age of shingles). Small hail, driven by high winds, can also cause extensive damage, especially to trees and crops.

In the 1970s, national property losses from hail were about one-ninth of the total annual hail losses (with most hail loss occurring to crops). By 2000, property damage had caught up with crop hail losses, in part because of a decrease in agricultural data reports. In the early1980s, the Illinois State Water Survey and a major agricultural crop insurer

Mexicans Ismael and his son Ismael Jr clear ice from the roof of their house in a village in the outskirts of Mexico City after an unusual hail storm blanketed the area with up to 25 centimeters (almost one foot) of ice on March 6, 2000. The highly unusual overnight storm left residents with a scene more akin to the Alps than Mexico.

Hail losses in "Hail Alley" for the 10-year period 1997-2006

State	# Reports	Property losses	Agricultural losses	Deaths	Injuries
Iowa	264	$121.2 M	$3.8 M	0	1
Kansas	517	$172.9 M	$23.5 M	0	6
Colorado	214	$170.1 M	$18.1 M	0	5
Nebraska	396	$297.0 M	$111.4 M	0	54
Oklahoma	351	$99.4 M	$250 K	0	2
Texas	1071	$802.0 M	$70.4 M	1	56
10-year TOTAL	**2813**	**$1.66 B**	**$227.45 M**	**1**	**124**

There is a significant variation in agricultural loss figures between Oklahoma and Nebraska. Hail injury statistics also vary greatly.

stopped collaborating on hail climatology research and the database for agricultural damage from hail suffered a climatological meltdown. Newspapers normally don't report small hail losses and Storm Data, the main US severe weather climatology document, lacks the framework for digging to find crop damage information. This is linked, in part, to how the publication is prepared and how the data for it is obtained. As a result, Storm Data still carries the disproportionate property-to-crop loss ratio that was reported back in the 1970s.

Although damage estimates due to hail (and other weather hazards) in the US suffer from serious limitations, they remain the best in the world. The data are the most easily accessible, most detailed, and highly reviewed. But scientists like Nolan Doesken, state climatologist of Colorado, bemoan the fact that efforts to determine hail climatology remain limited, at best, anywhere.

Yet, hail reports (and the amount of property damage) are often linked to population. Separate state-wide hail climatologies for Pennsylvania, North Carolina, and Colorado bear this out. Hail damage maps clearly show the locations of larger urban centers.

Above is a summary for 10 years (1997-2006) of large hail (greater than two inches (five cm) in diameter) data from the US online Storm Data publication for states in "Hail Alley". It is easy to see how agricultural data loss reporting varies from state to state and how reported agricultural damage remains only about 10 percent of other types of property damage. Reported injuries also show a marked variation from state to state; in Nebraska, for example, there were several instances in which large hail broke car windows, injuring occupants. Such events are rarely reported in other states. Much of the property damage is the result of relatively rare hailstorms (some with individual property losses in the tens of millions of dollars).

Hail swaths

Damage from hail is often confined to "hail swaths", the path in which hail is deposited by a moving thunderstorm. Sometimes, the swath is short and very narrow, only affecting part of a community or a field of corn. At other times it can be as much as 10 miles (16 km) wide and a hundred miles (160 km) long. It all depends upon the character of the parent thunderstorm and how fast the storm moves. At any one location, it is rare for hail to fall for more than a few minutes. In fact, a study of hailstorms in the Fort Collins, Colorado, area showed that the median hail fall duration was about six minutes.

However, storms with larger hailstorms tended to last longer. Of the 10 percent of all storms that lasted longer than 20 minutes, many were the most severe. Still, in some

Marble to golf ball sized hail up to 12 inches (30.5 cm) deep brings parking lot traffic north of Orlando, Florida, to a near standstill on March 6, 1992. Until another major hailstorm struck the Orlando area on March 25, 1992 (less than three weeks later), this storm was the worst in the area.

locations, even in places like Florida, hail can pile up to depths of many inches with wind-blown drifts measured in feet. If snow removal equipment is available, it is often taken out of storage to clear roads.

While severe thunderstorms often bring hail, it is usually the longer-lived storms that bring the largest hailstones. These storms have the strongest updrafts, which allow hail to grow and remain suspended the longest. The hail region then moves along with the thunderstorm dropping hail in a relatively narrow path. When the updraft weakens sufficiently or the hail gets too large to remain suspended, large hail falls occur.

A major hailstorm (hail up to the size of softballs) caused more than a million dollars damage to the US Geological Survey's EROS Data Center (EDC) in Sioux Falls, South Dakota on July 13, 1997. The storm destroyed an array of 512 solar panels (which heated 60% of the water used in the Center's photo labs), shattered many skylights and broke concrete roof paving tiles.

For documenting a hail event, it is better to collect, measure and photograph the hail. Keeping the hail in a sealed freezer bag in the freezer will keep the hail longer than

Cold air moving southeast collided with warm moist air from the Gulf of Mexico on May 13 and 14, 2005. Strong storms with hail and gusty winds pounded the area from the panhandle of Texas to the Ohio valley and Mid-Atlantic. The more intense convective cells (dark red) sit over north-central Texas in the image on the left, where there were several hail reports. The image on the right shows a 3-D perspective view of the squall line along the Oklahoma-Arkansas border.

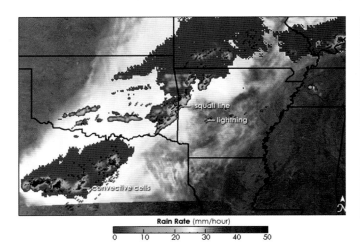

Rain Rate (mm/hour)

0   10   20   30   40   50

GLOBAL HAIL EVENTS

The following is a listing of just a few noteworthy (not necessarily the worst) hailstorm events around the world.

The earliest known severe hailstorm in the United Kingdom occurred at Wellesbourne in Warwickshire in May 1141 and contained hail not quite as large as a golf ball (by today's standards).

An artist's impression of the Wellesbourne hailstorm in May 1141.

April 4, 1977 (near New Hope, Georgia): A Southern Airways DC-9 jet crashed after being struck by large hail. The hail and associated heavy precipitation caused the engines to flame out. Seventy people (including eight on the ground) died; 21 people (including the flight attendants) survived.

July 12, 1984 (Munich, Germany): One of the most costly hailstorm events with damage estimated at $1-$5 billion.

April 30, 1988 (India): The deadliest hailstorm on record killed 246 people and 1,600 domesticated animals.

July 11, 1990 (Denver, Colorado): Softball-sized hail destroyed roofs and cars, causing more than $600 million in total damage.

September 7, 1991 (Calgary, Alberta, Canada): Hailstorm caused $400 million in property damage.

March 25, 1992 (Orlando, Florida area): Hailstones up to four inches (10 cm) in diameter resulted in more than $60 million in damage. This is still the costliest Florida hailstorm on record.

May 5, 1995 (Dallas, Texas): Costliest US hailstorm. Damage totaled $2 billion. More than 500 people injured.

April 14, 1999 (Sydney, Australia): Hailstorm causes $1.6 billion in damage, making it the costliest hailstorm to strike a populated city in the country. The hail damaged some 22,000 homes and more than 60,000 vehicles. Aircraft damage at Sydney Airport was extensive.

June 27, 1999 (Scottsbluff, Nebraska area): Large hail caused extensive damage to crops and property. Twenty-five people were injured as large hail broke car windshields. The Scottsbluff zoo reported injuries to many animals, particularly birds.

March 28, 2000 (Lake Worth, Texas): Last known hail fatality in the United States occurred when a 19-year-old male was killed by grapefruit-sized hail while trying to move a new car to shelter.

June 12, 2002 (Kearney, Nebraska area): Hailstorm caused $2 million in agricultural losses. More than 22,000 acres of corn, soybeans and alfalfa were severely damaged or destroyed.

July 19, 2002 (Henan Province, People's Republic of China): 25 people were killed and hundreds injured during a hailstorm.

June 22, 2003 (Aurora, Nebraska): Largest hailstone on record in the United States falls. It had a 7-inch (17.5-cm) diameter and a circumference of 18.75 inches (48cm).

In addition to simply damaging and destroying what it strikes, hail can also collect, either by hail fall or transit by water, and clog storm drains. Not surprisingly, during some hail storms urban street flooding may occur. Hail, much like its wintry cousin, sleet, can create slippery road conditions. In Maryland, one hailstorm, followed by bright sunshine, caused a multiple-car accident on an Interstate highway.

MEASURING HAIL

Most people estimate hail size by comparing it to things with which they may be familiar. Sports terms such as golf ball, softball, and even cricket ball (UK) are often used to size hail. In the US, coinage sizes (e.g. nickel-sized) serve the same purpose. Terms like "hen's egg", "marble", and "pea" are less descriptive as these objects can come in a variety of sizes and are often based on a person's past experience. Both the UK and US hail scales are shown below. Note that diameter is used to describe hail sizes in both.

Hail size and diameter in relation to TORRO Hailstorm Intensity Scale

Size code	Maximum Diameter mm	Description
0	5-9	Pea
1	10-15	Mothball
2	16-20	Marble, grape
3	21-30	Walnut
4	31-40	Pigeon's egg > squash ball
5	41-50	Golf ball > Pullet's egg
6	51-60	Hen's egg
7	61-75	Tennis ball > cricket ball
8	76-90	Large orange > Soft ball
9	91-100	Grapefruit
10	>100	Melon

Hail the size of golf balls and eggs. Since eggs come in various sizes, they do not make for reliable comparisons with hail.

leaving it unsealed. That's because most freezers today are "frost-free" and that means that sublimation form the hailstones will occur quickly.

Using a ruler provides measurable comparisons. However, you have to be quick to retrieve the hail before it melts, but even more careful while racing outdoors in a thunderstorm to collect samples. During the record-breaking hailstorm in the Aurora, Nebraska area, people actually wore hard hats as they collected hailstones. Clearly there is also the danger of lightning!

While rulers are often sufficient, using a caliper provides more precise measurements of hail.

Other ways to measure hail

Radar data provides useful clues that hail may be present in a thunderstorm. Radar works because reflected radar energy is returned to the radar site. Higher reflectivity values often suggest the presence of hail because more dense objects reflect more energy. Hailstones, especially those covered with a shell of liquid water, reflect radar energy even more. However, estimating hail sizes from radar is difficult because a large amount of small wet hailstones may be more highly reflective than a few large "dry" ones.

Another measurement approach involves placing "hail pads" outside when a thunderstorm approaches. Hail pads are styrofoam sheets wrapped tightly in heavy-duty aluminum foil. If hail falls onto a pad it will create an impact crater. Typically, the larger the hail, the larger the crater it leaves behind. By using a standard material and documenting the hail pad (the pads are sent to a central location for analysis),

Hail pads are now in widespread use across the United States. The new hail pads (left) are placed on the ground and the indentations (right) can be examined after the hailstorm has occurred.

meteorologists and others who look at hail damage don't have to "guess" at hail size. Hail impacts different materials differently – hail impact on a new roof may be substantially different that on one that is 25 years old.

As of early 2007, hail pad networks have been established in 18 US states thanks to the combined efforts of the National Weather Service (NWS) and others. Now 3,500 observers strong, the Community Collaborative Rain, Hail and Snow Network (CoCoRaHS) program is collecting and sharing precipitation and hail data in real-time. Although, the likelihood of hail occurring where a hail pad is located is small, the network has documented scores of hail events during the past few years. Several of these have actually helped the NWS issue more timely severe weather warnings (hail of 3/4 inch (2 cm) diameter – penny size or larger – is one criteria for issuing a severe thunderstorm warning).

Today, with digital cameras at hand, many people help to document hail occurrences. In a large number of cases, people have placed objects nearby the hail to make for easier sizing comparisons. Yet, except in rare cases, the overall sampling of hail sizes and distribution remains limited.

The major hailstorm that struck Munich, Germany, in 1984 had relatively few documented hail reports. For the Edmonton, Alberta, tornado in 1987, the area to the left of the tornado's track was very heavily sampled and hail size distribution mapping was easy to do. The story (see pages 154-155) about Northfield, Minnesota, shows what a community sampling effort can provide.

Making an impression

Hailstorms can "impact" storm chasers, too, especially when they get caught in the wrong part of the thunderstorm circulation pattern. One such event befell veteran Canadian storm chaser George Kourounis on May 12, 2005. While chasing tornadoes in the Texas high plains, Kourounis and other chasers were pummeled by giant hail. Their vehicles suffered broken windows and lots of hail dents. There were no injuries, but Kourounis noted that it was probably his "most intense chase day ever!"

Airplanes that fly into thunderstorms are also at risk from hail. A Southern Airways jet crashed in Georgia in 1977 after large hail and heavy rain were ingested into its engines. On August 10, 2006, a Capital Cargo International Airlines Flight from Calgary, Canada, to Minneapolis, Minnesota, encountered large hail at around 30,000 feet. The hail damaged many components on the plane's leading edge including the windshield, nose cone, cowling on the two engines, and leading edge of the right wing. Still, the damage was mostly cosmetic, according to some of the mechanics that worked on the aircraft.

Veteran storm chaser George Kourounis met up with some large hail on May 12, 2005. Or should I say that the hail met Kourounis and his vehicle?

Tennis ball sized hail damaged the nose cone and forward-facing parts of this aircraft over Alberta, Canada in August 2006.

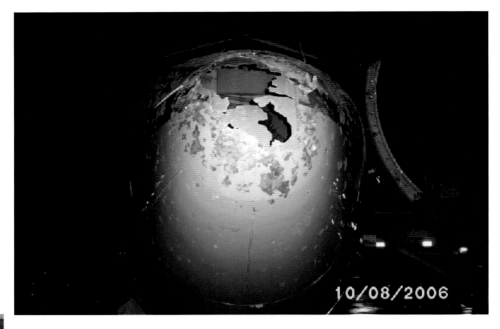

This Easyjet aircraft flew into a hailstorm over Switzerland on August 15, 2003.

During the past 10 years there have been several other hail incidents involving aircraft. These include a Brazilian Airlines Airbus (March 28, 2006) that suffered severe structural damage; a British-Midland International Airbus (May 27, 2003) in which large hail punctured the nose cone over Germany; an Easyjet 737 (August 15, 2003) that had its windshield shattered and nose cone pummeled by large hail outside of Geneva, Switzerland; and a DC-9 operated by Scandinavian Airlines (July 21, 2001) that even had its onboard radar damaged while on approach to Helsinki, Finland. In all instances, however, the planes landed safely without reported injuries.

A more serious incident occurred near Johannesburg, South Africa (April 22, 1999), when a Million Air Charter Boeing 727 flew into large hail. Although the plane and its 66 occupants landed safely, the aircraft was declared a total loss.

Then, on February 27, 2007, large hail (up to the size of golf balls) pummeled the Space Shuttle Atlantis on the launch pad at Cape Kennedy Space Center, Florida. The hail put more than 7,000 dings into the foam-covered fuel tank and caused a launch delay of more than three months while the damage was repaired and the spacecraft inspected for safety. According to Wayne Hale (perfect last name), the shuttle's program manager, "...the storm did the worst damage we've ever seen from hail."

A major hailstorm caused significant damage to the Space Shuttle Atlantis as it sat on its launch pad at Cape Kennedy, Florida on February 27, 2007.

Hail suppression

Because hail can be so devastating to crops, scientists have been studying hail suppression since World War II. They attempt to accomplish this by "seeding" (or inserting) silver iodide into thunderstorms. Silver iodide, when acting in consort with existing condensation nuclei, is supposed to overload the storm so that there can be many more small hailstones but fewer larger ones. The concept keys on the fact that smaller stones are less likely to damage crops and that smaller stones can melt more during their descent from a thunderstorm cloud.

Even with strict evaluation criteria, scientists remain hard-pressed to document success. Natural variability, varying storm characteristics and sampling procedures, and other factors can all mask statistical outcomes. David Atlas, a noted hail researcher, once observed during a hail project review that if just one hail event were moved from the seeded sample (seeding successfully limited large hail formation) to the unseeded one (seeding was unsuccessful), it would have totally reversed the results.

Hail suppression research, at least at the Federal and state levels in the US, has more or less ended. There are still some ongoing hail suppression projects underway in several agricultural regions around the world.

Climatic implications

Although the largest hailstone on record in the US fell in 2003, there is nothing in hail data that indicates hail is becoming more frequent or more severe. As with all other weather hazards, there are simply more people and more things at risk. When damage occurs, the value of losses is greater due to higher valuations.

Several hail studies, including one conducted at the National Weather Service Office in Lake Charles, Louisiana, indicate that the occurrence of a few significant storm systems (large damaging hail is relatively rare in any local area), the year-to-year variability, multi-year trends, and reporting procedures, all dominate hail statistics. As noted in earlier chapters, when significant hail events occur, they tend to "make the news". In our "instant breaking news" environment, this gives the impression too that storminess is getting worse.

Consider 1995, a year in which several strong storm systems affected the Louisiana area. There were 76 hail reports (hail at least 3/4 inch (2 cm) in diameter) in the counties for which the Lake Charles office issued severe weather warnings. That was double the previous record that occurred in 1991. This situation does not support a long-term trend, but rather the occurrence of certain types of large-scale storm patterns.

This large hailstone fell from a tornadic supercell northeast of Breckenridge, Texas. The storm left a path miles wide littered with three to four and a half inch stones. The hail fell from a dark cloud base to the south of a developing mesocyclone. No rain or thunder occurred in the immediate area while the stones were coming down.

SUPER-CELL THUNDERSTORM HAILS ON NORTHFIELD, MINNESOTA

On the morning of August 24, 2006, a super-cell thunderstorm raced southeastward across southern Minnesota. The storm was one of several that day which resulted in more than 150 reports of hail, and some two dozen reports each of tornadoes and high winds across the Dakotas, Minnesota and extreme western Wisconsin.

The storm that moved to the south of Minneapolis struck many communities but Northfield was especially hard-hit. Fortunately community residents were quick to document the event. Much of their story is told in the following images.

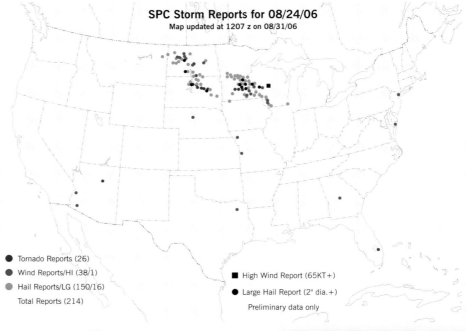

SPC Storm Reports for 08/24/06
Map updated at 1207 z on 08/31/06

● Tornado Reports (26)
● Wind Reports/HI (38/1)
● Hail Reports/LG (150/16)
Total Reports (214)

■ High Wind Report (65KT+)
● Large Hail Report (2" dia.+)
Preliminary data only

Gillian Wigley of Northfield, Minnesota, displays her hail collection. Some of the hailstones shown here are more than three inches (7.5 cm) in diameter.

Hundreds (if not thousands) of cars in Northfield were "punctuated" by hail. This VW bug (left) shows what hail can do. According to Griff Wigley, a resident, "The local car dealership (above) a couple blocks from us, got hit hard. Row after row after row of cars had windows smashed and dents."

Large and small hail litter this yard in Northfield. Since hail of many sizes is present in a severe thunderstorm, it is not uncommon for people to see such size mixtures on the ground.

Greenhouses, like this one at St Olaf College, are no match for large hail.

While large hail tends to break or dent solid objects (such as car windows and cars), it can also leave craters in softer objects (like wet soil). Here is an example of such cratering.

Due to cold downdrafts and the coldness of the hail itself, it is not unusual for significant hail events to be followed by dense fog.

-20 -10 0 10 20 30 40 50 60 70

Reflectivity image of the super-cell thunderstorm that produced the large hail in Northfield. Magenta coloration (highest radar reflectivity values) is where the largest hail is most likely occurring. The circled region is a "three-body scatter spike", typical of storms producing large hail. The "spike" is just one of many signatures meteorologists look for when evaluating severe thunderstorms. Notice that the spike extends radially outward from the radar through the hailstorm. Image time was 11.17 am CDT on August 24, 2006.

> "There is still an alarming lack of public support and education about the need for proper maintenance and repair of dams. Unless a dam fails, dam safety is not usually in the public view..."
> **American Society of Civil Engineers**

Flooding in Budapest, Hungary in 2006. In the foreground is a road flooded by the river Danube. Only a traffic sign and the most famous and oldest bridge (Chain Bridge) can be seen.

When one mentions the word flood, many people picture a raging river with water spreading across a floodplain. Others would see what appears to be a large lake where one did not exist before. Regardless, highways, homes and autos would be at least partially under water.

Watersheds

There was extensive damage caused by these floods in Massachusetts in spring 2006.

Floods are most likely the result of heavy rainfall across a large area, called a watershed. A watershed, or water collection area, can be likened to a bathtub. The water collects and is then moved along to another watershed downstream by a river or water channel. In the case of the bathtub, the drain handles this function. Eventually, the water drains into successively larger rivers or streams and eventually into an even larger water body (lake, bay or ocean, for example).

If more water enters the bathtub (watershed) than leaves from the drain, the water will still try to move along. But it "backs up". In doing so, it gets deeper, eventually overflowing the sides of the tub (river banks) and creates its own floodplain on your bathroom floor.

River floods happen all the time due to a long duration of heavy rainfall events (and other conditions, such as rainfall and snowmelt in

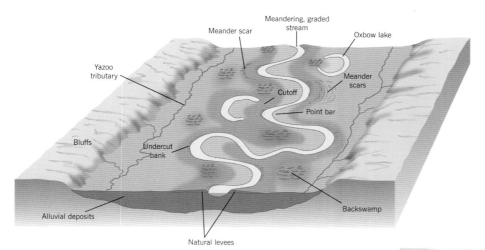

Meander scar
Meandering, graded stream
Oxbow lake
Yazoo tributary
Meander scars
Cutoff
Point bar
Bluffs
Undercut bank
Alluvial deposits
Natural levees
Backswamp

A diagram of a typical stream and flood plain.

combination). Here, water levels rise, more or less gradually, and then the waters spill out of their banks into the river's floodplain (the flat or nearly flat area near the river that experiences periodic flooding). As a result, the floodplain may become ideal agricultural land due to the addition of nutrient-rich soils, carried as sediment by the waterways, during the flooding event or it may become the deposition zone for rocks and boulders. The overflow also slows the river, hence lessening erosion downstream

Floodplains

Most river flooding is confined to the floodplain. If it weren't for cities, communities and businesses that have developed along waterways, these floods would not be a problem. Once human habitats are at risk, there are usually modifications made to the waterway, for example, straightening its channel, or its floodplain. In Johnstown, Pennsylvania, the site of three 100-year floods in the past 100 years, the city encroached upon the river, narrowing the river's banks to gain habitable land. In the process, it exposed itself to flooding risks.

Floodplains are more expansive along older, slow-moving, large rivers, for example, the Nile, Mississippi and Ganges. Here, meanders in the river, coupled with the overall slow speed of the flowing water, favor the deposition of silt and soil. Floodplains are often most extensive near the mouths of these rivers. Here, as rivers flow into larger water bodies, the river channel breaks up into smaller channels that spread apart from each other. As the water speed slows further, sand and silt in the river water can no longer be carried by the river. These materials are deposited and deltas – triangular shaped regions – often form.

Floods

River floods can also develop when heavy rain falls on melting snow pack, when rain falls on frozen ground and/or when rivers flow poleward into a region where the river remains frozen. The first scenario increases run-off through the addition of water from melting snow, as well as the rainfall itself. The second situation showcases what happens when rain falls on impervious surfaces (surfaces that do not allow water to permeate or enter the ground – e.g. frozen ground, concrete sidewalks and roofs).

This flood plain in Iceland clearly demonstrates the build up of gravel and other deposits that can occur.

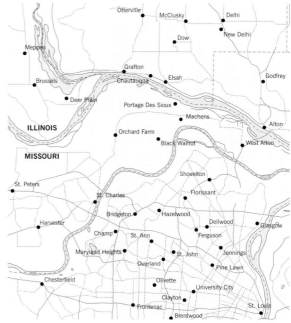

With many towns and cities built along the water's edge, flooding is a high risk event in many places as can be seen on page 158.

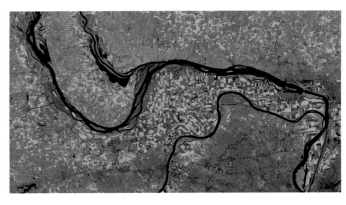
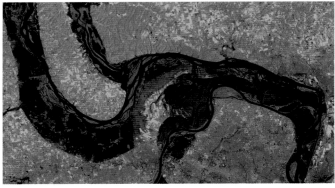

Above right is a satellite image for the area to the northwest of the St Louis, Missouri, metropolitan area during the Great Mississippi River Flood of 1993 (July 29). Notice how the normal river channels (left) are dwarfed by the flooded floodplain surrounding it.

The final scenario involves water backup either from water running into the ice itself and/or a collection of pieces of ice that have broken off from the frozen waterway and created an ice jam. If all three occur together, and if flood-producing forces are present, the flooding can be devastating.

When rain falls on any type of impervious surface the water has no choice but to run off. Eventually this water runoff will make it to streams or rivers either directly or through storm drains. If the watershed has steep terrain or is small, and rainfall is heavy, the waterway can respond almost immediately to the inflow and rise very quickly. This is when "flash floods" (floods that literally happen "in a flash") can occur.

Flash floods can also happen when thunderstorms align along a waterway or its watershed, each adding to the water inflow. When train echoes occur anyway along the waterway, the combined effect of the rainfall and runoff can create an overload.

A NASA ASTER image of Mississippi River delta. Notice how the river breaks up into smaller waterways and spreads apart as it enters the Gulf of Mexico.

If the waterway is filled with obstructions, for example, downed trees, rocks, or bridge underpasses, or if the channel becomes geologically or otherwise constrained, the water can back up and/or rise even more.

The Big Thompson Canyon flash flood

The flash flood that struck Colorado's Big Thompson Canyon on the evening of July 31, 1976, provided graphic testimony to this type of event.

In a meteorologically charged atmosphere (including high moisture, high instability, upslope surface winds and light winds at high altitude) a series of three slowly moving thunderstorms passed across the central portion of this narrow canyon area on the east slopes of the Rocky Mountains. Estimates place some 10-14 inches (25-35 cm) of rainfall in about three hours over a 70-square mile (181 sq km) region of the 800-square mile (2071 sq km) Big Thompson River watershed. The water was immediately channeled into a narrow waterway (the Big Thompson River usually runs only a few feet deep as it descends 2,500 feet (7,620 meters) down a 25 mile (40 km) narrow canyon area). The flood raced downhill with such ferocity that it swept away huge boulders, as well as homes, roadways, and bridges. The event took the lives of 144 people. Damage was more than $30 million.

The Red River of the North (which affects Minnesota and North Dakota) is especially susceptible to all three types of river floods. It flows northward from the extreme northeast corner of South Dakota along the North Dakota-Minnesota border and into Canada where it empties into Lake Winnipeg.

However, in addition to the factors described in the main text, the Red River has some other unique characteristics that add to its annual flooding cycle. Although most floods that affect the area are "minor", devastating floods happen about once every 10 years. In 1997, for example, the crest of the waters at Fargo, North Dakota reached almost 40 feet (12 meters). A typical flood stage (the height of the river at which flooding occurs) in the area is 18 feet (5.5 meters).

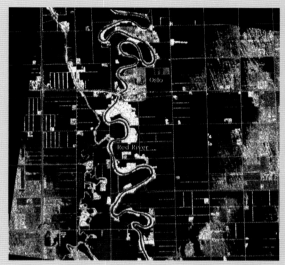

The Red River, meandering through North Dakota, is prone to flooding.

A look at the satellite image (left) shows the sinuous path of the Red River as its winds through the heart of Oslo.

Due to its direction of flow, the spring thaw begins near the headwaters of the river (where the river starts) and moves downstream with time. This forces melting snow and runoff water to move downstream and merge with other streams and rivers that are also discharging water from melting snow. If the melting from some or even many of the waterways is in synch, the more northern parts of the valley are at especially high risk of flooding. (However, the process might be disrupted by periodic cold fronts that slow the melting process.)

The Red River flows through what was once Glacial Lake Aggasiz, a giant, flat-bottomed lake that covered much of central Canada, southward to southern Minnesota. Aggasiz was formed as the North American Ice Sheet (that extended as far south as Kansas) retreated northward about 12,000 years ago. During the next 5,000 years it changed its character as the glacial ice advanced and retreated. In the process, the Red River incised a shallow, sinuous valley across the southern portion of what was once a giant lake bed.

Related to this physiographic factor is the young age of the Red River. In its present form, the Red River is about 9,300 years old and far too young geologically to have carved a significant valley-floodplain system. Since the land is so flat, when the river floods onto this plain, aerial coverage of the waters can become quite extensive. This is because the broad expanse of the lake plain actually becomes the "floodplain" to this river.

Gradient refers to the slope of a river. In the region just north of Fargo, the gradient of the Red River averages five inches (12.5 cm) vertically per mile of length. In the region near the North Dakota-Canadian border, however, the gradient drops to 1.5 inches (4 cm) of vertical drop per mile. During floods, as the Red River nears the Canadian border, it tends to pool due to lack of slope, the region becoming essentially a massive, shallow lake. As you can see, floods don't just happen. There are many factors – meteorological, hydrological, geological, and geographical – that have to come into play.

Confluence of Red and Sheyenne Rivers near Argusville, north of Fargo on April 14, 2001. In the foreground lies the meandering path of the Red River. The water level of the Red River in Fargo at the time of this image as 36.7 feet (11 meters) (flood stage is 18 feet (5.5 meters)).

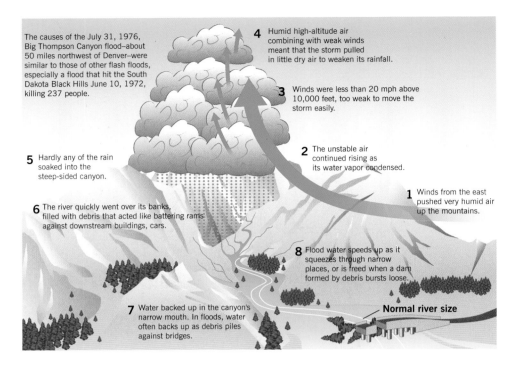

The causes of the July 31, 1976, Big Thompson Canyon flood–about 50 miles northwest of Denver–were similar to those of other flash floods, especially a flood that hit the South Dakota Black Hills June 10, 1972, killing 237 people.

4 Humid high-altitude air combining with weak winds meant that the storm pulled in little dry air to weaken its rainfall.

3 Winds were less than 20 mph above 10,000 feet, too weak to move the storm easily.

2 The unstable air continued rising as its water vapor condensed.

1 Winds from the east pushed very humid air up the mountains.

5 Hardly any of the rain soaked into the steep-sided canyon.

6 The river quickly went over its banks, filled with debris that acted like battering rams against downstream buildings, cars.

8 Flood water speeds up as it squeezes through narrow places, or is freed when a dam formed by debris bursts loose.

Normal river size

7 Water backed up in the canyon's narrow mouth. In floods, water often backs up as debris piles against bridges.

The following excerpt from a disaster report prepared by the University Center for Atmospheric Research (UCAR) expresses the magnitude and the extreme character of the event:

"Dozens of usually small tributary streams in the basin contributed large amounts of runoff to the Big Thompson River. In Dark Gulch and in at least two other small basins, discharge exceeded previously recorded discharge rates for basins less than four square miles (10sq km) in Colorado. However, Noels Draw, which lies just to the south of Dark Gulch, experienced a much lower maximum discharge of 2,050 cubic feet (50 cubic meters) per second from the 3.37-square mile (9 sq km) basin. This difference in run-off is probably related to velocity and direction of storm movement past the opposite-facing valley slopes and demonstrates the extreme concentration of the rainfall intensities experienced during this storm event.

"Further down in the drainage basin, above Drake, the peak discharge at the Big Thompson was 3.8 times the estimated 100-year flood discharge for that site. East of Drake, witnesses told of a wall of water 6-8 feet (2-2.5 meters) high in the narrows of the Big Thompson Canyon and in its north fork. The flood crest moved through the 7.7-mile (12km) stretch between Drake and the canyon mouth in about 30 minutes for an average travel rate of 15 miles per hour (24 km per hour)."

Levees

A levee is series of dam-like structures along a waterway instead of across it. If dams can fail, so can levees. Levees help to control water within a natural floodplain or other constrained area. While this normally protects areas outside the levee from flooding, it causes higher water levels between the levee and the river. Should the levee fail (i.e. collapse or breach), escaping flood waters can flow onto nearby areas with greater and more destructive force.

Levees are not a new creation. They were built along the Nile River in Egypt some 3,000 years ago. They were also part of other ancient civilizations including China and Mesopotamia. Today, levee systems can be found along many US rivers, including the

DAMS

This photo was taken less than two minutes after the Baldwin Hill Dam, Los Angeles, California, burst in December 1963 and shows water thundering down the hillside into the heavily populated area. Almost 300 million gallons of water poured down the hill causing an estimated $10 million damage and killing three people. A warning of nearly four hours was credited with preventing it from becoming one of the big disasters of all time in an area with 16,500 residents.

Dam failures (caused by construction flaws or leaks followed by erosion) can generate a flash flood downstream because so much water enters the waterway so quickly. Often small earthen dams fail this way. However, when leaks, erosion and collapse of the large Teton Dam in Idaho on June 5, 1976, occurred, it precipitated current efforts at assessing and improving dam safety. Several periods of heavy rains set the stage for the failure of Kelly Barnes Lake Dam in Toccoa, Georgia on November 7, 1977. The resulting flash flood killed 39, many students at Toccoa Falls Bible College, just downstream from the dam. Because the dam failed after the rainfall had ended, it became another "unexpected" event.

Historically, some of the largest disasters in the United States have resulted from dam failures.

On May 31, 1889, 2,209 lives were lost when the South Fork Dam failed above Johnstown, Pennsylvania following heavy rains. In 1977, six dams failed around Johnstown discharging six times the amount of water released in 1899. Almost 12 inches (30 cm) of rain fell in 10 hours as "train echoes" associated with a nearly stationary line of thunderstorms moved across the area. In 1928, the failure of the St Francis Dam (northeast of Los Angeles, California) killed more than 600. During the 1970s, the Buffalo Creek Dam in West Virginia failed and in 2003, the Silver Lake Dam in Michigan failed. Many of these dam failures and resulting flash floods happened at night.

These failures are dwarfed by the destruction resulting from a series of dam failures and intentional destructions (62 in all) along China's Ru River (far to the south of Bejing) in August, 1975 when Typhoon Nina struck the region. Nina interacted with a cold front and brought incredible rainfall to the province. During a one-hour period, an incredible amount of rain, almost 7.5 inches (19 cm), fell. Daily rainfall reached almost 42 inches (106 cm) in places. The magnitude of this one-in-2,000-years event comes into focus when one learns that the average annual rainfall for the area was only 31 inches.

According to the Hydrology Department of Henan Province, approximately 26,000 people died directly from the flooding and another 145,000 died during subsequent epidemics and famine. In all, 11 million residents were affected by the flooding. Almost six million buildings were destroyed.

The United States alone has some 76,000 dams which provide for one or more of the following functions: domestic, agricultural and/or industrial water supply storage; flood control; power production; recreation; streamflow maintenance in dry periods; and industrial cooling. Dams, many even taller and holding back more water than US dams, exist around the world.

The two largest dams are in Tadjikstan but Switzerland, Italy, Mexico, India, and Colombia also have large dams. The Hoover Dam, perhaps the most well-known US dam, is not even in the top 10 of the world's largest dams.

According to the World Commission on Dams (a three-year United Nations program that has now ended), there are some 40,000 large dams worldwide. According to the Australian National Committee on Large Dams Incorporated, Australia has about 500 large dams. China has now built the world's largest hydroelectric dam (the Three Gorges), one that is five times the size of the Hoover Dam. Globally, there are an untold number of smaller dams. There is also an International Committee on Large Dams, founded in 1928, that provides a forum for sharing knowledge and experience in dam engineering among its 85 member Nations.

This cross section of a New Orleans levee shows how the city had been protected against flooding but even these precautions could not prevent the deaths of around 1,500 inhabitants in 2005.

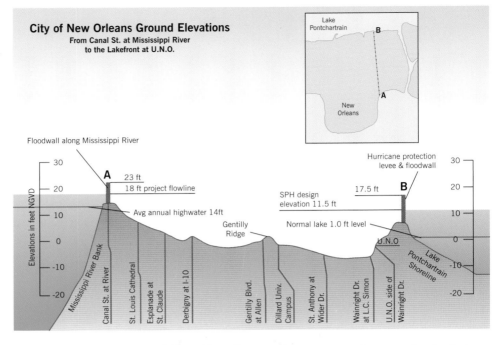

City of New Orleans Ground Elevations
From Canal St. at Mississippi River
to the Lakefront at U.N.O.

Mississippi, Sacramento and Ohio, their European counterparts, for example, the Loire, Volga and Rhine, and in Australia, India and some parts of central and southern Africa.

Levees can fail from erosion of soil material, overtopping, and resulting erosion, shifting of base earth materials, and even by being struck by large objects (such as boats) floating in the waterway. Of course, there is always the specter that the levee was not built properly or maintained well.

High waters that cause levee failure can come from river flooding, storm surge from a hurricane, and localized heavy rainfall atop either or both.

Hurricane Katrina and levee failure

Perhaps the most well-recognized levee failure occurred when Hurricane Katrina struck New Orleans, Louisiana in late August, 2005. Here, almost anything that could have gone wrong happened almost at the same time. Some levees were topped (and overflowed), some were struck by floating objects and either started to leak or collapsed, and many others simply breached. As water poured through openings in the levee system, about 80 percent of the city was inundated by flood water. The floods killed about 1,500 people and nearly destroyed the city.

A break is seen in a levee in an aerial view of damage from Hurricane Katrina in New Orleans, Louisiana, Tuesday morning, August 30, 2005.

However, meteorologists and others knew for years that New Orleans was vulnerable to such a disaster. Built on low-lying land and reclaimed marshlands, the city only existed because of a network of levees that frame its fragile infrastructure. The cross section (see above) of one levee in the city shows just how much of the city lies below sea level.

These levees have evolved over time since the early 1700s and have varying degrees of current

The Thames flood barrier was designed to prevent London from flooding. Opened in May 1984, the gates between the piers are raised when water levels are forecast to be dangerously high.

engineering factored into them. Other levee projects designed to further protect the area have been slow to happen due to fiscal and legal constraints. One levee, the Lake Pontchartrain Hurricane Barrier, was proposed following Hurricane Betsy in 1965. It was

THE MEASURE OF A FLOOD

There are many ways to measure a flood. Typically a flood is often defined as either the height of the flood crest or its impact (economic or human toll). Someone is always quick to assign a return period for the event (e.g, a 1-in-a-100-year event). But, there are other ways to assess floods.

Three major floods have affected the Missouri River in Kansas City, Missouri since 1844. The events are summarized in the table below:

Major floods affecting the Missouri River at Kansas City

Date	Crest (height above sea level)	Discharge (cubic feet per second
July 27, 1993	755.27 feet (230.2m meters)	541,000 (15,000 cubic meters)
July 16, 1844	754.40 feet (229.94 meters)	625,000 (17.5 cubic meters)
July 14, 1951	752.60 feet (229.40 meters)	573,000 (16 cubic meters)

Image at Westport Landing on the Missouri River showing signs commemorating historical floods.

Although the floods are ranked according to crest to match the signage at Westport Landing, there is not a direct relationship between crest and discharge (water actually flowing past a place).

Since all three of these flood events were in the summer, they had to have been related to excessive thunderstorm rainfall over a fairly large area.

Because of urbanization, there is no way currently to "normalize" historical floods by correcting for changes in impervious surface coverage, stream channel modifications and the like. In many urban areas, natural stream areas have been replaced by underground storm drains or concrete channels. While these help to remove water quickly from urban areas, they add water quickly to streams without first allowing some of the run-off to be absorbed into the ground.

never completed because a federal judge stopped the project on environmental and procedural grounds. Whether good or bad, environmentally responsible or not, the project will surely get a second hearing after Katrina's visit.

Global lessons

The Netherlands (which literally mean "lowlands") is a small country lying on the North Sea just to the north of France. For hundreds of years, the country has used dikes (another name for levee) and sea walls to keep their country from being gobbled up by the North Sea and periodic flooding events. The surge barriers along the coast in the Netherlands are much like sea walls in the United States (e.g. Galveston). However, the barriers in the Netherlands contain flood gates that help to control water flow on either side of the gate. Such a gated surge wall was envisioned for parts of the New Orleans area but was never realized.

Statistics

In 2006, according to the World Health Organization's Center for Research on the Epidemiology of Disasters (CRED), 226 out of 395 global disasters were flood related. To be considered a disaster, the event must have met at least one of the following: 10 or more fatalities; 100 or more people affected; a state of emergency declared; or a call for national assistance. Most of these criteria require that the event either strike a populated area and/or strike an area that is unable to fend for itself. Both of these can slant the occurrence of such events toward under- or less-developed nations. This perspective doesn't negate any impacts of the event(s), it merely places things into a known sociological, political, and economic framework.

In the United States, flooding is one of the largest causes of deaths, surpassed only by heat waves in certain years (see below). Yet, in terms of billion dollar disasters, floods only accounted for 17 percent (12 of 70) of such events during the period 1980-2006. If one factors in hurricanes and tropical storms (which have a large flooding component), there were 36 flooding events (51 percent) in this group.

However, the worst floods on record belong to China. In 1332, flooding on the Hwang He River drowned over seven million people, and a further 10 million died from the resulting famine. The worst recent flood was also on the Hwang He in 1887 where an estimated six million people were killed. Even the floods associated with the Ru Dam failure in China (1975, 85,000 fatalities) and the Johnstown, Pennsylvania (US) flood in 1889 pale by comparison.

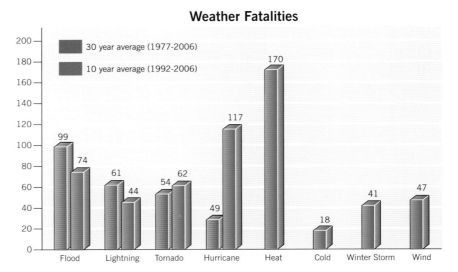

Weather Fatalities

Compilation of US hazard-related fatalities. Except for heat waves, flooding and events with a major flooding component (i.e. hurricanes) top the list. Many of the deaths due to Hurricane Katrina were included under hurricanes, not flooding. Any other fatalities linked to Katrina were not included under any specific hazard category. This is because the bodies were found long after the storm and local officials were not able to clearly assign a contributing cause of death.

Damage losses from floods are staggering, yet due to organizational, political, and social factors, the numbers are at best approximations. US losses (adjusted for construction costs) are now approaching $10 billion annually. In Australia, annual losses are about US $400 million.

A recently issued report from the United Nations University states, "Floods presently impact an estimated 520+ million people per year worldwide…the greatest potential flood hazard is in Asia…between 1987 and 1997, 44 percent of all flood disasters worldwide affected Asia, claiming 228,000 lives (roughly 93 percent of all flood-related deaths worldwide). Economic losses in the region in that decade totaled US $136 billion…" Although many of the Asian floods impact on less-developed areas, the numbers of people impacted raises the losses. The result is that Asian annual flood losses exceed US losses by about 60 percent.

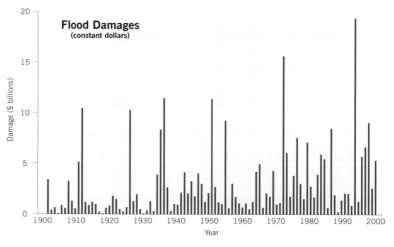

This graph clearly shows the escalating costs of repairing damage caused by flooding.

Climate implications

From a climatic perspective, climate change or global warming, one has to ask, "Is flooding getting worse?" From a death toll perspective, the answer would have to be resounding no. Deaths from flooding overall are not escalating. In fact, in many places, due to improved warnings, education, and community preparations, they are lower.

In terms of economic impacts, one would have to say yes. Losses from flooding continue to grow especially in vulnerable areas and in highly populated ones. With a push toward living in the coastal zone (see page 36) and along waterways, there is a rapidly growing population, and its homes, businesses, and belongings (however limited per household) is at risk. Thus, whenever a flood occurs losses are magnified by a population factor.

Flood frequency and severity

But are flooding events, and record-breaking floods, increasing? To answer this, I looked at record flood crests for four waterways (see the table on page 166); The Potomac River at Little Falls (just to the northwest of Washington, DC), the Mississippi River (Memphis, Tennessee), Greens Bayou (Houston, Texas) and the Concho River (San Angelo, Texas). Floods since 1970 are highlighted.

Other than for the very small Greens Bayou watershed, most top 10 flood events occurred before 1970 and all of the top several record flood events over 60 years ago.

What is readily apparent is how different the two Texas sites are. One has recorded nine record floods since 1970 while the other has none. The two sites are only 363 highway miles (585 km) apart.

Australia's Bureau of Meteorology has posted a listing of 10 major flood events in that country. Aside from the two events in the 1990s and one in the 1970s, all other events shown are pre-1956, with half occurring between 1916 and 1940. This looks much like the individual river statistics for the US shown above.

Since the cause of most flooding often remains too much rain too quickly, I looked to see if heavy rainfall events were increasing in magnitude and frequency.

Table showing top 10 floods for four US waterways

Potomac River, MD (data 1930-2007; drainage area 11,560 sq miles)	Greens Bayou, TX (data 1952-2007; drainage area 69 sq miles)	Mississippi River, TN (data 1872-2007; drainage area 930,000 sq miles)	Concho River, TX (data 1906-2007; drainage area 5,542 sq miles)
1 28.10 ft on 03/19/1936	**1** 67.81 ft on 06/09/2001	**1** 48.70 ft on 02/10/1937	**1** 47.50 ft on 08/06/1906
2 26.88 ft on 10/17/1942	**2** 66.04 ft on 06/27/1989	**2** 45.80 ft on 04/23/1927	**2** 46.60 ft on 09/17/1936
3 23.30 ft on 04/28/1937	**3** 66.00 ft on 10/25/1984	**3** 40.76 ft on 03/14/1997	**3** 40.70 ft on 10/03/1959
4 22.03 ft on 06/24/1972	**4** 65.89 ft on 05/18/1989	**4** 40.50 ft on 05/08/1973	**4** 39.80 ft on 05/09/1957
5 19.29 ft on 01/21/1996	**5** 65.75 ft on 09/21/1961	**5** 40.50 ft on 02/22/1950	**5** 36.80 ft on 04/26/1922
6 17.99 ft on 11/07/1985	**6** 65.17 ft on 06/06/2001	**6** 40.30 ft on 03/07/1975	**6** 35.90 ft on 07/23/1938
7 17.84 ft on 09/08/1996	**7** 64.68 ft on 08/19/1983	**7** 40.20 ft on 05/22/1961	**7** 32.65 ft on 10/13/1930
8 17.60 ft on 08/20/1955	**8** 64.61 ft on 11/18/2003	**8** 39.30 ft on 04/20/1936	**8** 31.76 ft on 09/26/1946
9 15.62 ft on 10/30/1937	**9** 64.15 ft on 03/04/1992	**9** 39.20 ft on 05/15/1983	**9** 31.20 ft on 05/31/1925
10 15.25 ft on 05/14/1932	**10** 63.00 ft on 10/16/2006	**10** 39.20 ft on 04/06/1945	**10** 30.88 ft on 08/23/1942

Top 10 floods on 4 US waterways. Floods since 1970 are shaded.

Heavy rainfall frequency and records?

There is little in the rainfall statistics that suggests that this is happening. In fact, a brief review of daily rainfall records at Moline, Illinois (a single location selected at random) shows the following.

Daily rainfall records show decline from the 1870s to the 1890s and then a gradual rise to a peak in the 1970s. Even with a large number of daily records in the 1990s, there is a clear decline in the number of records since. In fact, even when normalized to a full 2000-09 decade, the 1970s will still have had about 60 percent more daily records. If rainfall extremes are supposed to be increasing due to climate change, one would not expect such a dramatic decline in record breaking events.

According to a summary prepared by the Council on Environmental Quality (CEQ), the period since about 1970 has been much wetter than earlier decadal periods. Dry periods dominated in the early-mid 1930s again around 1950.

The same CEQ report notes that in an average year, about nine percent of the contiguous US is unusually dry and a comparable area unusually wet. Variability is quite significant, however. In 1983, more than one third of the country was unusually wet; in the Dust Bowl year of 1934, almost half the Nation was unusually dry.

The dry period across the US in the 1930s does not seem to appear in Moline's daily rainfall records.

I also looked Moline's peak daily rainfall by months. During the 135-year period, there was definitely a tendency for more daily records to be broken in more recent years. However, since the 1970s, only one monthly "extreme" record has fallen. Thus, even though the most recent 35 years or so has been wetter, extreme events have not occurred very often.

Daily rainfall records sorted by decades. Although there have been many more records in recent years, there are still a large number of records from more than 100 years ago.

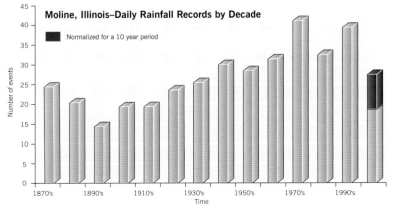

Moline, Illinois–Daily Rainfall Records by Decade

Floods

The intent of this assessment for a single rainfall-reporting site was simply to highlight some of the many different types of statistical and/or graphical measures that are possible. Depending upon which one is used, one may arrive at significantly different conclusions. I wonder what similar single station analyses (using relatively long-term rainfall records) from other parts of the world will show!

My challenge to global warming scientists and those who fund their research is to have each determine one data set or statistic that supports global warming and one that does not. This approach (while not completely unbiased) will help better define the problem than having scientists pre-disposed to supporting one side of the issue. Since flooding and drought are supposedly the by-products of the warming trend, rainfall would be a good place to start in terms of assessing changing conditions.

It is not just the volume of flood water that causes destruction, it is also the debris that is collected along the way. Branches have clearly dislodged a supporting pillar under this Austrian bridge, making it unsafe.

It can be more than climate change

While rainfall variation can be a factor worth looking at, population change, urbanization and watershed size are more likely at work. Around the time of the oil embargo in the 1970s, Houston's petro-chemical industry experienced a boom. More people meant more homes, more roads and more impervious surfaces. On a watershed that is only 69 square miles (179 sq km), even a small percentage change in the type of surface can yield large changes in runoff. That is most likely what has happened at Greens Bayou.

The period of data record can also play a role. All of the flood events on Concho River happened before 1960. Data for Greens Bayou only became available in 1952.

River or stream management is also a factor. In 1951, the OC Fisher Dam was completed on the north fork of the Concho River just to the west of San Angelo. This clearly helped to control river flooding even though two floods occurred after the dam was built.

A spring high tide floods a street in Richmond (near London) leaving cars waterlogged.

The other two sites drain larger watersheds and thus require more significant, larger-scale rainfall events in order to produce flooding. The Potomac River flooding in 1972 was tied to Tropical Storm Agnes; the flooding in January 1996 was linked to heavy rainfall and warm temperatures that followed a major snowstorm.

"...desert dust is virtually omnipresent within the atmosphere..."
Saharan Mineral Dust Experiment (SAMUM)

Looking down an interstate during a dust storm. Tumbleweed can clearly be seen blowing across the road.

Most people have never experienced a dust storm, the kind in which a wall of dust suddenly engulfs everything in its path. I accidentally came across one in Monument Valley, Arizona about 20 years ago. It was a small dust storm, generated by localized thunderstorm winds. As I watched the dust from a distance, the winds suddenly gusted and picked up a cloud of fine red desert dust. In an instant, I was in the cloud. Then, just as quickly, the wind slackened and the visibility immediately improved. The pulsing between being in a dust cloud and having good visibility continued for a few minutes. Then the storm was over.

Global deserts

My experience pales when compared to the periodic dust storms that develop over the sub-tropical desert region that extends from North Africa across the Middle East into interior Asia (Northern Hemisphere) and across Australia (Southern Hemisphere). The North African region's Sahara is known by many as the "classic" desert. However, the Atacama Desert in northern Chile may rank as the driest.

Deserts are often quite large (since they are forced by large-scale, long-term, precipitation deficits). The Sahara Desert covers almost 3.5 million square miles (9 million sq km). Deserts in Australia, combined, total around 1.3 million square miles (3.4 million sq km). The Gobi Desert (interior Asia) covers about half a million square miles (1.3 million sq km). But there are many smaller deserts around the world that also serve as breeding grounds for dust storms.

This south-looking, late-afternoon view shows one of the best examples in the shuttle photo database of a dust storm over the Sahara on May 13, 1992. A series of gust fronts, caused by dissipating thunderstorms, have picked up dust along the outflow boundaries while small cumulus clouds have formed over the most vigorously ascending parts of the dust front, enhancing the visual effect.

PARTIAL LIST OF THE WORLD'S DESERTS AND THEIR ATTRIBUTES

Desert	Location	Size (sq miles)	Topography
SUBTROPICAL DESERTS			
Sahara	Morocco, Western Sahara, Algeria,Tunisia, Libya, Egypt, Mauritania, Mali, Niger, Chad, Ethiopia, Eritrea, Somalia	3.5 million	70% gravel plains, sand and dunes. Contrary to popular belief the desert is only 30% sand. The world's largest non-polar desert gets its name from the Arabic word Sahra', meaning desert.
Arabian	Saudi Arabia, Kuwait, Qatar, United Arab Emirates, Oman, Yemen	1 million	Gravel plains, rocky highlands; one-fourth is the Rub al-Khali ("Empty Quarter"), the world's largest expanse of unbroken sand.
Kalahari	Botswana, South Africa, Namibia	220,000	Sand sheets and longitudinal dunes.
Gibson	Australia (southern portion of the Western Desert)	120,000	Sandhills, gravel, grass. These three regions of desert are collectively referred to as the Great Western Desert – otherwise known as "the Outback." Contains Ayers Rock, or Uluru, one of the world's largest monoliths.
Great Sandy	Australia (northern portion of the Western Desert)	150,000	
Great Victoria	Australia (southernmost portion of the Western Desert)	250,000	
Simpson and Stewart Stony	Australia (eastern half of the continent)	56,000	Simpson's straight, parallel sand dunes are the longest in the world – up to 125 miles. Encompasses the Stewart Stony Desert, named for the Australian explorer.
Mojave	US: Arizona, Colorado, Nevada, Utah, California	54,000	Mountain chains, dry alkaline lake beds and calcium carbonate dunes.
Sonoran	US: Arizona, California; Mexico	120,000	Basins and plains bordered by mountain ridges; home to the Saguaro cactus.
Chihuahuan	Mexico, southwestern US	175,000	Shrub desert; largest in North America.
Thar	India, Pakistan	175,000	Rocky sand and sand dunes.
COOL COASTAL DESERTS			
Namib	Angola, Namibia, South Africa	13,000	Gravel plains.
Atacama	Chile	54,000	Salt basins, sand, lava; world's driest desert.
COLD WINTER DESERTS			
Great Basin	US: Nevada, Oregon, Utah	190,000	Mountain ridges, valleys, 1% sand dunes.
Colorado Plateau	US: Arizona, Colorado, New Mexico, Utah, Wyoming	130,000	Sedimentary rock, mesas, and plateaus – includes the Grand Canyon and is also called the "Painted Desert" because of the spectacular colors in its rocks and canyons.
Patagonian	Argentina	260,000	Gravel plains, plateaus and basalt sheets.
Kara-Kum	Uzbekistan, Turkmenistan	135,000	90% gray layered sand – name means "black sand".
Kyzyl-Kum	Uzbekistan, Turkmenistan, Kazakhstan	115,000	Sands, rock – name means "red sand".
Iranian	Iran	100,000	Salt, gravel and rock.
Taklamakan	China	105,000	Sand, dunes and gravel.
Gobi	China, Mongolia	500,000	Stony, sandy soil and steppes (dry grasslands).
POLAR			
Arctic	US, Canada, Greenland, Iceland, Norway, Sweden, Finland, Russia	5.4 million	Snow, glaciers and tundra.
Antarctic	Antarctica	5.5 million	Ice, snow and bedrock.

THE DRIEST PLACE ON EARTH

The Atacama Desert (northern Chile) is caused by upwelling, a rain-shadow effect and its location in the subtropical high-pressure zone. As a result, it is the driest region on Earth (with an annual average rainfall measured in millimeters, at best). Some parts of this desert have not received a drop of rain in more than 100 years. According to Dr Tibor Dunai from Edinburgh University, UK, an analysis of some dry river beds in the Atacama indicate that they have not had water running through them for 120,000 years.

Dunai also found loose sediment surfaces that would have been washed away by any desert rainfall and these are older than 20 million years. This is much older than other such hyper-arid regions, such as the Dry Valleys of Antarctica (10-11 million years) and the Namib Desert in Africa (five million years).

Related research by a team of scientists from Louisiana State University, NASA, the Universidad Nacional Autonoma de Mexico and other research organizations, has discovered that the Atacama is amazingly similar to the surface of Mars. Both lack any forms of life and organic material. Dr Fred Rainey, an associate professor at Louisiana State University (LSU), noted: "The Atacama is the only place on Earth (from which) I've taken soil samples to grow microorganisms back at the lab and nothing whatsoever grew." This sets the stage for using the Atacama to test instruments and hypotheses about potential life on Mars.

The Atacama Desert in Chile is almost completely barren.

According to scientists, sand can be qualitatively defined as any particle that is light enough to be moved by the wind but too heavy to be held in suspension in the air. Very fine particles that can be held in suspension are therefore classified as silt or dust, while heavier particles unaffected by wind are classified as pebbles or gravel.

While deserts are typically found globally in the 15 to 35 degree latitude band (see map on page 171), the zone of the subtropical high-pressure systems (where rainfall is naturally limited), they can occur elsewhere; the interior regions of large continents (far removed from oceanic water sources) and places in the rain shadow (downwind side) of mountain ranges are other favored locales. Areas along the west coasts of middle latitude continents are also prime desert zones. Here, prevailing winds bring colder bottom waters to the surface (a process called upwelling). The colder surface waters stabilize the atmosphere, lessening rainfall.

Although most of us have the image of desert regions as being hot, some deserts are actually cold. The Gobi (interior Asia), on an elevated plateau and the Antarctic continent are perhaps the best examples of this type of desert.

Deserts are typically viewed as wastelands. Yet, some can provide ideal locations for raising saltwater fish, growing certain saltwater grasses and harvesting solar energy.

Moving sand and dust…

Just because somebody lives in or near a dry or desert region, it doesn't mean that person will experience a dust storm. Other necessary ingredients must come into play. These include dry, properly sized dust, sand or soil particles; winds strong enough to move larger particles horizontally (a process known as saltation); a surface that allows the wind to work most effectively; and sufficient surface heating or instability to allow convection to assist in the vertical transportation of smaller dust particles. In fact, depending upon

DESERTS OF THE WORLD

Deserts have many faces. Some desert regions may have large sand dunes; others may just have flat barren landscapes; still others may just be barren rocky areas.

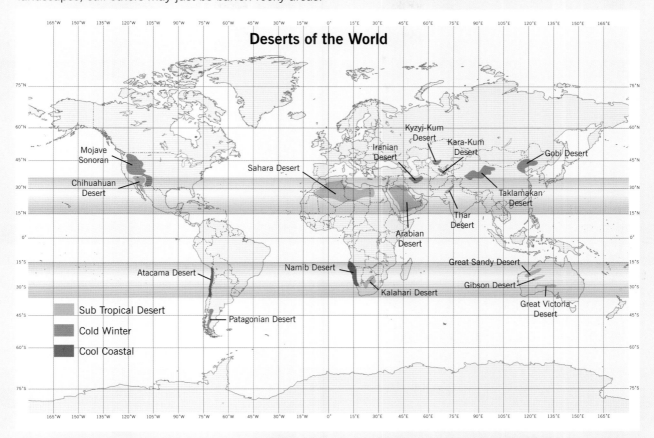

Deserts of the World

Mojave
Sonoran

Chihuahuan
Desert

Sahara Desert

Iranian
Desert

Kyzyj-Kum
Desert

Kara-Kum
Desert

Gobi Desert

Taklamakan
Desert

Thar
Desert

Arabian
Desert

Atacama Desert

Namib Desert

Great Sandy Desert

Gibson Desert

Kalahari Desert

Great Victoria
Desert

Patagonian Desert

Sub Tropical Desert

Cold Winter

Cool Coastal

Three different types of desert landscape: Gobi (above left); Sahara (above); and Mojave (left)

An example of wind erosion in the Arches National Park on the Colorado Plateau.

the size of the sand particles, sustained winds of only about 10-20 miles per hour (16-32 km/hr) are needed to start the sand movement process. Once the process begins, sand grains can bounce and/or dislodge other grains causing them to move. The heavier sand grains don't bounce or move very much above the ground. You may have experienced this on a windy day at the beach. The moving sand easily strikes your lower legs, but nothing much higher.

The movement of larger sand particles then dislodges smaller dust particles that may have been trapped beneath the larger particles or simply attached to other smaller particles by electrostatic forces.

In some desert regions, the moving sand and dust act like sandpaper on natural rock formations. Based on how the wind (and the sand/dust) is channeled, the rocks can be sculpted into very artistic patterns.

Moving sand can also collect in interesting formations. If you have walked along a sandy beach, you can also see how larger sand grains collect as they bump into and move around various obstacles (e.g. driftwood, dune grasses, or litter). In fact, these types of obstacles actually help to anchor the sand into dunes or hills. However, dunes can also form without such anchoring.

Although we are usually most familiar with dunes along sandy beaches, dunes are actually more widespread and larger in inland desert regions. Dunes can also be found on

A more comprehensive effort to build and maintain dunes occurred in coastal Delaware. Here, whole sections of dunes were restored and replanted with natural dune grasses.

other planets in our solar system (e.g. Mars, Venus and one of Saturn's moons, Titan).

Dunes along coastal beaches help to protect the land and human structures that lie on their land-facing side from ocean waves. Many coastal dunes are natural, including the types of plants that grow on them, such as sea oats, poison ivy, and American dune grass. With their deep or complex root systems, these plants help to anchor the dune.

Dunes, similar to these, were built along the North Carolina Highway 12 by the US government in the 1930s as part of Civilian Conservation Corps projects. The CCC also planted the dunes. Now that development has disrupted the normal seasonal transit of sand along coastal regions and human intervention is required to keep the dune structure intact. Sand fences are being built and maintained by some homeowners. Since each homeowner takes care of only a small part of the entire dune structure, this is a more haphazard process, where the integrity of the whole dune becomes questionable.

This house on Casey Key Beach, Florida is actually built on the dunes themselves.

Unfortunately, in some places, homes are built too close to dunes. In other places, such as northwest Florida, the homes and hotels are actually built on the ocean side of the dune or the dune is actually leveled to enhance beach access. In coastal North Carolina, the lack of dune structure has actually caused periodic erosion of a coastal highway and the loss of numerous coastal homes during hurricanes and/or winter storms. In 2006-07, the North Carolina legislature passed new laws that allowed for further development in coastal areas along beaches that were replenished after storm-related erosion.

Growing deserts

In some places, changes in land use and/or changes in rainfall allow desert regions to expand their coverage. As this process, known as desertification, takes place, sandy areas expand either through sand movement and/or the loss of overlying soils (which exposes the sand underneath). Near Freeport, Maine (Maine is not a natural desert region), the exposure of an underlying sandy area due to some bad farming practices (e.g. failure to rotate crops and excessive land clearing) in the 1800s has helped to create a growing sand area (now nearly 40 acres in size) that is overtaking nearby forest regions. Known as the "Desert of Maine," it provides visual testimony to how desertification can occur.

While it may be a picturesque setting, there are dangers involved in building houses so close to the ocean.

Similar poor framing practices, coupled with a very dry period, led to the "Dust Bowl" in the southern plains of the United States during the 1930s and a similar situation in Australia about 10 years later. The African continent is well-known for its desertification.

In some locations, such as India and Israel, plantings and irrigation have helped to reclaim desert areas and/or slow the advance of sand.

Recipe for a dust storm

While dunes show how wind can move sand and cause it to collect in local areas, it is the longer-range transportation of smaller dust and soil particles that is more important on a global scale. While expanding deserts and loss of agricultural areas force populations to relocate, it is the long-distance transportation of sand in the form of

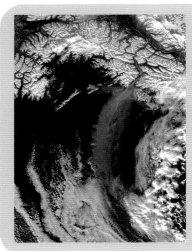

AN UNUSUAL SOURCE OF DUST

While deserts and semi-arid regions are the primary sources of dust, glacial deposits, known as loess, are another. Loess is a fine-grained silt formed as glacial ice grinds and crushes underlying and embedded rocks. When coupled with strong downslope valley winds (known as Chinooks or foëhns), dust plumes can blow out to sea from river valleys.

In this NASA Aqua satellite image, a large loess-laden plume extends from the Copper River Valley into the Gulf of Alaska. Here the winds and their dusty cargo blew down the valley, constrained by the valley walls. Once the dust escaped the valley channel, it was free to more easily disperse horizontally and the plume quickly widened.

dust storms that impacts on people's health, affects transportation, influences ecosystems health and much more.

Although dust particles are often about one tenth of the diameter of sand grains (or smaller), it takes winds strong enough to move sand before dust can break free from the ground. Once the sand grains start to move, they can bounce into smaller dust particles, dislodging them and sending them into the air. During daylight hours, when convective processes are at work, natural vertical motions and turbulent mixing of air can lift the smaller particles to high altitudes (10,000 feet (3,000 meters) or more). By late afternoon and evening, vertical motions diminish and so do many dust storms.

Once dust is airborne it can precipitate, creating a dust cloud near the ground, even without wind. This happened in the Dallas-Fort Worth area on April 2, 1982. While high winds picked up dust further west, winds were calm to light in the Dallas area. Dust from higher altitudes just settled across the north Texas area as the day time turbulent mixing and rising air currents weakened.

If the dust is carried far enough from its source region, it can interact with thunderstorms that develop ahead of the dust cloud. When this occurs, any rain that falls can become

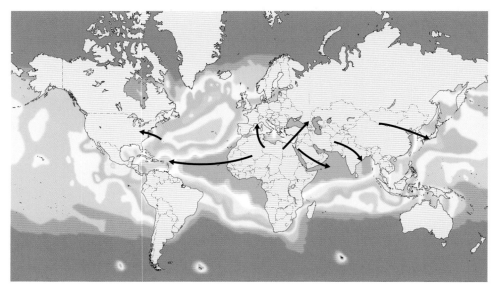

Worldwide distribution of dust over ocean areas. Notice the path that extends from western Africa to the Caribbean Sea (Trade Winds region). Another area extends from the Middle East into the Arabian Sea. Nearby land areas are, of course, affected.

Lake Chad dust storm with individual point sources to the northeast.

heavily dust laden. Even if rainfall is light, the muddy raindrop residue provides testimony to the dust storm nearby.

Contrary to expectations, it is not always the "wall of wind" that races across the desert landscape and lifts huge dust clouds. Rather, localized high wind regions (possibly due to channeling though rocky region gaps) and/or localized dust/sand sources allow for the formation of individual dust plumes. If enough of these are present, they can merge into a larger dust cloud.

Because winds often increase significantly with altitude, once the dust escapes from near the ground, it often travels much faster than surface winds would indicate. Once suspended, the finer dust particles can take a long time to precipitate out from turbulent air. Not surprisingly, some dust storm clouds can travel for thousands of miles. African dust clouds can bring dusty skies to the southeast United States. Some Saharan dust storms can bring dust as far north as north-central Europe. One recent Chinese dust storm made it across the Pacific and brought dust to Washington State.

Although it doesn't lead to dust storms, even normal activities or situations (e.g. driving vehicles across dirt roads, yard work, construction projects) can add significant amounts of dust particles into the air. Dust, often as small as 2.5 microns in diameter is considered to be an air pollutant in many places. (A micron is one millionth of a meter (and a meter is slightly larger than three feet)). Since dust is viewed as a potential air quality hazard, we'll look at dust again in Chapter 17.

Dust storm transportation

Since the Sahara is so large, it is often the source region for numerous dust storms. Many of these develop as strong winds from thunderstorm systems and race across the region. Others simply develop when localized stronger winds surge in the "Trade Wind" easterlies. These winds then pick up the dust and carry it westward over the Atlantic Ocean often affecting the Canary Islands. One of the main source regions for this

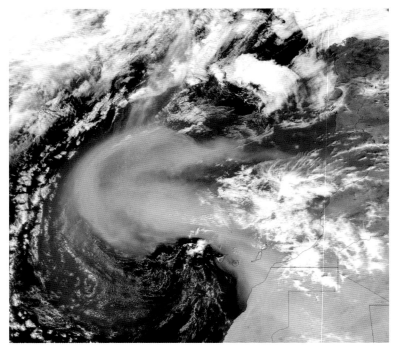

An African dust storm heads out into the Atlantic Ocean on February 26, 2000. The storm originated over the western Sahara Desert and moved westward thanks to the "Trade Wind" easterlies. The elongated shape of this dust cloud indicates a wind maxima extending to the west with a high-pressure circulation to its north and a low-pressure center to its south.

transport is the Bodele Depression in Chad where many dried lakes (known as playas) provide individual dust source regions. Often the dust cloud will have a curved bow-shaped western edge (like a thunderstorm outflow arc cloud) or it will look like a fountainhead with a surge region racing ahead of the overall dust plume.

African dust storms of this type are most likely to occur in the Northern Hemisphere summer (June to August).

Impacts

Saharan dust clouds can be carried across the Atlantic, bringing dust to the Caribbean and even to the eastern United States. When the skies in the Caribbean and Florida turn hazy during summer, it is almost always the result of suspended African dust. Scientists from the University of California – Davis reported about three such African dust incursions into the eastern United States annually between 1992 and 1995; on the average, the dust persisted for about 10 days and far exceeded the normal airborne dust amounts found in typical dusty central Plains states.

Dust may also trigger respiratory health problems in humans. According to Gene Shinn, senior geologist at the US Geological Survey (USGS) Center for Coastal Geology in St Petersburg, Florida, levels of asthma on the islands of Barbados and Trinidad are among the highest in the world. "The incidence of asthma on Barbados and nearby Trinidad, as documented by the Caribbean Allergy and Respiratory Association (CARA), has increased 17-fold since 1973," he said. "And that was the first year that the dust concentration graph for Barbados showed a big spike."

The St Petersburg Times reported that the number of Americans with asthma has increased 154 percent in the 20 years since 1980. Hillsborough County, Florida, which includes the Tampa Bay region, has one of the state's highest asthma rates, with more than seven percent of students reporting asthma symptoms, compared to fewer than three percent just four years earlier.

According to the research of Dr Joe Prospero, University of Miami, about half the particles breathed in South Florida during the summer months originate in Africa. Asthma epidemics in areas that are relatively free of industry correlates with the increased arrival of African dust.

In addition to causing respiratory distress to some people, the dust can settle over Caribbean waters. African dust can contain high concentrations of a fungus known to be lethal to fan corals. In recent years, fungal infection has killed more than 90 percent of the Caribbean Sea fans, a soft form of coral, and this may be related to mounting quantities of dust that have been emanating from the Sahara in the last four decades. Prospero also reported periodic outbreaks of African varieties of blight in Caribbean crops and at least one event in which swarms of emaciated, but live, African locusts dropped out of the sky in the Caribbean. On the other hand, nutrients in the mineralized African dust may be helping plants in the Amazon to grow.

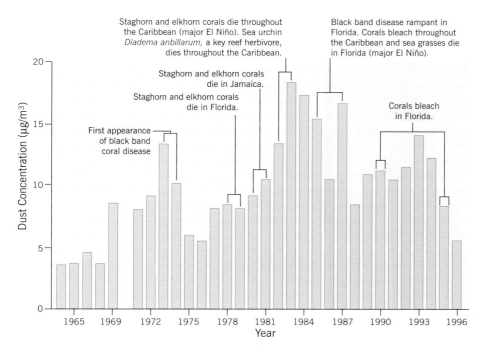

Average Annual Barbados Mineral Dust and Benchmark Caribbean Events

The graph shows the overall increase in African dust reaching the Caribbean island of Barbados since 1965. Notice the peak years for the dust deposition were 1983 and 1987. These were also years of extensive environmental change on Caribbean coral reefs.

Graph annotations:

- Staghorn and elkhorn corals die throughout the Caribbean (major El Niño). Sea urchin *Diadema anbillarum,* a key reef herbivore, dies throughout the Caribbean.
- Staghorn and elkhorn corals die in Jamaica.
- Staghorn and elkhorn corals die in Florida.
- First appearance of black band coral disease
- Black band disease rampant in Florida. Corals bleach throughout the Caribbean and sea grasses die in Florida (major El Niño).
- Corals bleach in Florida.

Graph axis labels: Dust Concentration (µg/m³); Year; 1965, 1969, 1972, 1975, 1978, 1981, 1984, 1987, 1990, 1993, 1996

Recent studies indicate a strong, albeit preliminary, link between Atlantic Ocean areas affected by airborne African dust and lowered sea surface temperatures. This is likely to be associated with the blocking of ocean warming due to the nature of the dust – it reflects sunlight back to space and also radiates energy out to space. With high concentrations of airborne dust in 2006, the number of Atlantic Ocean hurricanes fell dramatically from the very active year before (when there was a low dust concentration).

Further, the amount of dust leaving the Sahel region of Africa is closely linked to rainfall in that region. Prior to 1970, when rainfall in the Sahel was closer to average, less dust was removed from the region and carried over the Atlantic.

Still, African dust in the Caribbean region is nothing new. The presence of red iron- and clay-rich soils on the Caribbean's limestone-based islands are known to be of Saharan dust origin. Geochemical evidence of Saharan dust parent material for soils developed on Quaternary limestones of Caribbean and Western Atlantic islands. In short, the Trade Wind easterlies have been at work for a long, long time.

African dust moves northward

Sometimes dust from Africa is blown northward, usually in advance of a middle-latitude storm system that moves across central Europe. This is most commonly observed from late fall into early spring. As the desert air is carried across the Mediterranean Sea, it picks up moisture.

These dust-laden winds are known by varied names in the region:
- Scirocco and Sirocco are Italian names.
- Siroko in Greece.
- In Croatia and Montenegro, the wind is known as a Jugo.
- In France it is called the Marin.
- In Libya, it is called the Ghibli.
- In southwest Spain, the Levenche is also associated with heavy rains.

Black band disease a bacterial slime that spreads across coral, can be devastating. Growing radially outward from the whitish (dead coral skeleton) the disease has spread rapidly since first being detected around 1970. The disease has been linked to dust from Africa.

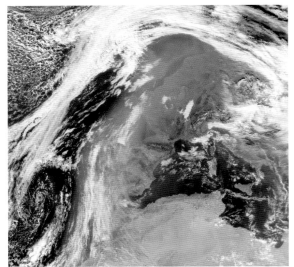

In these SeaWiFS images, Saharan dust covers large parts of Europe. In the top image, captured on October 30, 2001, the dust cloud extends as far north as the United Kingdom and parts of the Benelux region. In the bottom image, captured on April 16, 2003, the dust cloud extends into the Scandinavian region.

Middle Eastern dust

When a middle latitude low-pressure system passes to the north of the Middle East, there can be two dust storm scenarios. The first is similar to the European one in which south or southwesterly winds bring dust into the region. A second, more common setting is in the wake of the passage of the low-pressure system.

As the trailing cold front passes through the region, strong northwesterly winds, known as the Shamal blow into the region. Enhanced by mixing higher velocity air to the Earth's surface from higher altitudes, the Shamal easily moves sand and lifts dust. The satellite image sequence shown below shows how quickly the dust can be picked up and carried large distances. Even a day after the initial image pair, the dust cloud remains over Saudi Arabia, Kuwait, and southern Iraq.

Although primarily a colder season phenomenon, it can occur throughout the year.

Australian dust storms

Much of Australia is arid or semi-arid which means that dust transport is a fairly common event. Dust storms are, for the most part, restricted to the drier inland areas of Australia, where rainfall is limited to about 16 inches (400 mm) per year. Occasionally, during widespread drought, dust storms can even affect coastal districts. Dust storms are controlled by prevailing wind patterns and El Niño conditions (higher drought frequency); they also shift latitudinally with the seasons. In northernmost regions, dust is carried westward on prevailing easterly winds; to the south, westerly winds carry sand to more populated regions along the east and southeast coasts.

The extended dry period of the 1930s and 1940s generated many severe dust storms. During the summer of 1944-45, Adelaide was shrouded in dust that was so thick that street lighting had to be turned on. Uncomfortable as dust storms may be, the high wind and dry conditions that spawn them also contribute to the stripping of topsoil from Australia's arable land.

One of the most spectacular dust storms was the one that swept across Melbourne in early February 1983, late in the severe El Niño drought of 1982-83. Although coined a dust storm, it was actually a cloud of soil and topsoil.

In fact, the clearing of native vegetation and intensive farming and grazing in the past 200 years have contributed greatly to increased wind erosion. Most of the recent severe wind erosion events have occurred in the agricultural areas of southern Australia.

The total amount of dust emitted annually from the Australian continent has not been measured, but estimates place it on the order of several million tons.

India and Pakistan

In spring, just before the wet monsoon arrives, dust storms are common on the Indian sub-continent. Often these involve the transportation of sand from the Thar Desert region in western India and eastern Pakistan. The Thar is a small desert, only some 92,000 square miles (238,000 sq km) in area.

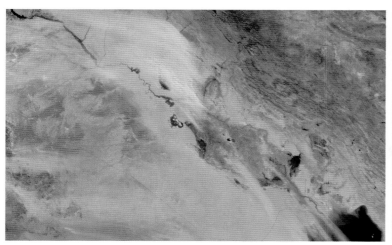

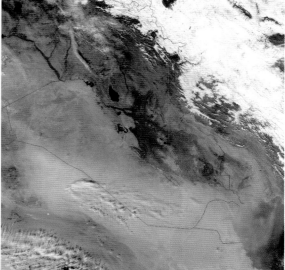

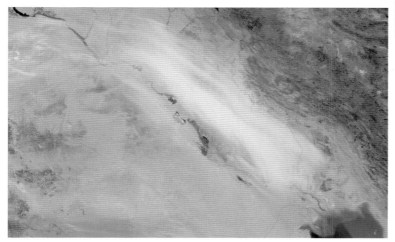

As middle-latitude cyclones pass to the north of the area, strong southerly to westerly winds easily pick up dust from the Thar. The dust is then carried across the more populated regions of the country. If the dust heads to the north, it is often blocked by the Himalyas and forced to move to the west and/or east depending upon wind patterns. The image below shows the cyclonic circulation responsible for a dust storm moving across Pakistan, on June 9, 2003. Dust from this storm has been carried to sufficiently high altitudes to be evident above some of the Himalayan snow cover.

The fine reddish, talc-like dust associated with these storms covers the landscape and as people often report, "…no amount of wet towels placed around windows and door frames keeps the dust out of homes."

China
The Taklimakan Desert in interior western China is only slightly larger than India's Thar Desert. However, because it is so far removed from water sources and surrounded by higher terrain, this desert (located between 37 degrees to 40 degrees north latitude) is the source of large amounts of atmospheric dust. In fact, dust from this desert has been found in North America, Greenland, and even atop the French Alps.

On a more local scale, dust from the Taklimakan has exacerbated China's already poor air quality. According to NASA scientists, northwestern China averaged a major dust storm every 31 years during the period from about 300AD to 1949. Since 1990, the same

Top right: A Middle Eastern dust storm in January 2005.

Top left: Iraqi dust storm August 7-8, 2005. Some dust is present on August 7, but the smaller dust clouds merge into a larger dust storm (bottom left) that covers the eastern half of Iraq, part of Iran and the northern part of the Persian Gulf.

Bottom right: A massive sand storm cloud is close to enveloping a military camp as it rolls over Al Asad, Iraq, just before nightfall on April 27, 2005.

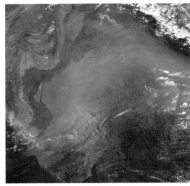

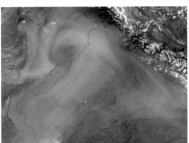

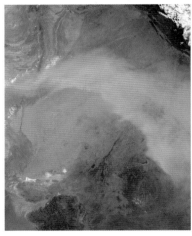

Three views of Indian sub-continent dust storms from NASA's Terra and Aqua satellites. Notice how the dust takes different paths depending upon the wind pattern. The Himalayan Mountains often block the northward transportation of dust shunting it to the east and/or the west. All three of these dust storm events occurred just prior to the onset of the wet summer monsoon.

region has experienced a major dust storm almost every single year. Although dust storms are most commonly observed in March and April, they can occur year round.

The Taklimakan is not the only source of dust in China. The Gobi Desert, some five times larger, is another major source of dust storms. One storm on March 10, 2006 saw a major dust storm across China linked to the passage of a middle-latitude cyclone. The dust cloud shrouded Bejing, as well as much of eastern China, and even parts of the Korean Peninsula. As can sometimes happen when dust mixes with precipitation, this dust apparently caused dusty snow in South Korea.

Other dust sources

While the Earth itself is the source of most airborne minerals (dust), dust is also added as meteors enter the Earth's atmosphere. Although meteors are not a concentrated dust source, some estimates indicate that some 330 million pounds of dust are added to our atmosphere annually by these "space invaders". Obviously, larger meteors (such as the one that allegedly caused the demise of the dinosaurs some 65 million years ago) contribute much more. As recently as August 30, 2005, an asteroid, about 33 feet (10 meters) across, entered the Earth's atmosphere over Antarctica. In moments, it deposited 2.2 million pounds (1 million kg) of dust into the upper levels of the atmosphere.

While these numbers may appear to be very large, note that the mass of the earth is 1,016 (that is, 10 million billion) times greater.

Dust from asteroids, comets, and other space phenomena is a concern for any space vehicles and for astronauts working outside spacecraft. Even relatively low speed impacts can damage or puncture protective layers.

Climatic impacts

In assessing climate variations, it makes sense to look at how the size of deserts change with time. Based on recent climate reports, deserts should expand and people should be displaced. Yet, a study of the Sahara Desert (1980-97) showed no systematic trend in desert size change, even though there were significant year-to-year variations (based on satellite data and where eight inches (200mm) of rain fell each year). This suggests that weather and climate factors may not be the most significant factor in how people are affected by desert conditions.

Dust in the atmosphere can have significant climatic and ecological implications, but research into these is in its early stages. Preliminary findings suggest these have to be studied further and incorporated into computer-based climatic models. Some of the findings (both positive and negative) include:

- Dust can block incoming solar energy and lower sea surface temperature; this may have an effect on the number and strength of hurricanes.
- Dust, laden with fungus, can devastate fan coral populations.
- Dust can provide nutrients which may enhance tree-canopy bromeliad growth in the Amazon River rainforest.
- Dust from Asia provides essential nutrients for Hawaiian rainforest growth.
- Dust may be a more significant atmospheric pollutant than previously believed.
- Because many desert sands are quartz-based, and since many desert areas are among the sunniest locations in the world, they could become a reliable, carbon dioxide-free source of solar energy. It is estimated that if the Sahara were completely covered over by solar collectors, it could provide enough electricity to exceed current global needs.

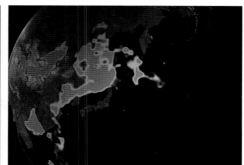

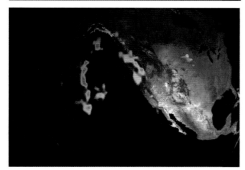

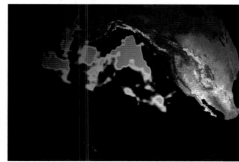

Perhaps the smallest dust storms are those created by dust devils. Likened to miniature tornadoes, these small atmospheric swirls typically have wind speeds far below their tornadic cousins and also lack any type of cloud linkage. In fact, they form most easily on days with limited cloud coverage, allowing sunlight to easily heat the Earth's surface. Common in deserts, they can also form in semi-arid areas and over super-heated asphalt parking lots. Dust devils can last for several minutes or can form and dissipate in an instant. In addition to being forced by solar heating, dust devils can also spin up when surface winds are strong.

For the last 25 years, NASA's satellite-based Total Ozone Mapping Spectrometer (TOMS) instruments have been looking at ozone and making daily maps of the ozone content of the atmosphere across the globe, showing scientists the evolution of the ozone hole from 1979 to today. A side benefit of the research was that the instrument was also found to be able to detect and track dust. Here is a sequence showing how a dust cloud from China traveled across the North Pacific reaching the northwest US coast in about one week (April 2001). The brownish area that can be tracked across the Pacific represents the highest concentrations of dust.

Below: Although it may not be everyone's idea of a typical desert, the Antarctic is classified as a polar desert.

Dust storms may look like tornadoes, but they lack the companion cloud. They are also formed primarily from strong surface heating and associated low-level, rising air currents.

"Drought is a normal, recurrent feature of climate and it occurs almost everywhere."
National Drought Mitigation Center (US)

Right: South Louisiana was having its worst drought on record in early 2006. Here is a picture of Bayou Savage, where portions are completely dried up.

Top right: The Germasogeia Dam in Cyprus was well below its normal levels when this picture was taken.

Bottom left: The Murray River in South Australia showing the effects of the prolonged drought in 2007.

Bottom right: Drought impacted corn in Nebraska with ears close to ground being eaten by deer and pheasant. The crop was simply not feasible to harvest.

Droughts occur when there is too little rainfall – at least that's the type of drought most people recognize. Television meteorologists almost always showcase this situation by showing a mapped drought area and discussing significant rainfall shortages over an extended period of time (several seasons to many years). Droughts are often slow to begin, long in duration and usually slow to end. Floods of all types begin much faster, and often end much faster, than droughts. Furthermore, droughts do not have a clearly defined beginning or ending.

Snow droughts may be mentioned on television weather reports, not only because they impact on skiing, but because snow pack in some places can also provide water supplies in drier months of the year.

Don Wilhite, head of the United States' National Drought Mitigation Center, is quick to point out that there are three other types of droughts in addition to meteorological drought. Wilhite notes that, "…what distinguishes the other types of droughts is that they are less associated with precipitation deficiencies and more associated with water shortages in other components of the hydrologic system." The hydrologic system is not just the classical water cycle (evaporation, condensation, precipitation, and

The Water Cycle

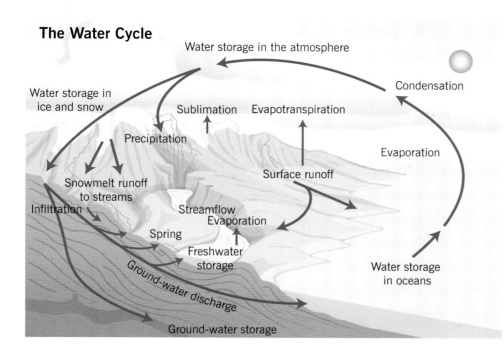

This illustration shows the various elements of the water cycle. In addition to the classical evaporation, condensation and precipitation, notice all of the ways water "accumulates" on Earth.

accumulation) that we learned about in school. Rather, it is the much more complex water system that includes competing water uses, water management, population demands, and much more. In short, no proper assessment of drought can be made by only viewing rainfall.

With each US household using an average of 94,000 gallons of water every year, US agencies like the Environmental Protection Agency (EPA), have begun promoting public awareness of the need for conservation.

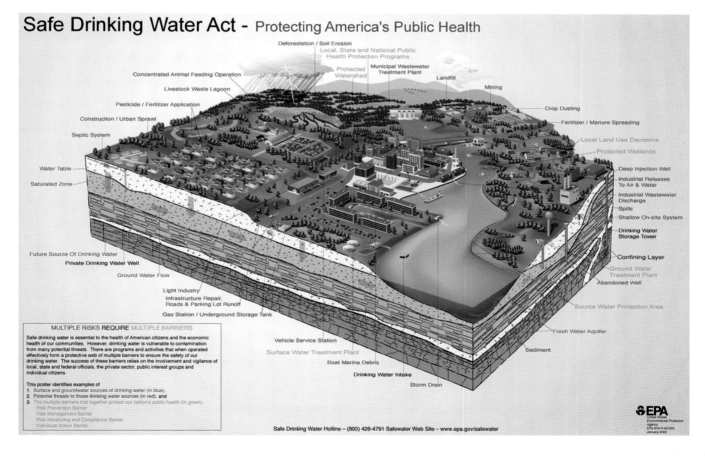

Safe Drinking Water Act - Protecting America's Public Health

MULTIPLE RISKS **REQUIRE** MULTIPLE BARRIERS

Safe drinking water is essential to the health of American citizens and the economic health of our communities. However, drinking water is vulnerable to contamination from many potential threats. There are programs and activities that when operated effectively form a protective web of multiple barriers to ensure the safety of our drinking water. The success of these barriers relies on the involvement and vigilance of local, state and federal officials, the private sector, public interest groups and individual citizens.

This poster identifies examples of
1. Surface and groundwater sources of drinking water (in blue),
2. Potential threats to those drinking water sources (in red), and
3. The multiple barriers that together protect our nation's public health (in green).
   Risk Prevention Barrier
   Risk Management Barrier
   Risk Monitoring and Compliance Barrier
   Individual Action Barrier

Safe Drinking Water Hotline ~ (800) 426-4791 Safewater Web Site ~ www.epa.gov/safewater

EPA United States Environmental Protection Agency EPA 816-H-02-003 January 2002

The Hoover Dam on the Colorado River, viewed from Lake Mead, was constructed in the 1930s and has provided much-needed reserves in times of drought.

Hydrological drought is involved with shortages of surface and subsurface water supplies. This can include a lowered ground water table and/or lowered water levels in aquifers (see page 186). Sandwiched between these layers, the water remains trapped. The "Ogallala Aquifer" that extends from Canada southward to Texas is perhaps one of the most famous.

Agricultural drought is associated with soil moisture deficits and evapotranspiration. Lack of soil moisture limits the water supply to plants. Evapotranspiration addresses how much water leaves plants (it can be likened to how a plant "perspires"). Even if rainfall is adequate, high evapotranspiration rates can produce drought-like effects. Think about the plants in your garden or in a nearby park. During periods of high temperature (daytime), you may see the plants wilt. After dark, the plant seems to perk up, even without the addition of water. The wilting is a defensive measure that actually helps reduce evaporation from plant leaves. If it persists too long, then the lack of water flow to the leaves will cause the plant to die.

There is even socio-economic drought. This relates to how much water is available and how much demand there is for the water. Competing demands for a limited water supply can create drought-like conditions. Consider the western United States where there is a relatively limited water supply, but high demands from people, the agricultural community, environmental requirements (e.g. maintaining minimum river flows), recreational users and, in some cases, hydro-electric power generating facilities. Depending upon how the water supplies are legally appropriated, one user may experience a drought, while another has plentiful water supplies.

It is likely that any drought experienced around the world today involves a combination of two or more of these drought types. For example, in parts of Africa, rainfall has diminished during the past several decades. This would lead one to think about meteorological drought. A growing population with its increased water demands has caused surface and subsurface water supplies to be severely taxed, even to the point of having large lakes dry up. When agricultural impacts and military conflicts are added, the resulting situation encompasses all four types of droughts.

However, almost everything eventually comes back to meteorological drought. Once a rainfall shortage occurs, other drought conditions often come quickly into play.

Meteorological drought

Look to any location on Earth and you will be able to find something called "average annual rainfall" or "average water year rainfall". This involves taking all the rainfall data and averaging them over a certain period (usually the most recent three decades). While most locations use annual rainfall, others with specific wet and dry periods often use a date other than January 1 to start annual rainfall tallies – California, for example, starts its rainfall year in July, during its dry season.

While "average" provides useful information, median rainfall (the number that splits the 30 years of data into two sets, each containing the same number of yearly values), tells more about the distribution of wet and dry years. There are other statistical measures; range (difference between highest and lowest) and mode (rainfall band that is most often observed) that are the most commonly used.

Perhaps even more telling than the numbers themselves is a graph showing the actual annual rainfall distribution. Although average annual rainfall for downtown Los Angeles

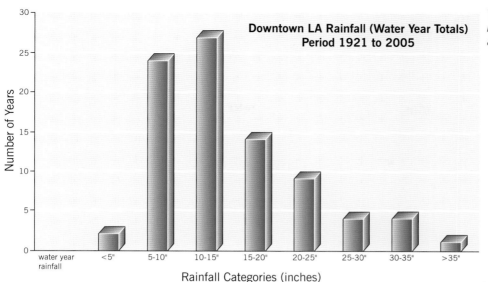

Water year (July 1– June 30) rainfall for Los Angeles in an 85-year period.

is slightly less than 15 inches (38 cm), notice that 53 years have 15 inches (38 cm) of rainfall or less, while only 32 have more than 15 inches (38 cm). This means that the average is higher than the median; this is because a few years had very large rainfalls (e.g. more than 25 inches (63.5 cm)).

During this 85-year period, the population of the Los Angeles area soared from about one million to more than 16 million (or a 1,600 percent increase). Average rainfall can be considered to have remained "constant." Due to limited region-wide water supplies and the fact that cloud seeding is not a realistic approach to increasing rainfall, the region has had to become very proactive in its water conservation practices (see page 195).

While annual rainfall provides a useful measure of rainfall, distribution of precipitation throughout the year (especially as it relates to the growing season) is even more important. Some locales (e.g. southern California and Florida) have very distinct wet and dry seasons; other places rain all year, but have significant seasonal variations, while other places (e.g. New England) have a more uniform annual rainfall distribution.

In areas with what would be considered ample rainfall, significant departures (especially toward the low side) can occur, sometimes for many months in a row. In low rainfall areas or in regions with distinct wet and dry seasons, rainfall reliability (e.g. length of the rainy season or when the season starts) becomes the critical factor.

Droughts can end when rainfall returns, though, often a return to "average" isn't enough. Consider the situation in which torrential rains arrive after a prolonged dry period. This quickly brings rainfall totals back to "average" and may lead people to think that the drought has ended. However, when heavy rain falls on dry ground, much of it runs off. This means that groundwater supplies may not be recharged. If the return to "average" rainfall period ends, drought conditions can quickly reappear. For a drought to end, it takes a long period of steady rainfall to fully replenish surface and (at least some) subsurface water.

Large and small scale factors can cause meteorological droughts. Typically droughts are linked to a persistent upper-level high pressure system or ridge that brings with it an inversion (layer of warm air aloft) and sinking air. Both of these are unfavorable for the

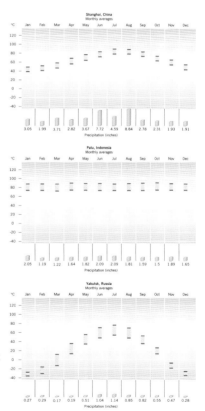

Climographs for three Asian locations aligned along a longitude line. Monthly rainfall is fairly uniform at Palu, but there are distinct summertime maxima at the other two locations.

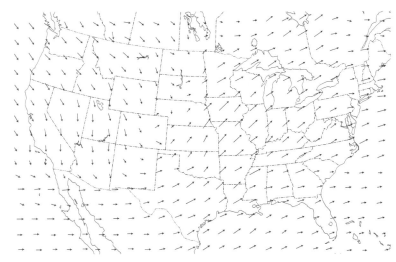

formation of clouds and precipitation. The satellite image and superimposed upper-level winds show how this type of pattern affected both California and Florida in late 2006.

Some scientists have argued that El Niño (the warming of ocean waters off the northwest coast of South America), Atlantic Ocean sea surface temperatures, and other similar large-scale factors, have affected upper-level wind and pressure patterns. Others argue that global warming has caused these drought patterns. Since the patterns experienced in current droughts are no different than patterns that have occurred at other times in the past 100 years, it is hard to accept global warming as the cause.

Type of upper level wind flow pattern that, if persistent, can lead to drought conditions in the western and southeastern United States. Data shown are for November 30, 2006, during the period in which overall weather conditions led to the onset and intensification of a drought in both areas.

Since large scale weather patterns help to force dry and wet regimes, usually when a drought ends in one place (weather pattern shift), it starts to develop in another. Compare these drought maps (below) for the US and Canada spanning a two-year period. In March 2005, drought conditions were present in the Pacific Northwest; a year later they had migrated to the southwestern Plains and much of Mexico; by March 2007, the pattern had migrated westward to the southwestern states. It is possible to trace the movement of this area as a large clockwise spiral. While other drought areas have formed and dissipated during this period, none was as pronounced as the one affecting the southwest in early 2007.

Similarly, opposite are four maps from Australia showing different types of drought (e.g. long-term, short-term).

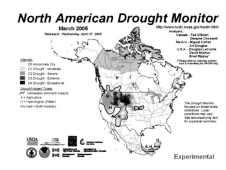

Drought maps for the United States and Canada from March 2005 to March 2007.

Hydrological drought

Hydrological drought is involved with shortages of surface and subsurface water supplies. Surface water is often easy to assess because lake levels often approximate the water table (the closeness to the surface of water stored in the soil). The higher the water level in a lake, the higher the water table (unless the lake is kept artificially full). As lake levels drop, beaches may appear or water-loving plants may be exposed to dry ground. Similarly, when lakes fill to excess, the reverse occurs.

Subsurface water includes water levels in aquifers. Aquifers are underground water storage areas formed when water (often from distant source regions) collects underground in layers trapped between impervious rock or soil layers. Some aquifers

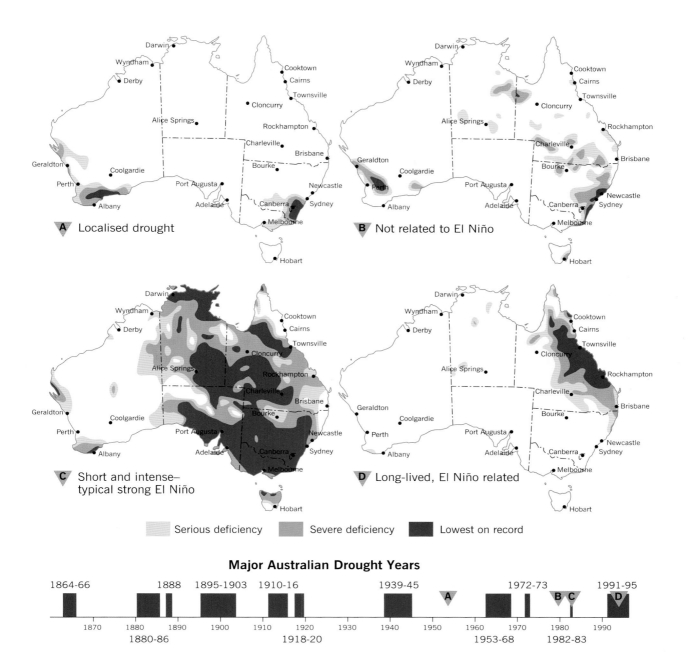

A Localised drought

B Not related to El Niño

C Short and intense—typical strong El Niño

D Long-lived, El Niño related

Serious deficiency Severe deficiency Lowest on record

Major Australian Drought Years

1864-66   1888   1895-1903   1910-16   1939-45   1972-73   B C   1991-95
1870   1880   1890   1900   1910   1920   1930   1940   1950   1960   1970   1980   1990
1880-86   1918-20   1953-68   1982-83

may be quite small; others can extend for tens of thousands of square miles. It is even possible to have aquifers stacked atop one another.

The Ogallala Aquifer System (174,000 square miles (450,500 sq km) that extends from Canada southward to Texas, through the High Plains region of the United States, is perhaps one of the most well known. Formed 10 million years ago, the aquifer system covers parts of eight states. Due to erosion, recharge of the aquifer from mountain snowmelt no longer occurs; the only recharge is from horizontal transport within the aquifer and limited rainfall. Until efforts were undertaken to lessen depletion of the aquifer's "fossilized waters" for irrigation (see circles on page 189), people in the area believed that the aquifer's water supply was infinite.

Although Australia's Bureau of Meteorology notes that there is an average 18-year return period for major drought, the period ranges from four to 38 years.

EXPLAINING DROUGHT MAPS

Drought classifications in the US are typically addressed on the national level (although there is not yet a global standard for defining them). But it is up to state and local jurisdictions to address the drought. Only when the drought takes on significant impacts does the federal government again enter the picture. Here disaster declarations and response actions may follow.

National Dryness Categories used in the US. (Note that the classification system involves several types of indices and some human interpretation.)

Australia's Bureau of Meteorology focuses on accumulated rainfall over three successive months as its measure of potential drought. If the 3-month rainfall (6-month for arid regions) lies within the lowest 10 percent on record, the Bureau issues a Drought Watch. Meteorologists also factor in the expected rainfall pattern for that time of year.

Category	D0	D1	D2	D3	D4
Description	Abnormally Dry	Moderate Drought	Severe Drought	Extreme Drought	Exceptional Drought
Possible Impacts	Going into drought: short-term dryness slowing planting, growth of crops or pastures. Coming out of drought: some lingering water deficits; pastures or crops not fully recovered	Some damage to crops, pastures; streams, reservoirs, or wells low, some water shortages developing or imminent; voluntary water-use restrictions requested	Crop or pasture losses likely; water shortages common; water restrictions imposed	Major crop/pasture losses; widespread water shortages or restrictions-	Exceptional and widespread crop/pasture losses; shortages of water in reservoirs, streams, and wells creating water emergencies
Palmer Drought Index	-1.0 to -1.9	-2.0 to -2.9	-3.0 to -3.9	4.0 to -4.9	5.0 or less
CPC Soil Moisture Model (Percentiles)	21-30	11-20	6-10	3-5	0-2
USGS Weekly Streamflow (Percentiles)	21-30	11-20	6-10	3-5	0-2
Standardized Precipitation Index (SPI)	-0.5 to -0.7	-0.8 to -1.2	-1.3 to -1.5	1.6 to -1.9	2.0 or less
Objective Short and Long-term Drought Indicator Blends (Percentiles)	21-30	11-20	6-10	3-5	0-2

Drought Watches fall into two rainfall deficiency categories:

A severe rainfall deficiency exists in a district when rainfall for three months or more is in the lowest five percent of historical records.

A serious deficiency lies in the next lowest five percent (i.e. lowest five percent to 10 percent of historical records) for a three-month or longer period.

The Drought Watch is discontinued once "plentiful" rainfall returns. Plentiful is defined as well above average rainfall for one month, or above-average rainfall over a three-month period.

Although the US and Australian systems are different, both reflect a realization that drought is a significant and extreme weather and climatic event.

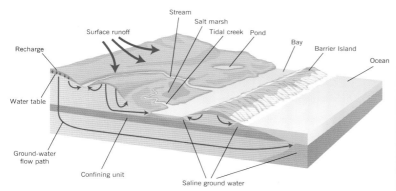

Ground-water flow paths in an idealized coastal watershed along the Atlantic coast. Fresh ground water is bounded by saline ground water beneath the bay and ocean. Fresh ground water discharges to coastal streams, ponds, salt marshes, and tidal creeks and directly to the bay and ocean.

South Florida communities often include small lakes (for wildlife habitat and water storage). This image was taken at the end of Florida's 2006-07 dry season. Notice how low the water level is compared to normal.

Florida's population may have the same mindset. The figure on page 190 shows how withdrawals from Florida aquifers have escalated as the population has grown.

Other aquifers filled with "fossil water" during ancient geologic times (but no longer being recharged) include the Nubian aquifer in North Africa and another in Libya. Libya has mined its aquifer to bring drinkable water to its coastal urban population and is now turning on the spigots for agricultural use.

During periods of hydrological drought, water flow in rivers diminishes. This can have serious implications, especially along coastal areas where there is already a delicate balance between fresh and salt water (much like within the aquifer). Here, reduced river flows mean that the "salt front" (the interface between fresh and salt water, used much like its counterpart weather front) can advance inland.

First, this affects salinity in bays, estuaries, and marshlands. Depending upon their biological make-up, some plants and animals have a hard time adjusting to large changes in salinity. If the drought persists, the front can even march up large rivers. This routinely happens during summer when less reliable rainfall and increased demand lessens river flow.

In late 1964, it happened to an extreme. The front made its furthest inland advance on record, reaching Philadelphia, PA and Camden, NJ. Due to water needs, the Camden water pumping station actually drew in river water, water that had a higher than average salinity. Although the salt-water invasion did not affect human consumption of water during this situation, it demonstrated how vulnerable coastal water supplies are when affected by droughts and possibly by sea level rise attributed to global warming.

Aerial image showing an array of irrigation circles. At the center of each circle is a pump which brings ground water to the surface. This water is used for plants and allows agriculture to prosper in normally dry regions.

Agricultural drought

In regions with reliable and/or ample rainfall, agricultural drought results when typical rainfall patterns are disrupted by an extended dry period (often with high heat). In these cases, available surface and subsurface water resources are used up far more quickly than they can be replenished. Soils literally bake and crops fail.

In regions termed arid and semi-arid the situation is more tenuous. Here, agriculture is normally "on the edge". Even the slightest disruption can spell the loss of crops, emaciation of livestock, or worse.

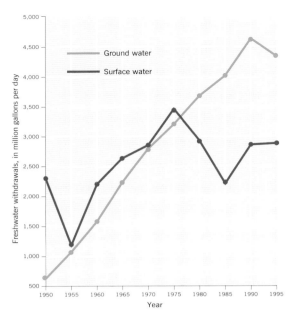

Arid and semi-arid regions are defined by their precipitation-evaporation (P/E) ratios (0.5 for arid and 0.5 to 1.0 for semi-arid). In short, there is more water loss from plants than can be expected from rainfall alone. In addition to a lower net precipitation rate, rainfall events often occur in the form of rare, short-duration, high-intensity storms (some even causing flash floods).

Given the nature of the soil and sand in the area, the ground often becomes baked, with dried lakes and riverbeds showing the "cracked earth" character. Such a hardened surface does not allow for infiltration of the high intensity rainfall; instead, runoff dominates. Thus, even though rainfall may be at, or above, average values, the usefulness of the rainfall may only be marginal. Unless there is a way to capture runoff (e.g. lakes, reservoirs), the lost water will not be available for agriculture and other needs.

Agricultural droughts are common along the edges of large desert regions (especially in Africa) where rainfall is limited even under the best conditions.

Socio-economic drought

Bring competing demands into play, especially in areas with marginal water supplies, and you have the makings of a major socio-economic drought. In places ravaged by war, overpopulation, or other societal situations, the impacts can be even more devastating. In socio-economic drought, a myriad of justifiable needs compete for water.

Consider the Bonneville Dam on the Columbia River in Oregon. Here a large dam was built in 1938 to ensure generation of hydroelectric power. The project was a huge success, producing hundreds of kilowatts of power each day. The project also improved navigation on the Columbia River and created a lake that enhanced recreational activities. The lake provided water supplies for agricultural and river flow management needs.

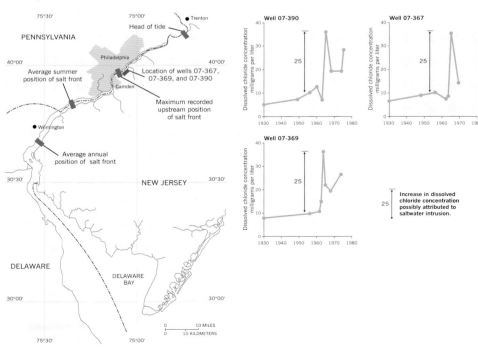

Right: Location of the salt front in the Delaware River and Estuary for average annual, average summer and maximum recorded upstream positions.

Far right: Dissolved-chloride concentrations of water from selected water-supply wells near the Delaware River in Camden, New Jersey, 1930-80.

However, there were trade-offs. The dam disrupted spawning activities of some species of fish (e.g. salmon and steelhead trout) and land used for other purposes had to be flooded to create the lake (reservoir) for the dam.

When a drought struck the area, there was major competition for water. Boaters and others involved in recreational activities wanted the water kept in the reservoir while farmers wanted to divert some for their crops. Environmentalists were concerned about water flow in the river and urban water managers wanted to ensure enough reserves for human and business use.

In coastal areas with major drainage canals, dam-like control structures have been constructed. These have gates that can be opened or closed to control runoff and the inland movement of saltwater up the canals.

Were the Columbia River situation unique, there would be no story. However, even on rivers without dams, competition almost always surfaces during a drought. There are a lot of droughts and a lot of rivers!

In some places, drought (coupled with war and/or economic hardship) forces major societal impacts. This is a common situation in the semi-arid region south of the Sahara Desert. Here, frequently occurring droughts force equally frequent large-scale migrations. Once people move, the hosting country, with already stressed resources, faces a refugee situation including population control, care and feeding, and more. The United Nations, sometimes individual countries, and many international relief organizations often step in to help. Other than war alone, droughts are the largest contributor to refugee problems.

Historically, droughts have also contributed to migration in the United States. The Great Dust Bowl of the 1930s forced large numbers of farmers to leave the Plains region (see historical droughts on page 192). Many headed west to California.

Around 1300, the Anasazi Indians, a Native American culture that thrived in what is now Arizona, vanished. Were they attacked by nearby tribes or did they migrate in search of water supplies? While there are now several theories (or combinations of potential causes) for this, many believe that the effect of drought and possibly water control by nearby tribes played major roles.

Water wars?

Over the last 40 years or so, oil has been touted as one of the driving forces behind wars. That may have been true in the past, but "water wars" are coming.

Even if rainfall were distributed in a reliable manner, the routine competition described above could lead to conflicts. In the United States, Australia, Europe, and other places with established legal systems, there are ways for citizens to address these conflicts. In other parts of the world, other measures come into play.

The Anasazi White House cliff dwelling in Canyon De Chelly, Arizona.

On an even larger scale, when droughts, poor framing practices, overgrazing and overpopulation combine, food and water supplies can rapidly dwindle. The result can be famine and the resulting displacement of large populations. With local military conflicts and/or political instability added, a temporary displacement can quickly become a refugee situation. This, in turn, can exacerbate political and social instability in the host country. International relief is almost always needed. In short, what starts out as meteorological drought grows and its impact can spread far beyond the original drought area.

OTHER AFFECTS OF DROUGHT

As you have seen, drought can influence many aspects of life. However, there are still more. If forest or grass fires develop, then the smoke from these can affect those with respiratory illnesses. Fires can also disrupt ground and air travel and use up lake and reservoir water in firefighting efforts. Burned areas also become vulnerable to erosion and mudslides once rains return. If water levels drop in reservoirs, there can be cutbacks in hydroelectric power generation (which can affect business and residences). Environmental impacts can far exceed those already noted in the fresh-salt water interface area (e.g. spawning and nesting areas can be lost, dissolved oxygen levels in still water areas can drop to levels lethal to some species). On the other hand, at least temporarily, mosquito populations are kept down. However, once rains return, mosquito eggs (which can lay dormant for years) will hatch.

Similarly, bad feeling can be heightened by religious and/or sectarian violence in an arid or semi-arid region (consider the area from North Africa through the Middle East into Asia). Some Middle Eastern leaders have already noted that conflicts over water could supersede any peace treaties in place.

Water and the law

In the arid western United States, water is recognized as a prize commodity and hence there are rights and responsibilities surrounding how it can be accessed from waterways. In writing Water Rights Law – Prior Appropriation, Anne J Castle of Holland and Hart, LLP, noted that rights to water are established by the actual use of the water, and maintained by continued use and need. They are much like real property rights and can be sold and mortgaged in the same manner. All of this is independent of the place at which the water comes from and where it is used. To further confuse the issue, each state has its own legal precedents and may have different rules for surface and subsurface water.

International waters are muddier. In most places, rivers and streams cross multiple national borders, and local political and religious interpretations often act to deprive downstream users of water resources.

Where dams are built to provide multiple benefits, there are often significant trade-offs to be made. Consider the Aswan Dam in Egypt. In a lecture before the Geneva conference on Environment and Quality of Life in June 1994, Adel Darwish noted that, "…Nasser refused to listen to any argument but politics when he built the Aswan Dam – while the dam is providing multiple benefits to farmers and generating electricity – it had other devastating effects…the lake the dam created covered priceless archeological sites, destroyed valuable ecosystems and fishing grounds, eroded beaches and damaged nutrient and sediment balances, and uprooted 100,000 people (who had called the area home for some 5,000 years…)."

Dam building around the world raises similar concerns. Yet, with burgeoning populations and a need for water, nations have little choice but to invest in one or more methods of increasing water availability.

Worst droughts

From an historical perspective, we have weather records for about the past 300 years or less. From a geological perspective, this is an extremely limited sampling period given the age of the planet. Still, it is possible to list some significant droughts. Consider the following candidates arranged from the US eastward.

US "Dust Bowl" Drought (1930s)

Poor agricultural practices (which pushed land marginally suited for agriculture into use) coupled with a multi-year drought, led to this American (and Canadian) disaster. Beginning with 14 documented dust storms in 1932, the drought worsened, with 38 dust storms occurring in 1933. By 1934, an estimated 100 million acres of soil and topsoil was lost to the winds. On April 14, 1935, a day that has come to be known as "Black Sunday," a massive dust storm blew soil from the Plains to the US East Coast. Within months, Congress passed soil conservation legislation.

The Dust Bowl displaced hundreds of thousands of people. Many in the US migrated to California. Canadian farmers migrated eastward to cities like Toronto.

Jamestown, Virginia (1607)

When colonists from England arrived in the New World in 1607, they had no idea that they had entered a region in the midst of a multi-year drought. At least that's the finding of a team of scientists from the University of Arkansas. Measuring tree ring widths from the trunks of bald cypress trees (which can have lives of 1,000 years or more), Matthew Therrell and his team found much narrower rings during both the 1606-12 and 1587-89 periods. This would make the two back-to-back droughts the worst in about 800 years. This likely severely taxed the colonist's ability to find and trade for food.

Brazilian Drought (2005)
Burning of tropical rain forests, warmer than average Atlantic Ocean temperatures, and the specter of global warming can all be blamed on this drought, the worst in between 50 and 103 years (depending upon how one measures the drought). Regardless, the drought dried rivers and curtailed river transportation. This in turn affected food, medicine, and fuel distribution, further impacting on people across the Amazon region. As a result, the governor of Amazonas State declared a drought crisis.

Once a picturesque scene, this dry river creek in Morocco is testament to the power of a drought.

Spanish Drought (2004–2007)
A three-year dry period (starting in fall 2004) plunged large parts of Spain into a serious drought. This has affected agriculture, tourism, swimming pool use, and prompted significant water conservation measures. There has also been an increase in forest and brush fires. Rationing was already underway in seven regions for agricultural uses; rationing for human use is possible, if dry weather persists.

Although reliable, long-term weather records have only been kept in Spain since 1947, the drought is the worst to strike the Iberian Peninsula in recent history. It has also prompted the first ever "drought warning."

Rainfall has been well below average for most areas during the three-year period, even breaking several "lowest ever" rainfall totals during some months or seasons. Meteorologists link the dry period to a change in wind patterns. A large high-pressure system across western Spain has brought drier high latitude air masses into the region.

African Drought (1968 to early 1980s)
Large regions of Africa, just to the south of the Sahara Desert, experienced a major drought during this period. Some coined this the worst drought of the 20th Century, although more drought conditions struck in the 1990s. In fact, this part of Africa, at the interface between arid areas (deserts) and semi-arid areas (grasslands), experiences many droughts. This is due to a set of factors that include periodic dry periods, overgrazing, increased agricultural use and wood-cutting. In fact, prior to this drought, the period 1950-68 was fairly wet across the region from Mauritania eastward to Chad.

KW Butzer, Department of Geography, University of Chicago, reviewed stream, dune and lake phenomena in several well-studied sectors of the Sahel and found that these painted a picture of ongoing, wild environmental change. Between 12,500 and 20,000 years ago, sand encroached on Sahel's semi-arid savannah. That was followed by a period (5,000-9,000 years ago) in which lakes and rivers had greater volumes of water. Butzer also found significant short- and long-term fluctuations during the past 2,500 years. Butzer concluded that the drought starting in 1968 clearly falls into "…the norm of at least six earlier, dry anomalies verified since 1400."

North Korean drought (1997–2001)
Information from this country is limited, but several dry periods have led to a serious drought. With some reservoirs almost dried up, water supplies for agriculture and animal raising have been curtailed. Hydroelectric power production has been compromised with some areas requiring significant energy cutbacks (e.g. no lights at night). Although

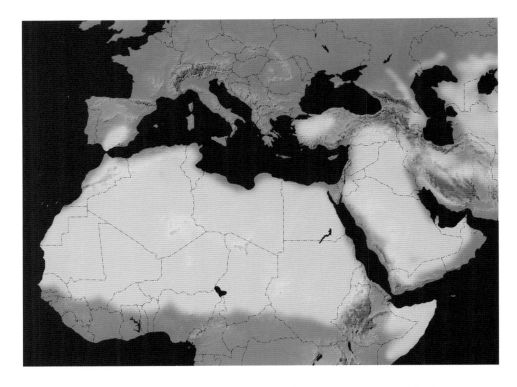

This map of Africa shows the countries at the southern edge of the Sahara that are most at risk of drought.

a Communist nation, North Korea had no choice but to accept donated food from other countries to avoid massive starvation.

Australian drought (2002–2007)

This multi-year drought has contributed to major agricultural losses, water shortages, and low levels of flow in streams. Many note that it is the worst since 1910 (when reliable weather record period began); others claim it is the worst in 1000 years (statistical). As with most Australian droughts, El Niño conditions appear to have influenced tropical rainfall patterns across the Pacific Ocean basin. There have been other major droughts in Australia during the past 140 years.

Solutions to drought

There are many activities that can be done to mitigate drought. These include prevention, preparation, and planning; changes in lifestyles or business use of water; and developing new sources of useable water or better capturing and storing of water that falls from the sky. Due to the extensiveness of this topic, I'll not even attempt to cover all the aspects here.

Monitoring

Clearly monitoring weather conditions is paramount. Changes in sea surface temperature patterns (e.g. El Niño), changes in upper-level circulation patterns, and careful monitoring or rainfall, river flow, and aquifer levels are important. Posting this information on the Internet ensures that scientists, water managers, and the public know what is unfolding.

Satellite monitoring of vegetative health, known as NDVI (Normalized Difference Vegetative Index) provides comparisons between "normal" or expected average vegetation and current conditions. Another satellite comparison, the EVI (Enhanced Vegetative Index), provides similar comparisons to NDVI but compensates for certain reflectivity and smoke interference.

Planning and conservation

Around the world, many countries, states, regions, and cities are developing drought plans. These include response actions, education, coordination, and planning.

In south Florida, for example, the South Florida Water Management District can mandate water restrictions; then it is up to local agencies and police to enforce them. Restrictions can include how and when water is used outside. From April 13 to May 4, 2007, when home and garden irrigation watering was cut back to a maximum of two times a week (drought emergency level 2), almost 4,000 citations were issued for violation of watering restrictions. Fines and penalties vary by county, but in Lee County (Fort Myers, FL), fines escalate from an initial warning through up to a $500 fine and/or up to 60 days in jail for a third offence.

In the Los Angeles area, conservation efforts since 1970 have offset a significant amount of population growth. Although population soared by 35 percent, water use only increased by seven percent. In a city whose population is now more than 10 million and whose annual average rainfall is only slightly more than 15 inches, this is a significant accomplishment.

Of course, by planting drought tolerant plants (xerigraphic landscaping) one can significantly reduce watering needs.

The changing colors of these three maps show the onset of drought in Australia during 2000–02. The more yellow or orange the color, the less healthy the vegetation.

Capturing and storing water

In Hawaii, water collected by aquifers provides much of the island chain's water supply. To ensure maximum replenishment, there are building restrictions in water catchment areas (located on the higher elevations of ancient volcanic cones). These are to ensure that catchment areas remain free of impervious (solid) surfaces. Builders are not happy that they cannot capture this valuable real estate for development. Similar types of catchment plans are being considered for places in northern Africa and the Middle East.

Linked to creating water supplies is water storage. While above ground reservoirs are the mainstay of storage today, evaporation losses can be enormous, especially in hot and/or dry climates. Tom

A sign posted on a lawn in Collier County, Florida advises residents about water restrictions.

Landscaping in Los Angeles has seen an increase in the number of cactus plants being utilised in an attempt to cut down on the amount of watering required and therefore help to conserve this precious commodity.

SAVING WATER

A typical US household uses about 120 times the amount of water shown here on a daily basis.

According to the American Water Works Association, an average US household uses 350 gallons (1.3 KL) of water per day. Depending on the region, households use from 20-62 percent of that total outdoors. Lawns, gardens, and pools dominate this use. The uses shown here do not include business uses (e.g. golf courses, community pools, and manufacturing).

While saving water outdoors may seem to be paramount, it can have serious impacts on maintaining home landscaping. So, indoor savings may be easier to realize. It is estimated that homeowners can reduce water use by 30 percent by installing more efficient water fixtures and regularly checking for leaks. Taking shorter showers and using front-loading washing machines can also help. See resources for more ways to save water.

Home water usage (numbers rounded to nearest whole percent)

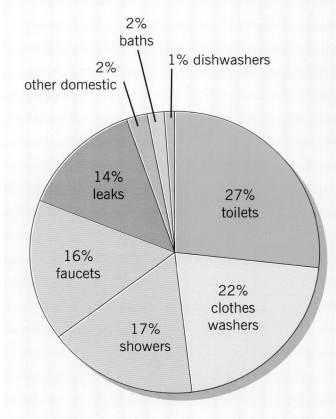

2% baths

1% dishwashers

2% other domestic

14% leaks

16% faucets

17% showers

27% toilets

22% clothes washers

This chart gives a detailed breakdown of how water is used in American households. Note the amount lost through leakage; in some parts of the UK this is claimed to be as high as 33 percent.

Brokowski, a professor of hydrology at the University of Texas, Dallas, and Wayland Anderson, a Denver engineer, studied three areas – Hays, KS; Guymon, OK; and Las Vegas, NM – and concluded that draining reservoirs and pumping the water back into aquifers not only prevents evaporation, but it also recharges the aquifer. Doing this, however, curtails recreational uses of the reservoir.

Making water

Recycling water is another way to capitalize on existing supplies instead of creating new ones. Many municipalities routinely process wastewater and then return it to nearby streams where it mixes with water that is eventually drawn into municipal water systems. It is also possible to process wastewater for non-drinkable uses such as irrigation (see sign). Florida and California lead in this practice, although a growing number of places around the world are planning to begin doing so. In many jurisdictions, car washes are required to capture and recycle water.

Desalination involves a choice of several ways to create fresh water supplies from salt water. The process, which removes salts and other minerals from the water, is expensive and uses a lot of energy. Still it is a viable source of fresh water in dry regions. Saudi Arabia has about 24 percent of the world's desalination capacity. Israel recently completed a major desalination facility that will provide around 13 percent of the country's domestic consumer demand.

Florida leads in US desalination, but Texas is developing its capabilities quickly. By 2004, Texas had 100 desalination plants. The desalination plant in Tampa, Florida, when completed, will only produce about one quarter of the drinkable water that the Israeli plant will. Still, it will be largest such plant in the US.

The first desalination plant in Australia opened near Perth (Western Australia) in November 2006. It is the largest of its kind in the Southern Hemisphere and is also the largest to be powered by renewable energy. When its expansion is completed, it should meet about 17 percent of Perth's water needs.

Checking for leaks

Homeowners are often urged to check for water leaks in their homes. However, recent reports show that older water transmission pipes in parts of South Florida are losing massive amounts of water. For example, in Miami-Dade County, pipe leakage costs more

HYDRO-ILLOGICAL CYCLE

IR Tannehill wrote in *Drought: Its Causes And Effects* (1947) that, "We welcome the first clear day after a rainy spell. Rainless days continue for a time and we are pleased to have a long spell of such fine weather. It keeps on and we are a little worried. A few days more and we are really in trouble. The first rainless day in a spell of fine weather contributes as much to the drought as the last, but no one knows how serious it will be until the last dry day is gone and the rains have come again."

Don Wilhite of the US National Drought Mitigation Center has adapted those words and various diagrams to create the image of the "hydro-illogical cycle" shown here. This graphic blends human action and inaction with the water cycle and clearly shows that we humans tend to act only when faced with a crisis.

Fortunately, many scientists, planners, government officials, and others no longer follow this approach. Instead inter-disciplinary teams often work together to solve complex problems. You've seen some examples in this chapter and elsewhere in this book. Nowhere will such collaboration be more important than in defining appropriate responses to the myriad of water-based extreme weather and climatic events affecting our planet.

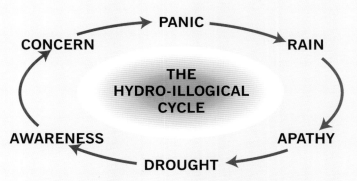

The hydro-illogical cycle clearly demonstrates the never-ending train of human thought. While people may be concerned about a drought at that particular time, they believe their fears are unfounded once rain eventually arrives.

Recycling water is one way in which the authorities are trying to prevent future shortages.

than 12 billion gallons of water; this is nearly a 10th of what the county moves through its pipes. These numbers do not account for any local water system leaks. Other public water utilities, especially those with older infrastructures, have similar water losses.

Although the cities and counties who operate such water systems pledge to spend huge sums on detecting and repairing leaks, even more funding and effort are probably needed. Given competing needs and the cost of retrofitting existing pipelines, the actions will likely remain in "catch-up" mode for many more years. All the while, valuable water is being lost.

Climate implications

Whether on long- or short-terms, precipitation can vary significantly. Most places in middle and low latitudes have histories punctuated by wet and dry periods and floods and droughts. About ten thousand years ago, the Sahara, for example, was savannah-like, with rivers, lakes, and plentiful rains. Rainfall then decreased, only to increase somewhat again around the 1600s. That rainy period was short-lived and desert conditions resumed quickly.

Although the savannah landscape is gone, the rains from that period (and earlier rainy periods) progressively percolated into the ground to be collected in aquifers. That is how some Sahara locations (e.g. Libya) have obtained much of their water supplies. Similarly, aquifers provide significant portions of water supplies in places like the high plains and Florida (US), and Perth (Western Australia).

Space-based research at NASA, tree-ring studies, and historical reconstructions have added evidence that global warming is not necessarily the cause of all of our recent weather and climate problems. Shorter-term fluctuations in ocean temperature patterns (including, but limited to El Niño) can create major shifts, especially in rainfall patterns. The ongoing drought in Australia can be linked, to a large extent, to an El Niño pattern (warmer waters in the eastern Pacific Oceana and cooler waters near Australia).

Droughts can also become self-sustaining (i.e. they can have a remote response and control over weather and climate even in other places). Scientists know that droughts can become localized and intensified based on soil moisture levels, especially during summer (lack of rain and high temperatures). When rain is scarce and soil dries, there is less evaporation, which leads to even less precipitation. With less evaporation to cool plants and use up the solar heat, temperatures rise. Both create a feedback process that reinforces the lack of rainfall.

However, as we have seen, droughts are not just weather-based. Water storage (surface and subsurface), land use and an array of socio-economic factors all have roles. Whether one subscribes to global warming as a cause, there is no doubt that human activity has affected large parts of the global water cycle.

With a world population that is soaring (1.7 billion in 1990; 5.9 billion in 2000; and an estimated 9.3 billion in 2050, according to the US Census Bureau), demands for water, agricultural commodities and land will be strained even further. Efforts to explore new ways of recycling water, creating new water supplies and conserving the water that is available are now becoming the norm around the world.

BOTTLED WATER

According to the Beverage Marketing Corporation, European countries lead in use of bottled water, with the US ranked 11th in per capita consumption. The US (top country consumer) and Mexico combined comprise almost 30 percent of the world market for bottled water. Several Middle Eastern countries (desert regions) show up in the top fifteen consumer nations. Although still small in numbers, many developing countries are starting to use bottled water as a low-cost alternative to unsafe tap-water supplies.

Bottled water sales are big revenue generators, too. In the US alone, sales in 2005 exceeded $10 billion (7.5 billion gallons sold); although global sales value was not available, consumption totaled some 43.3 millions of gallons. If the value per gallon matched that of the US, global sales would be on the order of $58 billion. No wonder that aside from carbonated soft drinks, water has become the beverage of choice in many counties.

Reasons for such increased consumption of bottled water include life-style changes (globally we are on the run more and more involved in activities like jogging), convenience, and concern for quality of tap water. Consumers believe that bottled water (or as the World Health Organization calls it – "packaged water") has fewer chemicals in it and comes from more pristine water supplies like springs and glaciers. Yet most of the bottled water sold today is simply treated water from various sources.

Although bottled to standards, bottled water is not necessarily better than regular tap water. Some bottled water is obtained from springs; most, however, originates as tap water and is then purified (which can include being distilled, undergoing demineralization through various means, or being acted upon by ozone as a disinfectant).

Marketers of bottled water tout its low-cost compared to other drinks. While this may be true of some coffee-based and alcoholic beverages, bottled water is more expensive than soft drinks at many US entertainment sites (e.g. sporting events or movies). At many fast food restaurants, combination meals do not include water as a choice. To have water, you have to buy the water in addition to the cost of the meal itself. Starting in early 2007, one fast food restaurant chain, Subway™, began offering bottled water as a low-cost upgrade to combination meals. Still water is being sold at a premium.

Finally, consider the impacts on the environment and on weather and climate of creating bottled water. According to the Container Recycling Institute, sales of the plastic most commonly used in water bottles, shot up to 1.63 million pounds (738,000 kilograms) in 1999, more than double the amount in 1990. Producing 2.2 pounds (1 kilogram) of this plastic requires almost 39 pounds (17.5 kilograms) of water and results in significant emissions of hydrocarbons, sulfur and nitrogen oxides, carbon monoxide, and carbon dioxide. This does not even address the transportation of the product and introduction of plastic waste into the trash cycle.

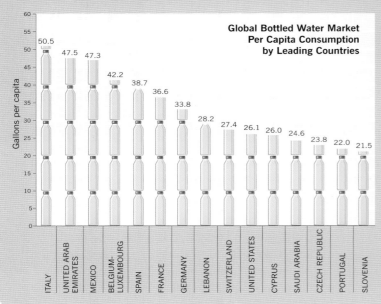

Global Bottled Water Market Per Capita Consumption by Leading Countries

Country	Gallons per capita
ITALY	50.5
UNITED ARAB EMIRATES	47.5
MEXICO	47.3
BELGIUM-LUXEMBOURG	42.2
SPAIN	38.7
FRANCE	36.6
GERMANY	33.8
LEBANON	28.2
SWITZERLAND	27.4
UNITED STATES	26.1
CYPRUS	26.0
SAUDI ARABIA	24.6
CZECH REPUBLIC	23.8
PORTUGAL	22.0
SLOVENIA	21.5

Perhaps surprisingly, it is Italy that leads the market in bottled water in this survey conducted between 2000-05. The UAE and Mexico complete the top three places while America lags in 10th and the UK is nowhere to be seen.

"There's been a shift, from many people having the sense that all (forest) fire was bad to almost as many people now seeing good fire and bad fire. With the caveat, so long as it's not in their backyards."
Barb Stewart, Fire Communications and Education Staff, National Park Service

Forest fires* are a natural part of life. Even before humans inhabited the planet, forests burned. It was nature's way of removing sickly vegetation and thinning forest stands, lessening insect populations, priming the forest floor for new growth and even opening the seed containers of some plant species. Without forest fires, forests would not be what they are today.

The world needs forests because they provide a means of transforming carbon dioxide into oxygen (through photosynthesis) and storing carbon in the wood. Yet, the world's forests continue to shrink in size through urbanization, the need for farming land and forest fires. This loss of forest is one of the factors contributing to a growth in atmospheric carbon dioxide and recent atmospheric warming.

Firefighters are dwarfed by the flames from this forest fire in Big Sur, California.

State forestry crews attempt to cut off the head of a wildfire in western Oregon. The fire started as a slash burn but east winds pushed sparks into brush piles, timber stands and newly planted trees.

While the United States, Canada, and Australia have extensive fire data bases, other nations are now starting to develop similar real time and historic data information sources.

Setting the stage for fires

There are many ingredients that can come together to create a forest or wild fire. As the number of ingredients grows, so, too, does the risk of having a more significant fire.

Poorly managed forest lands help to set the stage. In the United States, during most of the 20th Century, fires were deemed undesirable and were quickly extinguished. Land

*Although most readers may be familiar with the term forest fire, a more encompassing term "wild fire" is being used in many circles today. Wild fire includes all types of vegetative fires including forest, grassland and brush.

managers thought this was a good thing (because it saved habitats, trees, and homes). By preventing natural burns from occurring, forests become over-vegetated and full of kindling (woody matter that litters the forest floor). As a result, when a fire started, it became larger.

The large amounts of tinder have also allowed forest insects to gain a stronger foothold. Swarming pine bark beetles and their accompanying blue fungus vociferously decimate large stands of trees. The beetles burrow into the tree breaking the tree's moisture-nutrient circulatory system. The fungus negates the tree's natural defense against the beetles.

One of the most effective tools in the firefighters' arsenal when it comes to combating forest fires is to use an airplane to dump tons of water on the flames.

In 2006, in British Columbia alone, the pine beetle was estimated to have killed or infested some 23 million acres (13,000 square miles) of trees. Canadian Forest Service officials fear that more than three-quarters of the province's pine forests will be lost by 2013. The situation is similar from Alaska's Kenai Peninsula southward to the desert southwest region of the United States and to other parts of the world including Scandinavia.

Humans have also placed themselves at greater risk to fires. Homes are built in forested areas (for reasons that include peace, tranquility, and "becoming one with nature") or in expanding urban areas, and large numbers of people and their personal worth now reside in areas known to be fire prone. By building homes with flammable roofs (for example, cedar shake shingles for a "natural look"), and keeping trees in proximity to their home (no fire break or clear space), the homes have literally become part of the forest. Hence, when fires have occurred, homes are consumed as well. Even in non-heavily forested areas, homes are built where grass and brush fires can easily reach homes.

Of course, natural events also contribute. Droughts set the stage for fires by causing trees, shrubs, and grass to die and, by drying out, creating tinder. Indirectly, droughts allow disease and insects (such as the pine beetle and its associated blue fungus) to further destroy forests.

A brush fire threatens residents' homes in southern California.

Fires themselves are caused by many factors, including lightning, winds downing trees onto powerlines, contact with hot volcanic matter and human activity. It is human activity that accounts for the majority of fires. Some of these are purposely set (e.g. agricultural biomass burns; arson); others are caused by carelessness (cigarettes thrown from cars, unextinguished campfires); and some are the result of accidents (ashes from fireplace fires, automobile catalytic converters, explosions).

Fire weather
On a daily basis, it is atmospheric temperature, relative humidity and dew point, the strength of the winds, and the potential for "dry" lightning that are the main concerns of weather forecasters and fire fighters.

Above: The forest landscape clearly shows which trees are dying. Discolored foliage is a sign that these lodgepole pines have been attacked and killed by the mountain pine beetle.

Above right: With more homes being built in secluded forest locations, residents have put themselves at greater risk from forest fires.

"Dry" lightning is almost always confined to areas that are dry to begin with. Here there is enough heat and moisture to create convection (rising air currents) and the formation of cumulonimbus clouds and lightening. However, when precipitation falls from these clouds, it almost always partially or completely evaporates on the way to the ground. Thus the lightening happens without much, if any, rain.

This evaporation process helps to create more rapidly sinking air currents known as downdrafts and downbursts (see Chapter 6). If fires are already in progress, the downdrafts and downbursts create strong, rapidly shifting winds that can spread a fire quickly, cause it to jump fire lines, and possibly block escape routes for the firefighters.

Even if a fire was not ongoing, lightning can ignite one or more blazes and associated thunderstorm winds can quickly spread a fire. Since lighting fires can happen anywhere in forested lands, many lightning fires begin in remote locations, making detection and firefighting efforts more difficult.

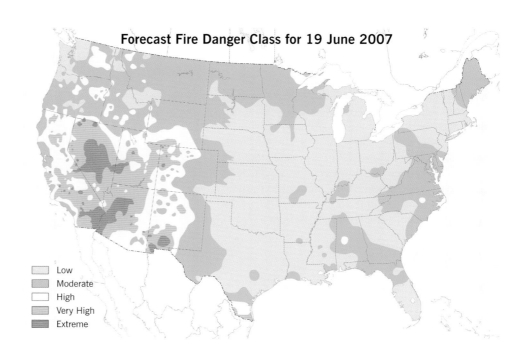

Forecast Fire Danger Class for 19 June 2007

Low
Moderate
High
Very High
Extreme

The Forecast Fire Danger map for the United States on June 19, 2007. Very high to extreme fire danger conditions existed in 8 of 11 western states.

Lightning starts fires because it is hot (sometimes five times hotter than the surface of the Sun or 50,000 °F (28,000 °C)). When heat like this is applied to dry materials, ignition can be immediate. Sometimes the lightning strikes and merely starts woody materials smouldering. It may take hours for a sudden gust of wind to fan the heated material and create a fire. Lightning can also cause powerline transformers to arc or spark. In a worst case scenario, lightning can ignite combustible materials like gasoline.

Humidity is important because moisture exchange is always taking place between fuels and the air. When the air is dry, moisture leaves forest fuels, making them even drier. When humidity readings increase, moisture is transferred back to the fuels, making them less likely to burn. This isn't much different than what happens when we leave some clothes outside during the afternoon, but don't take them inside until the next morning. The clothes may have dried during the day, but become damp again overnight.

Relative humidity increases at night (lower temperatures), as does atmospheric stability. This stability prevents the downward mixing of higher winds above the ground. Also, thunderstorms (and their associated winds) are less likely to occur after dark. Collectively, these factors help to lessen the spread of fires overnight and allow firefighters to make gains in fighting them.

In coastal areas (especially along the US west coast), onshore breezes often help to increase humidity values. Sometimes, these onshore winds even bring in fog (which further aides in transferring moisture to the fire fuels).

Fighting fires

Once a fire starts, firefighters and meteorologists need weather information around the fire site. Local winds and other conditions – some of which are created by the heat and drying power of the fire itself – are key to making more accurate forecasts and assessing fire behavior.

Satellite, radar and lightning detection data help in determining where thunderstorms are forming and how winds above the ground are blowing. To obtain local data, specially trained meteorologists at the fire site need to set up and monitor data from portable weather stations. These battery-powered instrument packages can be repositioned as needed.

HUMIDITY

Humidity is a measure of the amount of water vapor in the air. Absolute humidity (measured by the dew point) tells how much moisture is present absolutely. Air with a dew point of 60 °F (15 °C) contains more moisture than air with a dew point of 40 °F (4 °C).

Relative humidity describes the amount of water vapor present compared to the maximum amount of water vapor the air could hold at the current air temperature. Thus, air with a temperature of 40 °F (4 °C) and a dew point of 40 °F (4 °C) would have a relative humidity of 100 percent (the air holds the exact amount of water vapor it could possibly hold).

However, air at 60 °F (15 °C) with a dew point of 40 °F (4 °C) would have a relative humidity of only 48 percent. This means that the air would be holding only 48 percent of the water vapor possible at 60 °F (15 °C).

Typically, dew points will change little during the day except when there are air mass changes. Relative humidity undergoes a daily cycle. Relative humidity is typically highest during the early mornings (coolest time of day). That's when we see the formation of dew.

By afternoon, temperatures will have warmed and relative humidity values will have plummeted. During this warming period, any dew on the grass will have evaporated.

By late afternoon and evening, temperatures will have again fallen and relative humidity values will rise.

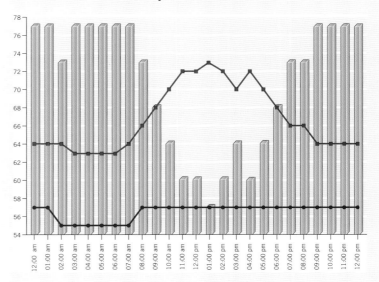

Temperature (red line), dew point (blue line) and relative humidity (bars) graph for San Diego, California on June 20, 2007. Notice that while the temperature undergoes a daily cycle, dew point values remain nearly unchanged. Thus, relative humidity has an inverse (opposite) relationship to the daily temperature graph.

Chuck Redman from the NOAA National Weather Service forecast office in Boise, Idaho, setting up the FireRAWS equipment near a wildfire.

Once the data is obtained, it can be used in high-resolution computer models. These models can be updated as often as new data arrives.

In addition, meteorologists and fire fighters can measure the moisture content of the forest tinder and obtain other measurements that help define firefighting measures and approaches. Although fire fighting remains a dangerous vocation, meteorologists are helping lessen the uncertainty involved with the weather.

Firefighters know that fires move with the terrain. In general, fires move uphill faster than downhill and they move uphill faster when terrain is more steeply sloped. Fires will move up narrow canyons faster due to channeled, heated upslope winds. The type of terrain and the type of fire will determine whether fires burn at the ground or move at tree top (or crown) level.

Following the same approach as storm chasers, firefighters want to be sure their escape routes do not get cut off. The same approach applies to evacuations involving local residents. Firefighters and local emergency service officials no longer hesitate at ordering these. Much like hurricane or flood evacuations, there's no time and often no place for pets and much in the way of personal belongings.

Before fires start

Over the past 100 years many fires were extinguished instead of being allowed to burn. Thus, flammable materials built up in forest lands. Further, species that needed periodic fires to reproduce were overrun by other species. As a result, our forest lands are now primed for fires.

In fact, almost every year, in almost every season, in one part of the United States or around the world, major forest fires make the headlines. The total area of forests now burnt annually in the United States alone is the size of the state of Vermont.

Researchers Jan Kucera and Yoshifumi Yasuoka, of the Institute of Industrial Sciences, The University of Tokyo, Japan, used satellite-based algorithms to assess forest fire burn areas during the warmer parts of the year in a relatively small portion of eastern China. Their 18-year study period spanned the period 1984-2001.

Kucera and Yasuoka found that the acreage burned has increased significantly during the period. Prior to 1992, there was but one year in which the area burned exceeded 7,000 square miles (18,130 square kilometers). From 1992-2001, five of the 10 years had burn areas that exceeded 8,000 square miles (20,720 square kilometers). The two most recent years in the study had burn areas of more than 12,750 square miles (33,022 square kilometers).

With forests, grasslands, and people at risk from wildfires, it is no longer an option to just let forests and other areas burn naturally. Thus, many countries now have major firefighting systems in place. The US system, based on local, state, and national levels, is primarily concentrated in the drier, western portions of the country. However, firefighting systems are in place in most states. Still, when Georgia and Florida, for example, were ravaged by forest and grassland fires in the spring of 2007, it took enormous resources to gain control of the fires.

There are several ways to prevent major fires. First is to allow smaller fires to burn and clean out forest tinder and undesirable vegetation. Another is to actively set small fires (called prescribed burns) which accomplish the same thing. Clearly for both, weather conditions must be "prefect" to ensure that a small fires does not become a large one.

Harvesting some trees (especially ones that have crowded larger tree stands), even from public lands, is another. However, this solution often comes with controversy as logging activities disrupt the forest and, at times, larger trees are incorrectly harvested.

When fire danger is high, campgrounds may be closed, outdoor burning prohibited, and watering restrictions implemented so water will be available for fire fighting efforts.

Lightning versus humans

Lightning and humans both cause forest fires and statistics tell the real story.

In the US, human-caused fires outnumber lightning-caused fires by 6:1, but lightning fires account for nearly double the burned acreage, mainly because of where they occur (more remote locations) and have longer detection times.

According to Australia's Department of Sustainability and Environment, lightning, intentional and combined agriculture-campfire, and forest and brush fires each account for 25 percent of the total number of annual blazes. This is based on a 20-year period ending in about 2000.

These graphs compare the number of fires and acreage burned due to the two main fire igniters. Humans start the greater number of fires, but lightning-caused fires consistently consume more acreage.

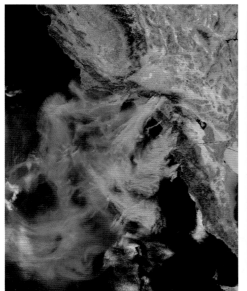

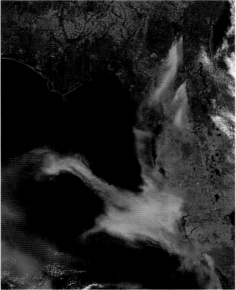

These satellite images show smoke plumes from major fires on both the west and east coasts of the US. Notice that initial plumes may be more or less straight lines, but changes in wind patterns, dispersion and other forces spread the plumes in many directions. I live in Naples, Florida and the smoke plumes covered our skies for more than 450 miles, with a cloud sufficiently thick to completely shroud the sun.

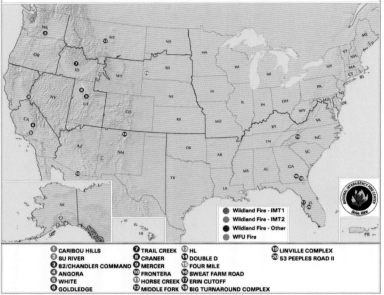

Large Incidents - June 26, 2007

Wildland Fire - IMT1
Wildland Fire - IMT2
Wildland Fire - Other
WFU Fire

① CARIBOU HILLS
② SU RIVER
③ 82/CHANDLER COMMAND
④ ANGORA
⑤ WHITE
⑥ GOLDLEDGE
⑦ TRAIL CREEK
⑧ CRANER
⑨ MERCER
⑩ FRONTERA
⑪ HORSE CREEK
⑫ MIDDLE FORK
⑬ HL
⑭ DOUBLE D
⑮ FOUR MILE
⑯ SWEAT FARM ROAD
⑰ ERIN CUTOFF
⑱ BIG TURNAROUND COMPLEX
⑲ LINVILLE COMPLEX
⑳ 53 PEEPLES ROAD II

The MODIS Active Fire Mapping Program presents a daily update of incidents around the United States.

Just by looking at this forest fire in Ecuador, it is easy to see how these incidents are responsible for so much carbon dioxide and particulate matter being introduced into the atmosphere.

Fire, smoke and more

When forest fires occur, it is more than the forests themselves that are at risk. First, forest fires worldwide disperse untold amounts of smoke and ash into the atmosphere. These affect visibility, deposit dust and ash (which can affect our homes and cars), and most importantly provide a major source of particles that can affect breathing (especially those with breathing impairments). The particles in forest fire smoke include large particles (which fall out fairly quickly) to small particles that can remain suspended in air for months.

In addition, it is estimated that forest fires worldwide emit at least 85 million tons of carbon dioxide, 13 million tons of carbon monoxide, 2.2 million tons of nitrogen oxides (a precursor of ground-level ozone), hydrocarbons (such as benzene), aldehydes (such as formaldehyde), and various trace minerals.

When the smoke is trapped by an atmospheric inversion (warm air layer atop a slightly cooler one), concentrations of smoke and pollutants can be become even more hazardous.

Record fires – United States

Major forest fires of today often burn tens of thousands of acres or more (see list). However, the worst forest fire in US history burned more than 1.5 million acres. That is comparable to about half an average year's fire burn across the US and is more than the area of the state of Delaware.

The Peshtigo Fire occurred on October 8, 1871, the same date as the Great Chicago Fire. Much as the Chicago fire involved human carelessness, so, too, did the Peshtigo Fire of eastern Wisconsin. The Peshtigo Fire was the culmination of forest clearing, piling of pine slash debris, an extended drought, careless fire control and the arrival of an intense, autumnal low pressure system with high winds. Clearly, the high winds did much to spread both fires.

The resulting fire burned forestlands, communities and farms on both sides of Green Bay. It killed between 1,000 and 1,500 people and destroyed about a dozen communities.

Although some allege that a meteor shower contributed to both blazes on that fateful evening, there is little evidence supporting that hypothesis.

Possibly the largest North American wildfire occurred in October 1825, burning from Maine through New Brunswick, Canada. A group of loggers ignited a fire in a drought area that soon burned out of their control. The fire burned three million acres of forest and killed more than 160 people.

Record fires – globally

Internationally, forest and wildland fires are just as noteworthy. Consider the following.

On February 16, 1983, the Ash Wednesday Fires (about 100 separate blazes) began in the Victoria region of southeast Australia. Fueled by a lengthy drought, low humidity

MAJOR US FOREST FIRES

State	Dates (start-control)	Description
Arizona	June 18-July 7, 2002	Rodeo-Chediski fire; two fires (one arson and one accident combined to burn 467,066 acres (1,890.15 km) of forest. Worst forest fire in state.
Wyoming	Summer-fall 1988	The Yellowstone Fire of 1988 was the largest forest fire known to affect Yellowstone National Park in recorded history. The fires burned for several months and by the time the winter snows extinguished the flames, 793,880 acres or roughly 36 percent of the park had been impacted by the fires.
California	October 25-29, 2003	15 devastating forest fires burned for two weeks, primarily in San Diego, Ventura, Riverside, and San Bernardino Counties. The fires killed 24 people, forced more than 80,000 people to evacuate their homes and burned 800,000 acres and destroyed more than 3,600 homes. More than 15,500 firefighters battled the blazes. The Cedar Fire in San Diego, which burned through 200,000 acres, was the largest fire in California's history.
New Mexico	April-May 2000	A prescribed fire started by the National Park Service raged out of control, destroying 235 structures and forcing evacuation of more than 20,000 people. The blaze consumed an estimated 47,000 acres and threatened Los Alamos National Laboratory.
Colorado	July 2-11, 1994	The South Canyon Fire was a relatively small fire (2,000 acres), but it still led to the deaths of 14 firefighters.
Maine	October 25-27, 1947	A forest fire destroyed part of Bar Harbor and damaged Acadia National Park. In all, 205,678 acres burned and 16 lives were lost.
Florida and Georgia	April and May 2007	On April 16, 2007, high winds blew through Okefenokee National Wildlife Refuge (one of the oldest and most well-preserved freshwater areas in America) causing a tree to fall on a power line and showering sparks on the drought-ridden land. By mid-May, this fast-moving wildfire quickly became not only Georgia's largest fire in recorded history, but Florida's as well.

levels, very dry conditions, and the passage of a strong cold front (with strong and shifting winds), the fires, once started by various causes, spread quickly. If it weren't for the fires reaching the ocean and running out of eucalyptus fuel, the losses would have been even more extreme. As it was, more than a million acres of forest and grasslands were burned. Seventy-five lives were lost (including 17 firefighters) and damage exceeded US $250 million.

Malaysian fires have become an annual ritual. Although primarily linked to El Niño conditions, the fires are now occurring even in years when fire risk should be lower. These fires are caused by many factors, including the burning of peat lands.

Russia's northern forestlands, the taiga, account for about half of the world's evergreen forests and 20 percent of the world's natural forests. In recent years, there has been a dramatic increase in forest fires and scientists and Russian officials speculate that this is beyond natural causes.

The results of a forest fire can be devastating to the natural natural landscape and its inhabitants (human, animal and plant)

Haze over Malaysia during fires that took place in the summer of 2001. Haze from these fires often affected other parts of Asia and the southwest Pacific region.

Part of the cause may involve Russian rules that allow for easier logging permits in burned areas than in natural areas. Thus, arsonists may be purposely setting fires to make the land more harvestable. The timber, now far more valuable than as natural forest, is then sold to China and other countries. However, Russian forests are at high latitude and this means that they are slow growing. Thus, fire losses here have a longer-term impact than in warmer areas of the world.

In the mid 1980s fire losses in the Siberian forests totaled around 8,000 square miles. In 2004 the losses had increased 10-fold. The area lost to fires in 2004 alone was more than that of the area of Idaho, the 11th largest US state.

In early July 1995, a fire started near Jerusalem, Israel. With high temperatures and strong winds, the fire spread quickly. Although it was also extinguished quickly, the fire still destroyed four to eight square miles (10-21 sq km) of natural and planted forest areas. This is a lot coverage for a country smaller than the size of New Jersey that is mostly desert. Two million trees succumbed to the blaze, thought to be the largest in the nation's history.

Climatic implications

Just looking at the growing incidence of forest and wild fires, one would be tempted to blame warming temperatures and more intense droughts (both part of the global warming scenario). However, the situation is really more the result of poor forest management practices. In short, forest policies have prevented periodic small to medium fires and in the process set the stage for more major, large scale burns.

One of the most telling examples involves that of western United States Sequoia groves. According to US Geologic Survey ecologist Dr Nate Stephenson, of the Western Ecological Research Center, sequoias have responded to "an ever-changing climate and fire regime." This has included all types of natural fires that partially, and sometimes totally, cleared forest areas and ground cover. Only with bare ground can sequoia seeds germinate.

The loss of fire in sequoia groves over the past century and a quarter has greatly affected the age distribution of sequoias. Instead of a wide range of ages, many groves today lack trees that are younger than 100 years.

And Stephenson notes, "Fire exclusion by humans has done more than the last three millennia of climate and fire regime changes… Essentially, when you exclude fire, sequoia reproduction crashes to zero."

On a longer time scale, fossil finds, including a major one in an Illinois coal mine, showcase how forests have changed with natural changes in climate and landform configuration. While working the mine, some miners noted and reported unusual markings on a tunnel ceiling. The result is a scientifically-reconstructed, 300 million year old fossil forest record, locked in coal layers, spanning several square miles.

Apparently submerged suddenly due to an earthquake, the forest vegetation was quickly covered by sediment and preserved. In addition to how fast the forest was placed in fossilized mode (and possible associated changes in weather and climate), the fossilized landscape offered insight into the entire forest ecosystem, not just the usual scattered fossilized remains. In fact, scientists now have an idea of what the forest may have looked like.

According to the UK-US team, the forest had a layered structure (much like modern forests) with a mix of plants. There was a sub-canopy of tree ferns and an assortment of shrubs and tree-sized horsetails that looked like giant asparagus. Towering above these were club mosses that stood more than 130-feet (40-meters) high. Although now extinct, this type of forest is quite different than forests we know today. This also indicates that the weather and climate of the region were much different than they are now.

Firefighters battle against a blaze in the Amazon.

Other research has also helped to place plants and animals in their proper place in our geologic history. Some fossils have been preserved in mud, coal, and even volcanic ash. Collectively, these provide part of the geologic timeline we need to understand if we are to understand our planet's long-term weather and climate and how plant species die and evolve.

Fires themselves not only destroy forest timber, they affect all aspects of the forest ecosystem. Animals can easily become casualties; burned forest land is vulnerable to flooding and mudslides since tree canopy and roots no longer slow down water flow; and ash and mud can affect species that live in forest area rivers. Smoke from forest fires adds to the amount of pollution in the atmosphere (even hundreds of miles from the fire source) and can affect the amount of incoming solar radiation, trap heat in the Earth's atmosphere and add massive amounts of carbon dioxide, ash, and other pollutants to the atmosphere.

With its combined vegetative loss, the forest area is less able to process carbon dioxide out of the atmosphere and replace it with oxygen (part of the atmospheric CO_2 – oxygen cycle) The loss of tree canopy affects how the surface of the Earth absorbs in coming solar energy. Burned areas are often hotter and drier than their forest counterparts. Almost all of these factors contribute to atmospheric warming. Thus while forest fires might not be caused by global warming, they may be contributing to it.

A wildfire in the Bitterroot National Forest in Montana, United States, on August 6, 2000, photographed by John McColgan, a fire behavior analyst at the Forest Service, an agency of the US Department of Agriculture.

"Extreme heat may be one of the most underrated and least understood of the deadly weather phenomena. In contrast to the visible, destructive and violent nature associated with "deadly weather", like floods, hurricanes, and tornadoes, a heat wave is a "silent disaster". Unlike violent weather events that cause extensive physical destruction and whose victims are easily discernible, the hazards of extreme heat are dramatically less apparent, especially at the onset."

NOAA Natural Disaster Survey Report on the Chicago Heat Wave, 1995

A farmer harvests his crops as the sun sets. Notice how dry and dusty the conditions are.

Heat waves are a usual summertime phenomenon. Some are short in duration where others can last for weeks. They are always accompanied by temperatures that are well above seasonal averages (daytime and night time alike) and, in many locales, high humidity is also present.

What are heat waves?

Heat waves are most easily determined by looking at the combined effects of temperature and humidity on the human body. Typically, this is done via the NOAA National Weather Service's Heat Index (also known as "apparent temperature"). When the index reaches a critical value(s), it means that for most/many people heat is actually transferred to the human body or that evaporative cooling is stymied. This means that human body temperatures can increase, leading to heat exhaustion and possibly death. The index does not address human activities being engaged in at the time, age, sex, disease and other such variables.

The Heat Index often varies significantly during the course of a 24-hour period due to the diurnal temperature cycle. Although wind and temperature combine to create a wintertime wind chill measure, wind is not a factor in computing the heat index.

Your experience with hotter weather of all types depends largely on what you are used to. If you live in the tropics, then weather that makes the average North American or European feel as though they are melting would probably be quite comfortable. On the other hand, those used to weather in cooler climates might find even non-tropical

Right: A fire hydrant is opened on a particularly hot day. It won't be long before children and adults alike are taking advantage of the cool water.

Far right: Air conditioning is one way in which people attempt to combat heat waves.

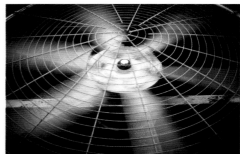

weather unbearable. We become used to our "home" climate (acclimatization), and any significant change (toward higher values or lower values) becomes uncomfortable, possibly stressful. The effects are partly physical, due to excessive heat affecting the way the body temperature is controlled, and partly psychological, because many people just do not like the sensation of feeling hot and sticky.

Definitions

Definitions of "heat wave" vary among countries and even within a country (depending upon who classifies the event). In the US, the National Weather Service issues a heat advisory when a daytime heat index is expected to reach or exceed 105 °F (40.6 °C), and a night time minimum ambient temperature of 80 °F (26.7 °C) is expected to persist for at least 48 hours. In Dallas County, Texas, the criterion used by the medical examiner's office to designate a heat wave is greater than or equal to three consecutive days of temperatures greater than or equal to 100 °F (37.8 °C).

In the Netherlands and neighboring countries, a heat wave is defined as a period of at least five consecutive days in which the maximum temperature in De Bilt – a small inland town in the Netherlands – exceeds 77 °F (25 °C), provided that on at least three days in this period the maximum temperature exceeds 86 °F (30 °C).

In Australia, where no formal heat wave definition exists, Ken Granger and Michael Berechree have managed to identify at least 18 heat wave events in since 1899. These are based on the use of a threshold for temperature that is within the top five percent of daily maximum temperatures for a continuous three-day period in the South East Queensland area. Thus, on average, there is a heat wave in the area every five to six

Heat Index Chart

Relative humidity (%)

Air Temperature (°F)	40	45	50	55	60	65	70	75	80	85	90	95	100
110	136												
108	130	137											
106	124	130	137										
104	119	124	131	137									
102	114	119	124	130	137								
100	109	114	118	124	129	136							
98	105	109	113	117	123	128	134						
96	101	104	108	112	116	121	126	132					
94	97	100	103	106	110	114	119	124	129	135			
92	94	96	99	101	105	108	112	116	121	126	131		
90	91	93	95	97	100	103	106	109	113	117	122	127	132
88	88	89	91	93	95	98	100	103	106	110	113	117	121
86	85	87	88	89	91	93	95	97	100	102	105	108	112
84	83	84	85	86	88	89	90	92	94	96	98	100	103
82	81	82	83	84	84	85	86	88	89	90	91	93	95
80	80	80	81	81	82	82	83	84	84	85	86	86	87

Heat Index (Apparent Temperature)

With prolonged exposure and/or physical activity:

Fatigue possible.

Sunstroke, muscle cramps, and/or heat exhaustion possible.

Sunstroke, muscle cramps, and/or heat exhaustion likely.

Heat stroke or sunstroke highly likely.

The Heat Index accounts for the combined effect of temperature and humidity on the body's ability to cool itself. At Heat Index values of 85 and higher, the risk of heat-stress symptoms – such as dizziness, excessive weakness, headaches, heavy perspiration, high body temperature, irregular heartbeat, loss of consciousness, muscle cramps, nausea, pale and clammy skin (sometimes skin that is red and dry), rapid pulse, rapid shallow breathing and severe mental changes – increases.

Although this thermometer on a New York City sidewalk reads 120° F (49 °C), the reading is somewhat erroneous. Air temperature was still only in the mid 90's (upper 30's °C) on August 2, 2006. Still people enjoy some ice cream to cool down.

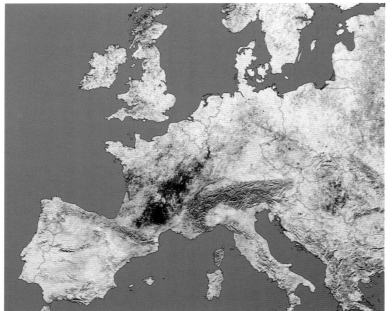

Temperature Anomaly (C)

-10   -5   0   +5   +10

NASA image showing satellite-based temperature departures during the great European heat wave of 2003 compared to those from the same period in 2001. While France bore the brunt of the heat wave, temperatures were also much higher across the UK, Germany, and northern Italy.

years. January (Southern Hemisphere summer) is the most common month in which to experience a heat wave.

Researchers around the globe are exploring other indices (many statistically-based) to better define heat waves. Much of this research is a direct result of the current focus on global warming.

A heat wave can also simply involve an extended period of hotter than average weather, as is common in the Mediterranean and southern US states in the summer; but, it can also be an exceptional period of heat which can have very serious and prolonged effects on human health and the economy. For instance, the effects of a heat wave when crops are just starting their growth or when they are ripening can destroy an entire harvest. In underdeveloped parts of the world, such conditions may lead to famine and starvation, as well as death of livestock. Heat waves can exacerbate the drying of forest areas, increasing the risk and occurrence of fires. They can lead to electricity brownouts (a temporary reduction of power – like the opposite of a power surge), water restrictions and even cancellations of some community and sporting events. In some situations, concrete and asphalt highways and even railroad tracks have buckled from the high temperatures created in heat waves.

Heat can be good or bad for those who grow grapes. As temperatures have warmed since the mini-Ice Age, grape growing has made a resurgence in the UK (grapes used to be grown there prior to 1300). And the length of the growing season has expanded in some grape growing regions. On the other hand, higher temperatures can lead to earlier ripening and require harvesting at night (when grapes are cooler). Some scientists suggest that wine growing regions will undergo a major poleward shift if global warming continues.

A sunflower field in South Dakota during a drought clearly shows the lack of crops to harvest.

Air transportation can also be affected by heat waves. While the main problem for aviation involves storms (and clearly none are around during a heat wave), high heat and humidity require longer runway distances for an airplane to gain sufficient take-off speed. That's because hotter air is less dense and provides less lift. While it is rare for flights to be cancelled to due high heat, airlines may have to adjust aircraft cargo weight and increase aircraft spacing between take-offs.

When heat waves occur

Heat waves occur when a large upper-level high-pressure system (and associated surface high pressure system) park themselves over an area for an extended period of time. These typically bring the following meteorological conditions; sinking air and an inversion (both unfavorable for clouds and precipitation), and either light surface winds or a wind flow from a moisture-rich source region. In the southwest US and in other desert areas, a so-called "heat low" may appear. This is a fictitious weather feature caused by how meteorologists recompute sea level pressure accounting for temperature variations.

A vineyard in Sussex, southern England. The length of the growing season has increased enabling grapes to again be grown at such a high latitude.

Lacking cloud cover and the tempering effect of evaporative cooling from plants and wet ground, the dry ground bakes and heats up even more. The longer the upper-level high stays in place, the more the feedback mechanism causes it to stay in place.

CHICAGO HEAT WAVE, 1995

These maps showcase the surface and upper level (30,000 foot (91,500 meters)) conditions associated with the Chicago heat wave in July 1995. Notice that the jet stream is displaced far into Canada while a large upper-level high-pressure system dominates the Midwest. At the surface warm, humid, southerly winds are moving into the Midwest from the Gulf of Mexico. Both upper level and surface wind patterns are clockwise (Northern Hemisphere). While every heat wave around the world is different, these are the types of weather conditions that are often associated with significant heat waves.

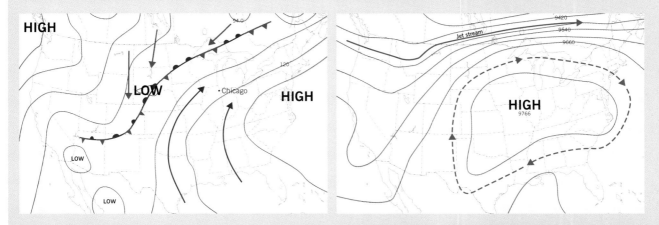

Upper air level (30,000 foot (91,500 meters)) and sea level pressure pattern during the peak of the July 1995 heat wave in Chicago. Notice that upper high parked near Chicago while surface winds were bringing in warm, moist air from the Gulf of Mexico.

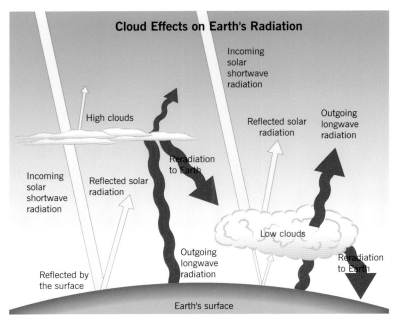

Cloud Effects on Earth's Radiation

High clouds

Incoming solar shortwave radiation

Reflected solar radiation

Reradiation to Earth

Incoming solar shortwave radiation

Reflected solar radiation

Reflected solar radiation

Outgoing longwave radiation

Low clouds

Outgoing longwave radiation

Reradiation to Earth

Reflected by the surface

Earth's surface

The complex role of clouds in the Earth's energy balance.

Meanwhile, near the ground, in many situations, atmospheric moisture values remain high (either from transportation from moisture-rich regions, such as the Gulf of Mexico, or the effect of increased evaporation from crops such as corn. The result is a heat wave replete with high temperatures and high humidity.

Measuring heat waves

Although heat waves can occur almost anywhere, they are most likely to occur outside of normally hot locales. Still, places like Phoenix, Arizona and New Delhi, India, can suffer heat extremes.

Globally, as we saw in Chapter 1 (page 10), every continent has its temperature extreme, at least in recent recorded history.

However, there are other extremes that can be added to individual daily high records. For example, according to the *Guinness Book Of World Records*, the hottest place in the world – in terms of average annual temperature – is Dallol, Ethiopia, at 94 °F (34.4 °C). But, this record was established not over a climatological time period of 30 years, but rather during a seven-year period (1960-66). The same book notes that temperatures in Death Valley rose above 120 °F (48 °C) on 43 consecutive days.

In Bullhead City, Arizona, a string of daily record national high temperatures occurred in 1998 that got meteorologists to think that something might be wrong with the weather observation equipment. A detailed review found the instruments to be okay, but that the weather equipment was incorrectly situated over a sandy, barren surface instead of a grassy area. Dry soil gains and loses heat more quickly than a vegetated surface.

While working for the National Weather Service in the San Francisco Bay area in the early 1980s, I uncovered a suspicious situation in the Merced-Modesto area (just east of San Francisco and west of Sacramento). One of the weather stations reported abnormally high temperatures, but only during the afternoon. Readings in the morning meshed with nearby locales. After days of instrument analysis, technicians found the error and corrected the faulty sensor. The records, however, still remain in the climatological record.

The bottom line is that any weather record must be based on some form of measured (temperature, rainfall) or estimated (EF Tornado scale) condition. And when the records have potentially contaminated data, and not any type of new measured data, it is impossible to correct them.

It's not the heat, it's the humidity

Hot weather alone does not make a heat wave. As we have already seen, it is often the combined effect of heat and humidity. That's because both act to lessen the body's ability to cool itself.

Humans have several mechanisms for keeping cool. We can radiate heat from our skin and we breath out hot air through our nose and mouth. But when the air temperature exceeds our core body temperature, we breathe in hotter air then we exhale. Similarly when the air temperature exceeds our body skin temperature (around 93 °F (39 °C)), we capture more heat then we lose. During the normal evaporative process, we expel large amounts of water vapor and perspire large amounts of salt-laden water. Unless the water and salt are resupplied to our bodies, our evaporative capabilities will diminish.

You can use the figure on page 211, at the start of this chapter, to evaluate how both temperature and humidity conspire against us during heat waves. Even with a low relative humidity, heat index values can soar. Arizona may be noted for its expression, "It's may be hot, but it's a dry heat." Nonetheless, 100 °F (37 °C), even with only a 40 percent relative humidity, still spells danger!

By the numbers

Although we think of hurricanes and floods as the great killers, heat waves are actually deadlier. They just arrive without much fanfare and often overstay their welcome. The numbers tell the story. For example, in the US alone, there were 2,190 deaths related to hot weather between 1992 and 2001, compared to 150 deaths from hurricanes and 880 deaths from flooding.

SUMMER IN THE CITY

"Summer In The City" is the title of a 1966 pop song, and it describes a situation very familiar to people living in areas where heat waves are common. Urban heat islands are hot spots, where the center of an urban area is markedly warmer than the surrounding suburbs, and these are hotter than the countryside around the city.

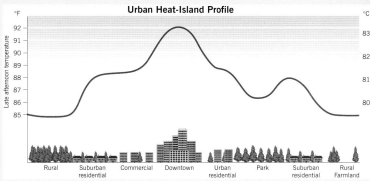

The temperature is normally higher over built up areas.

This phenomenon can be clearly seen in satellite photos using infrared photography. According to the Environmental Protection Agency in the UK, on hot summer days urban air can be 35.5-43 °F (2-6 °C) hotter than the surrounding countryside.

Landast satellite image pair showing the urban heat island for Atlanta, Georgia, on September 28, 2000. The upper image shows the daytime view as one might see it visually from space. Areas of trees or vegetation are the greenest. The lower image shows infrared (heat) view. Notice how easy it is to see the hottest areas (highways and most urbanized areas) in red and the coolest areas (trees) in yellow on the IR image.

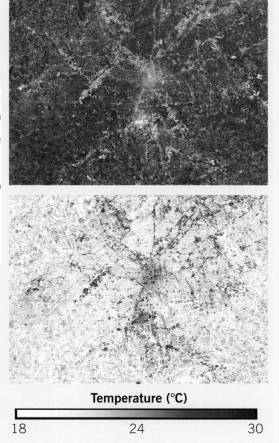

Temperature (°C)

18   24   30

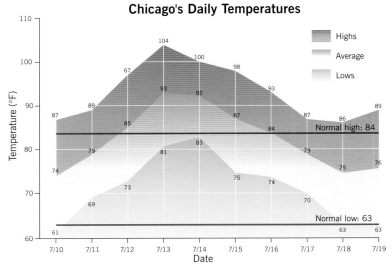

Chicago's Daily Temperatures

Legend: Highs, Average, Lows

Normal high: 84
Normal low: 63

Observed daily high, low, and average temperatures for Chicago, Illinois, during the mid-July 1995 heat wave. "Normal" or expected average high and low temperatures are also shown. Notice that while daytime highs were much above average values, night time lows approached values normally observed during the day.

The worst weather-related disaster to befall the US between 1980 and 2006 was the onslaught of Hurricane Katrina in August 2005 (losses of $125 billion and more than 1,800 deaths). But the second worst disaster was the great August 1988 Midwest heat wave. Losses in that event were estimated at more than $61 billion with an unbelievable 7,500 fatalities.

Other US disasters prior to these most recent events, include the Galveston Hurricane (1900) with some 6,000 fatalities, the Tri-State Tornado (1927) with nearly 700 deaths and the Dust Bowl (1930s) with an unfathomable economic and human toll. (The latter falls into the categories of both drought and heat wave, as we saw in Chapter 12).

In 1988, Chicago, Illinois, suffered from a heat wave – 77 people died. The event was not unusual in terms of temperature. High temperatures had occurred in the area many times before.

Another Chicago heat wave struck seven years later (1995). In this heat wave, as in so many others, night-time low temperatures approached the average expected daytime high temperature values. Still, during the three-day period from July 13-15, 1995, approximately 70 daily maximum temperature records were set at locations from the central and northern Great Plains to the Atlantic coast.

Chicago alone recorded 465 deaths during the period July 11-27 with Milwaukee, Wisconsin, reporting another 85. A NOAA Disaster Survey team found no evidence that large-scale climatic effects (e.g. El Niño or global warming) contributed to the heat wave. However, increased urbanization was believed to be a major contributing factor.

Social and societal issues contributed, as well. A disproportionate number of the fatalities were elderly, many living in small rooms at the highest level of an apartment building. Many lacked air conditioning. Those who had air conditioning suffered from periodic electricity outages and the lack of family members to check in on them. Once the heat wave began, hospitals were overwhelmed with victims. Lacking a coordinated city-wide plan, ambulances travelled from hospital to hospital trying to find an emergency room that could handle their patients.

The hottest ever temperature (in recent history) recorded in Chicago was 105 °F (45 °C) on July 24, 1934. Although low temperatures were around 80 °F (26 °C) on several nights during the 1995 heat wave, the record high minimum temperature in Chicago remains 85 °F (29.5 °C) (July 29, 1916). In fact, that record occurred during at least a five-day period in which record lows were above 80 °F (26 °C) every night and at least one record high temperature exceeded 100 °F (38 °C). The fact that these records remain on the books 90 years after they occurred is testimony to their extreme nature and demonstrates that we can't always credit global warming as the cause of recent heat waves.

While the Chicago heat wave was studied extensively on many fronts, there are many other significant heat waves. Consider the European heat wave of August 2003 that killed an estimated 50,000 people.

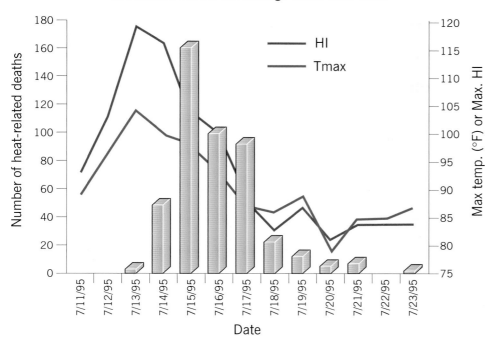

The human toll for the Chicago 1995 heat wave

There was a strong relationship between high temperatures (including the highest heat index (HI)) in Chicago during and following the summer 1995 heat wave. Note the lag in increasing deaths. It took time for the heat wave to claim its victims. Even after the heat wave relaxed its grip, a significant number of heat related fatalities continued to be reported.

Much of the heat was concentrated in France, where nearly 15,000 people died during a nine-day period. Night-time temperatures were well above average and much of France does not have air conditioning. The situation may have been compounded by the usual summer holiday in which large numbers of people leave major European cities. The elderly and infirm did not have this luxury and normal caretakers were not present at a time of great need.

Australia has a relatively long period of heat wave statistics. According to one researcher, heat waves kill more people than any other natural hazard experienced in Australia. During the period between 1803 and 1992 at least 4,287 people died as a direct result of heat waves. This was almost twice the number of fatalities attributed to either tropical cyclones or floods over much the same time-frame.

Climatological considerations

While heat waves are more in the news than ever before, there is little to suggest that they are the result of global warming. An ever-aging population, the separation (in distance) between family units, urbanization, and other societal factors may be playing significant roles in their impacts.

In addition, urbanization adds daytime and night-time heat. It does this because surfaces like asphalt and concrete absorb solar energy and reradiate it to the atmosphere. The effect is most evident at night. If you watch television weather reports, for example, it is often easy to see this in the temperature mapping.

Also, urban areas often have fewer trees than in rural areas. Trees help to lessen the absorption of solar energy. The urban heat island also fuels the statistics that indicate global warming. With urban centers reporting warmer temperatures, people are quick to note that the "climate must be warming". Yet, scientists recognize that about half the increase in average global temperatures is linked to the growing size and thermal impact of cities.

Boschi Pope, aged two, looks up as she drinks water from a fire hydrant to beat the heat on August 1, 2006 in the Harlem area of New York City.

While heat waves have been around forever, they have only become a meteorological watchword since about 1980. Not surprisingly, documentation of such events has soared since that time. To the untrained, this discontinuity in weather record keeping can be presented as proof that heat waves are becoming more widespread and severe. It is interesting to note that David Changnon, a climatologist at Northern Illinois University, has reported a marked increase in the number of high dew points in recent years. Dew points are the absolute measure of atmospheric water vapor. If one cools the air to the dew point, dew or condensation will occur. Relative humidity (RH), which is usually used to define atmospheric moisture, is just that, "relative". It compares the actual water vapor to how much water vapor the air could hold at a certain temperature.

The study began by analyzing hourly dew-point readings recorded from 1959-2000 at Chicago's O'Hare and nearby Rockford Airports. Changnon and his students found the top four high dew-point frequency years were 1983, 1987, 1995 and 1999.

Then the researchers analyzed the 10 most extreme heat waves in the region. This provided proof that the number of high dew-point hours was much greater after 1980. Additionally, the researchers discovered that over the 42-year period they studied, three different dew-point indices at both airports showed general increases over time. Those indices included:
• Hours each summer with high dew points reaching or exceeding 75 °F (24 °C).
• Summer days with at least one hour of high dew point.
• Summer days with 12 or more hours of high dew points.

According to Changnon, "Dew points that exceed 75 °F (24 °C) are considered rare in most regions of the United States, the exception being the Gulf Coast." Changnon and other meteorologists are well aware of the high dew point values often experienced in places like New Orleans and Houston and how this air mass affects the Midwest in of the United States during late spring and all summer. Since this source of water vapor close to the Earth's surface hasn't changed over time, Changnon ruled out the Gulf as being the source for the growing frequency in high humidity levels. He also noted that during the 1995 and 1999 heat waves, the dew-point levels were greater in the upper Midwest than in those areas between the Midwest and the Gulf Coast.

It was easy to rule out an urban heat island effect as a primary cause of humidity spikes. That was because similar trends were identified at both suburban and rural locales. Eliminating all other possible variables, Changnon concluded that for such high dew points to occur, evotranspiration from crops such as corn and soybeans must have been the cause.

With greater soil moisture (from increased rainfall) and a sharp increase in acreage in which corn and similar crops are planted, Changnon concluded that the greater number of high dew points evident in the 1995 and 1999 heat events, were preceded by average to above average precipitation across northern Illinois.

The heat wave of 1988 was anomalous. It was accompanied by a drought. "Corn yields dropped by nearly 50 percent. And although it was an extremely hot summer, with Chicago experiencing temperatures of 90 °F (32 °C) or greater on more than 40 days, very few high dew points occurred."

Scientists now believe that mortality statistics during heat waves can be linked to the so-called "harvesting" effect which merely advances the death date of marginally susceptible people. The drop in death rate following a heat wave tends to balance out the increase during the heat wave.

James O'Brien, Florida's state climatologist, was critical of colleagues who he thinks are too quick to link short-term and long-term weather to heat waves. O'Brien believes that the Midwest heat wave in 1988 was caused by high sea-surface temperatures in the tropical Pacific, not by global warming.

And, Philip Klotzbach, an atmospheric scientist at Colorado State University, is cautious, too, noting that, "Heat waves have happened for many years (i.e. the Dust Bowl in the 1930s), so to say that this one particular event is caused by global warming is really impossible."

Further, higher than average temperatures do not necessarily beget more heat waves and/or more severe heat waves. Still, Australian researchers and others have found a tendency for there to be more high temperature extremes than cold ones and for there to be more warmer nights. When very warm nights accompany daytime heat, there is less time for the human body to recover and be primed to deal with high heat the next day.

Chris Field, director of the Department of Global Ecology at the Carnegie Institution of Washington's branch at Stanford University, said scientists can't attribute singular weather events to global warming. But many studies conclude that heat waves tend to get hotter as the planet warms.

"This week's heat wave might or might not have occurred without global warming, but it is a good bet that heat waves will be hotter and more frequent in the warmer world," Field said.

Others, such as Michael Mann, a leading global warming expert at Pennsylvania State University, agree that climate change is "…stacking the deck…" and making heat waves more likely.

While some claim that numerical models portend a growing heat wave problem, the jury remains out. After all, in spring 2007, the northeast US was very cold and snow fell in parts of Colorado as late as late May.

The duck weed-covered Limehouse Cut, a canal in East London on August 8, 2006. Hot weather resulted in green duck weed stretching six miles along East London's canals. Duck weed, a flowering water plant, can double in size every day and the heat wave provided perfect growing conditions.

"This past summer and fall have been so cold and miserable that I have from despair kept no account of the weather. It could have been nothing but a repeatation (sic) of frost and drought."
From the diary of Adino Brackett, at the end of the "Year with no Summer" 1816.

The best way to combat the cold is to wrap yourself up well and always remember to wear a wear a hat and a scarf. Between 30 and 50% of body heat loss is from the head and neck

From late fall to early spring, cold air masses (large volumes of air with similar temperature and moisture characteristics) often make their way toward the Equator from polar source regions. Most of these arctic intrusions into the middle latitudes reach to between 30 and 35 degrees of the Equator before being repulsed by warmer winds. On occasion, an arctic air mass will reach to 20-25 degrees of the Equator. Much as heat waves strike areas outside and poleward of normally hot regions, cold waves typically affect areas just outside and Equatorward of normally cold regions. Cold waves can sometimes affect places used to cold weather.

Often the arrival of the cold air mass brings a shift in wind, possibly some snow or rain and lower temperatures. Sometimes air masses arrive with a flourish, bringing brisk winds and sharply colder temperatures. Occasionally air masses take on epic proportions and bring the potential for life-threatening conditions. Such extreme cold weather events typically displace an abnormally warm weather pattern. This makes the cold wave feel even colder than it would otherwise be.

While most people in middle and high latitudes (poleward of about 30 degrees latitude) think "sub-zero" when it comes to cold waves, it's another story in lower latitudes and in warmer climates. Florida, India and even southern California have a different standard when it comes to cold waves.

Rancher Eddie Linke breaks the ice on a waterhole in a pen so that his cattle will have water to drink in Colorado, USA.

Formation and movement

Air masses normally develop over more or less uniform source regions (see Chapter 4, page 40). For polar or Arctic air masses, source regions include Northern Hemisphere polar and Arctic regions, interior Canada, interior Russia, and Siberia. The Southern Hemisphere lacks such large land masses at high latitudes. At high latitudes, the ground is often snow-covered and nights are long. In addition, the daytime sun angle is low or the sun may not rise for weeks or months due to the tilt of the Earth's axis and Earth's orbit around the sun. These conditions allow for the Earth and its atmosphere to lose, overall, heat to space. The longer the air mass stays in place, the colder it usually becomes. Air masses are usually associated with high-pressure systems and cold air masses are often accompanied by very high barometric pressure readings.

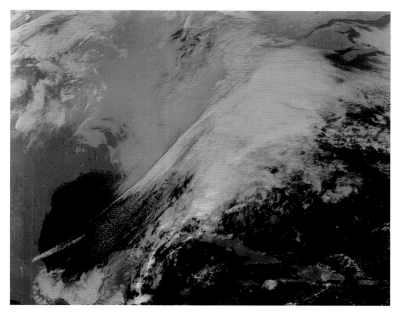

Once the cold air mass develops, upper level winds and transient storm systems may allow it to move toward warmer latitudes. As it does, the air mass may move over warmer ground or warmer water, undergo warming and lose its lifeblood. However, if the air mass moves over snow- or ice-covered ground or water (and arrives with little modification), it can earn the title of a cold wave. Not surprisingly, cold waves are often preceded by a major wintry precipitation event (if not in the immediate area, certainly poleward of it). If the high-pressure system remains over its new home for an extended period, its clear skies and light winds can allow for more pronounced and localized night time cooling. Such radiational cooling events are most significant when it comes to agricultural freezes associated with cold waves.

While cold waves often involve a single high-pressure system, sometimes, a cold wave is reinforced by the arrival of additional cold air masses. When this occurs, instead of a bitterly cold air mass arriving and depleting the supply of Arctic air, smaller pieces

When cold air masses move Equatorward, they clash with warmer and more humid air masses that are moving poleward. The result is often a storm, like this one forming along the US East Coast. The brighter, whiter clouds associated with the storm extend from eastern Mexico northward to southeastern Canada.

We experience seasons because the Earth rotates on an axis that's tilted in its orbit. That 23.5 degree tilt causes the different hemispheres to be at different angles to the sun at different times of the year.

June 21
Summer solstice in the northern hemisphere.

March 21
Vernal equinox. Spring in the northern hemisphere, autumn in the south.

Because of the angle, during winter the energy from the sun must travel through more atmosphere and reach the Northern Hemisphere latitudes. Also, a given amount of the sun's energy is spread over a larger area.

Sun

Earth's orbit

September 23
Autumnal equinox. Autumn in the northern hemisphere, spring in the south.

December 22
Winter solstice in the northern hemisphere.

Earth's seasons in the making. The tilt of the Earth's axis from vertical and its orbit around sun allows nearly all locations on our planet to experience differences in the receipt of solar energy. This includes changes in length of daylight and vertical position of the sun in the sky. Both affect how much solar energy arrives at a place. At high latitudes, part of the year involves complete darkness followed by complete daylight. Near the Equator, day length is close to 12 hours year-round. Other than poleward of the Arctic and Antarctic Circles, solar elevation can vary by 45 degrees (halfway from horizontal to directly overhead) between summer and winter.

Cold Waves

Snow-covered ground is highly reflective and keeps ground and air temperatures colder; without snow-cover, grass, soil and other surfaces can better absorb solar energy and warm the air above the ground.

of Arctic air arrive in series, keeping temperatures cold (but not as bitterly cold) for a longer period. That's what happened when a series of cold air masses dominated the weather in Lincoln, Nebraska, from November 6 to December 25, 2000. On 42 days out of the 50, period temperatures were below average. Not surprisingly, snowfall was above average, with the 11 inches (28 cm) on the ground on December 26, 2000 being the greatest since December 1983.

Blizzards (see also, Chapter 4) involve cold temperatures, wind, and snow, while ice storms can involve bitterly cold temperatures and dangerous ice accumulations. Because the dangers are many, these events are treated separately from stand-alone cold waves.

A cold wave has serious implications for humans, energy supplies, transportation, livestock, and agriculture. Clearly human health and safety are paramount concerns.

Bodily impacts

If accompanied by even light winds (e.g. 10-15 mph (16-24 km/h)), wind chills can become dangerously low during a cold wave. Wind chill temperature is related only to wind speed or motion past exposed skin and air temperature. Thus, for computing wind chill, skiing or jogging in cold weather would be much like standing still in windy conditions. Wind chill does not address individual metabolism, health, or the effects of humidity and/or precipitation. It is not a factor when it comes to inanimate objects (e.g. car radiators). Thus, water in a car's cooling system (even without antifreeze) will not freeze even if the air temperature is 40 °F (4.5 °C) and winds are blowing steadily at 45 mph (72 km/h) – creating a wind chill of 26 °F (-3 °C)!

If exposed to low wind chills long enough, skin can freeze and in the process die (frostbite). Having worn poorly insulated gloves once while shovelling snow from my driveway, I can attest to this. My fingers became hard to move and started to lose their normal pinkish color. Using an infrared thermometer, much like those used on weather satellites, I measured my fingertip temperature to be 30-40 °F (0-5 °C). In my case, scientific inquiry almost won out over my own safety.

Usually it is the extremities (ears, fingertips, toes, and nose) that can suffer frostbite most quickly. That's because these areas are furthest away from the heart and the source of heated blood, and because they often have large exposed surface areas compared to their individual volumes.

While frostbite is the main concern in the wind chill chart shown above, it is not the only one. Once the blood in these exposed areas is chilled, it returns to the heart. In the

NWS Windchill Chart

Temperature (°F)

Calm	40	35	30	25	20	15	10	5	0	-5	-10	-15	-20	-25	-30	-35	-40	-45
5	36	31	25	19	13	7	1	-5	-11	-16	-22	-28	-34	-40	-46	-52	-57	-63
10	34	27	21	15	9	3	-4	-10	-16	-22	-28	-35	-41	-47	-53	-59	-66	-72
15	32	25	19	13	6	0	-7	-13	-19	-26	-32	-39	-45	-51	-58	-64	-71	-77
20	30	24	17	11	4	-2	-9	-15	-22	-29	-35	-42	-48	-55	-61	-68	-74	-81
25	29	23	16	9	3	-4	-11	-17	-24	-31	-37	-44	-51	-58	-64	-71	-78	-84
30	28	22	15	8	1	-5	-12	-19	-26	-33	-39	-46	-53	-60	-67	-73	-80	-87
35	28	21	14	7	0	-7	-14	-21	-27	-34	-41	-48	-55	-62	-69	-76	-82	-89
40	27	20	13	6	-1	-8	-15	-22	-29	-36	-43	-50	-57	-64	-71	-78	-84	-91
45	26	19	12	5	-2	-9	-16	-23	-30	-37	-44	-51	-58	-65	-72	-79	-86	-93
50	26	19	12	4	-3	-10	-17	-24	-31	-38	-45	-52	-60	-67	-74	-81	-88	-95
55	25	18	11	4	-3	-11	-18	-25	-32	-39	-46	-54	-60	-68	-75	-82	-89	-97
60	25	17	10	3	-4	-11	-19	-26	-33	-40	-48	-55	-62	-69	-76	-84	-91	-98

(Wind Speed (mph) — vertical axis)

Frostbite times   30 minutes   10 minutes   5 minutes

$$\text{Wind Chill (°F)} = 35.74 + 0.6215T - 35.75(V^{0.16}) + 0.4275T(V^{0.16})$$

Where T = Air Temperature (°F) V = Wind Speed (mph)

Wind chill chart with formula for computing wind chill (updated in 2001). Color-coding signifies how quickly frostbite can occur when human skin is not protected from the cold. Wind chill requires relative wind flow past human skin of four mph (6 km/h) or more. This can involve the wind itself and/or human activities (e.g. walking, running).

process, it brings colder temperature into the body's core, lowering internal temperature. If this continues long enough, the body temperature may drop to levels termed as hypothermic. During hypothermia, the body tries to shut down unnecessary functions in order to conserve internal body integrity. This can result in the loss of mental acuity, hand and leg mobility, and more. If warmth doesn't follow shortly, even the core internal organs can shut down.

In cases of just cold weather, it is easy to classify how weather claims its victims. When cold weather, snow, ice, and/or wind combine, casualty reports may not clearly specify the actual cause of death (e.g. car accident or freezing). For this reason, winter weather casualty figures include all wintry-related causes of death.

Cold waves and water

The impact of cold waves goes far beyond human health. It can affect many aspects of everyday life.

A boy stands beside an icy wall as old and ruptured heating pipes release steam that quickly freezes into ice on January 28, 2001 in Spassk, Russia. Gripped by the worst winter in half a century, and the collapse of aged Soviet heat and electricity grids, residents of Russia's Far East and Siberia battled daily against brutal cold and bundled up even to sleep.

The weight of car and truck traffic coupled with the freeze-thaw cycle is responsible for the birth and growth of potholes. Unless repaired quickly and correctly, potholes can become very large. Potholes cause considerable damage to automobile steering systems each year.

Locally, cold waves can simply make life miserable. In lower latitude regions, where homes are not as well insulated as in more northern areas, water pipes in outside walls can freeze. So, too, can in-ground water pipes. Both are due, in part, to how water expands in volume as it chills below 39 °F (4 °C); the expansion is even more pronounced as water freezes.

The freeze-thaw cycle also plays havoc with road surfaces. Once water gets into small cracks and freezes, the force exerted by the expanding ice breaks up the road surface even more. Water can also get into the underpavement and cause pavement heaving. Over time, the same type of process (weathering) that helps to break up natural rock formations acts to break up manmade ones. Then, as additional freeze-thaw cycles occur, a small pothole (a hole or pit, especially one in a road surface) soon grows. Once formed, and with cars bouncing into the pothole and straining its fragile edges, the pothole can expand in area and depth quickly. Although all potholes do not develop from freeze-thaw effects, many do. (Other contributing factors are the effects of placing salt on roadways to help with snow and ice melting and a decline in routine road maintenance.)

The global impact of potholes is large. In the US alone, local, state and Federal government agencies spend billions on winter-time highway repair. In Michigan, for major highway repairs alone, the state spent an average of about $7.3 million annually between 2001 and 2003. North of the border, the City of Toronto, Canada, spent nearly $3 million on repairing 55,000 potholes in 2006. A recent report issued by the Asphalt Industry Alliance placed the annual cost of pothole repair in England and Wales at more than $110 million. The AIA also noted that it would take 11 years to repair all of the existing potholes in the two countries at current repair rates!

People spend equally large amounts collectively to maintain their car's alignment, repair front end damage, and replace blown out or prematurely worn tires from interactions

The amount of freezing of the Great Lakes depends upon wintry temperatures and the depth of various parts of the Great Lakes. Here, Lake Michigan's nearshore waters are frozen, but deeper mid-lake waters remain ice-free. Other lakes show cloud cover as well as ice.

with potholes. Whenever potholes of sufficient size are around, travel slows as people try to avoid them.

Boat and plane travel

While potholes can disrupt ground transportation, cold waves can also affect airplane and boat travel. An extended period of very cold temperatures can cause fresh water bodies to freeze. Each winter, portions of the five Great Lakes, between the US and Canada, freeze with the deeper and warmer lakes staying unfrozen the longest. In recent years, warmer winter temperatures have resulted in the lakes remaining unfrozen most of the winter.

In January 1977, the Potomac River near Washington, DC froze solid and people could actually ice skate. Since then the river has frozen on occasion but not to a sufficient depth.

In the 18th and early 19th Centuries, cold waves allowed the Thames River in England to freeze over. With the ice sufficiently thick, Ice Fairs were held on the frozen waterway. While this was fun for the locals, it caused a cessation of ship transit on the river. Partly due to warmer winters, partly due to the rebuilding of London Bridge (allowing for faster water on the river), and channelling (which allows for faster flow), the Thames has only frozen once, in 1963, over the past 200 years.

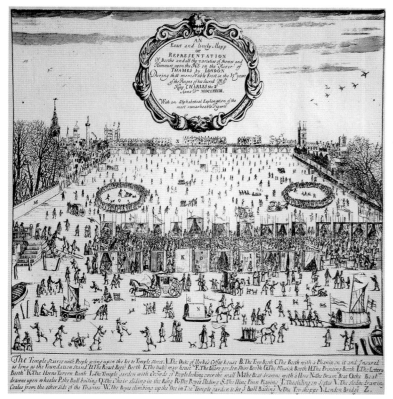

An artist's impression of the 1683 Frost Fair on the frozen River Thames in London.

Quebecers (residents of Quebec, Canada) celebrate the bravery of those who used to deliver mail and supplies along the ice-clogged St Lawrence River in winter, with an annual ice canoe race.

Ice Canoe racing in Quebec is another way to celebrate the annual freeze-thaw cycle of waterways.

Portions of major rivers in the Midwest, most rivers in Russia, and rivers in other high latitude regions also freeze each winter. Sometimes parts of large bays or estuaries (even though there may be brackish (partly salty water) will freeze. As noted earlier, the only recent cold wave that has allowed ice to exit the Mississippi River into the Gulf of Mexico occurred in 1899.

Other cold wave affects

As described in Chapter 4, when very cold air (even if not cold wave status) flows across warmer lake or ocean waters, clouds and snow squalls can develop.

Under windy conditions, waves and spray can coat piers, boats, and other near-shore objects with ice. If the spray interacts with sand and/or blowing sand, it can create frozen mounds of unstable sand and ice known as ice dunes.

Even without precipitation, under very cold and calm conditions, ice can form on aircraft wings and fuselages. The ice that forms, in a process known as deposition (the direct phase change of water vapor to ice crystals) must be removed by a chemical de-icing spray before the aircraft can safely take-off.

Some newer vegetable oil-based energy fuels may actually freeze or solidify at certain temperatures and are not suitable for use in places in which cold waves or simply cold

This image compares temperatures from January 1-24, 2006, to the average temperatures for that period from 2001-2005. Regions where temperatures were up to 10 °C colder (18 °F) are dark blue, while places where it was 10 °C warmer are shown in dark red. Places where temperatures matched the five-year average are white.

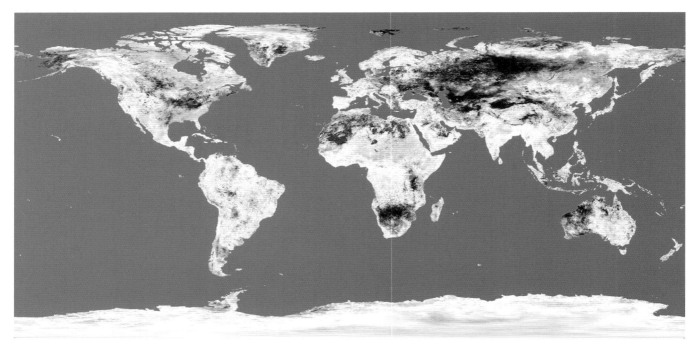

temperatures occur. Thus, some vegetable-based fuels have limited distribution regions, especially in the winter, or require additional modifications to ensure that they can be preheated before use.

Much as humans can succumb to wintry chill, so, too, can livestock. Because livestock often feed outdoors, any type of snow or ice cover that doesn't melt during a cold wave, or which glaciates, makes it impossible for the animals to feed. Such conditions can require that food be provided.

Water supplies can become frozen. Firefighters may have to contend with ice formation on hoses, trucks, and the ground on which they walk. Not surprisingly, home fires increase during cold waves as people turn to other sources of heat (e.g. kerosene stoves or electric heaters) that are more prone to creating a danger.

Produce, the agricultural products that we require for a healthy daily diet, and other agricultural commodities are particularly prone to damage by cold waves. Perhaps the January 2007 cold wave in California provides the most vivid testimony. During that cold wave agricultural losses exceeded $1.4 billion. However, any type of cold wave can lead to the need for protective measures and/or agricultural losses.

Environmentally, cold waves with their usual attending high-pressure system often block precipitation areas. This decreases wintry precipitation, which is unfavorable for adding snow cover to mountain glaciers. However, the cold weather lessens melting which allows for glacial ice retention. When bitterly cold weather strikes, anything out of doors (especially waterfowl, fish, and other vulnerable species) is at risk.

Ice covered Lieutenant Mike Hobin of the Quincy Fire Department's Rescue 1 unit stops to take a break after containing a fire in frigid weather in Quincy, Massachusetts, January 16, 2004. Firefighters had to contend with frozen hydrants, hampering efforts to fight the house fire. It was -6 °F (-21 °C) at 6 am when the blaze started.

Historic cold waves
Cold waves typically don't last as long as heat waves because they are linked to moving air masses. Heat waves thrive on stagnant conditions. Nonetheless, cold waves can be as deadly as heat waves.

Usually, somewhere over the Earth each winter, a cold wave of noteworthy proportions occurs. If the cold wave affects places with a vulnerable population and/or a significant agricultural base, the impacts become more newsworthy.

European and Asian cold wave of 2006
The cold wave that struck Europe in January 2006 resulted in scores of deaths from Russia into Central Europe. Temperatures in the region dropped into the −30 °F to -40 °F (-34 °C to -40 °C) range. The cold caused rail tracks to crack in Vienna, iced a key German canal, and brought heavy snow to Greece. The cold weather was blamed for around 120 deaths in parts of eastern Europe.

The cold wave struck northern India and nearby countries. The death toll (mainly involving the homeless) reached 160 in India and more than 100 in neighboring Pakistan.

Frost occurred in New Delhi for the first time in 70 years with temperatures dipping to 32.3 °F (0.2 °C) on January 9. The previous record occurred on January 16, 1935, when New Delhi reported 31 °F (-0.6 °C).

The effects of the week-long cold wave spilled over to Europe and Japan. With sea level pressure readings reaching 31.30 inches of mercury (1,060 hPa) within the high-pressure system, the resulting strong pressure gradient helped to push the colder air into Japan and parts of Eastern Europe. Strong northwesterly winds blowing across the warmer Sea of Japan allowed "lake effect" like snows to develop (see also page 21).

Since the pattern persisted for about a month, it wasn't surprising to find that Japan reported record high snow drifts of 13 feet (about 4 meters) by mid-January. On January 21, Tokyo received its heaviest snowfall 2.8 inches (7 cm) since late January 2001. The lengthy cold wave and associated snowy weather claimed more than 80 lives during the period.

Weather satellites monitored both temperature and snowfall effects. Satellites easily measure the Earth's surface and ocean temperatures through infrared sensors. This allows scientists to reconstruct ground temperature pattern maps and visible satellite images show cloud patterns easily.

It is interesting to note that one year later, in January 2007, many European nations reported their warmest January on record. In the rush to proclaim "global warming," this record-breaking cold wave seems to have been forgotten. Yet averaging the two extreme events, wintry weather was close to "average."

United States cold wave of December 1983

This cold wave made national news once it arrived in the southern states, but that event was preceded by incredible and lengthy cold in the north. For example, on 19 December 1983, Coronation, Alberta, reported its coldest day in 99 years as the temperature plunged to -42 °F (-41 °C).

In the US northern Plains, the average temperature for the month of December 1983 in Sioux Falls, South Dakota was 2.1 °F (-16 °C) (17.6 degrees below average). Readings were equally cold in nearby states.

Even in north Texas, cold air was firmly entrenched by December 18, 1983. In fact, starting on the 18th (and continuing until the 30th) temperatures remained below freezing in the Dallas-Fort Worth area for 295 consecutive hours.

Then, just before Christmas 1983, the cold air charged as far southward as the jet stream pattern transitioned. Prior to the cold wave, the jet stream had helped to keep the cold air locked across Canada and the northern United States (west to east or "zonal" flow); then, the jet stream took on a new character (greater north-south orientation known as "high amplitude"), allowing cold air to race southward. Since the cold air moved quickly southward over snow-covered ground, it did not have time to be modified. The result was a bitter cold wave reminder.

The cold wave seriously strained power supplies and disrupted travel across the south. It also caused an incredible amount of burst pipe damage. One reason for this was that homes had water pipes in outside walls without sufficient insulation for such unusual cold. Even in-ground water pipes burst. Damage in Texas alone was in the range of $50–100 million.

More than 125 cities across the eastern US reported record low temperatures for December 25th, and 34 of those cities reported all-time records for the month of December. For example, the temperature plunged to 1 °F (-18 °C) at Huntsville Alabama (its coldest temperature ever until another cold wave arrived in 1989).

Volcanic eruptions have been known to have a definite impact on global temperatures. This Mount St Helens eruption on July 22, 1980 sent pumice and ash six to 11 miles (10-18 kilometers) into the air and was visible in Seattle, Washington, 100 miles (160 kilometers) to the north.

The United States cold wave of February 1899

During the period January 26 to February 14, 1899, there was widespread and severe cold across the eastern United States. There were record lows in many places, including a state-wide Florida record low of -2 °F (-19 °C), on February 13th, in Tallahassee. That record remains on the books today. As the cold wave came to an end, an intense low pressure developed along the southeast U.S. coast and brought blizzard conditions to the Middle Atlantic states. Washington, DC reported snowfall of 25-30 inches (63-76 cm).

According to a description prepared by NOAA's National Climatic Data Center, the "Great Cold Wave of February 1899" actually featured two different cold waves, both of which set some all-time state records which exist to this day. The first cold wave around the 10th of the month, set the all time Ohio state record low temperature of -39 °F (-39 °C) at Milligan. A quote from the Monthly Weather Review of February 1899 remarks on the Great Cold Wave, "These cold waves established many new landmarks for future reference – whether we consider the instrumental readings or the physical phenomena resulting from the cold. The most striking of the latter perhaps was the flow of ice down the Mississippi River on the 17th, past New Orleans and into the Gulf of Mexico, an event never before witnessed within the memory of man. Ice an inch thick formed at the mouth of the Mississippi in East and Garden Island Bays, and the temperature fell to 10 °F (-12 °C) on the 13th."

More than 100 people died in the mid-February cold wave. However, untold numbers of poultry and domestic animals froze to death while game birds and fish perished in large numbers. There were food and fuel shortages in some larger cities. The cold was so severe that schools closed.

Northern Hemisphere cold wave of 1816

While cold waves seem to happen through classic meteorological formation processes, sometimes geologic processes enter the scene (see Chapter 19 for a more detailed explanation). Such was the case in the cold period during the early to mid 1810s, culminating in the "Year without a summer" (1816).

During that period, there were a number of major volcanic eruptions. In 1812, the Soufriére and St Vincent volcanoes erupted; in 1814, it was the Mayon and Luzon volcanoes in the Philippines; and Tambora in Indonesia erupted during 1815.

According to climate scientists, volcanic ash, suspended in the stratosphere (the atmospheric layer about 10-30 miles (16-21 km) above the Earth), caused increased reflection of incoming solar energy while not preventing the radiational loss of heat (which water and snow clouds do). This resulted in lowered Earth temperatures.

Recent volcanic eruptions have uploaded large amounts of ash into the atmosphere, but almost nothing compares to what Tambora did. The US Geological Survey indicates that the Tambora eruption ejected more than seven cubic miles (160 cubic km) of earth material into the atmosphere. This was almost double what Krakatoa ejected in 1883 and tens of times more than Mount St Helens sent airborne in 1980.

When coupled with other meteorological events, it is easy to see how such eruptions could lead to a cooling trend. In 1816, that trend devastated large regions of the Northern Hemisphere including New England and parts of northern Europe. The most notable effects centered on agriculture where the cold either destroyed crops or never even allowed some to start growing.

Yet, NOAA's NCDC in reconstructing weather records suggests that the year was cold in middle latitude areas of the Northern Hemisphere, but was actually above average elsewhere.

Climatic implications

As much as heat waves have us thinking of global warming, cold waves tip the scale in the other direction. Yet both, when viewed in the larger climatic scenario really lead to overall balance.

We've seen that the Eurasian cold wave of 2006 was tempered by above average readings across North America and how a cold wave in 2005 brought temperatures back to average in the Washington, DC area. In Europe, one of the coldest years on record in the 1800s was in 1816 (partially linked to volcanic eruptions). Yet, six years later, the continent experienced its second warmest year of the century. In 1834, Europe had its warmest year of the century only to be followed four years later by the coldest year (without volcanic eruptions).

Our perception that winters are not as cold as they used to be can be linked to short-term memory, relocation to warmer climates (more Equatorward regions and regions closer to coasts), and to media reports that key more on warmer than colder weather. While our ancestors and nomadic tribes migrated to warmer climes during winter, many of us today are making that transition as part of lifestyle changes.

Yet state-by-state US weather records do not suggest that record high and low temperatures are being seen any more or less often, respectively. In fact, using data compiled by NOAA and USA Today, most state temperature records from 50 or more years ago still remain.

What does seem to be evident is that weather patterns associated with both cold and heat waves seem to be more persistent (i.e. they are remaining "locked" in place for longer periods). There is nothing in the global climate model that suggests that weather patterns will persist longer.

Comparison of ash/dust cloud from selected volcanic eruptions. Mount St Helens' eruption in 1980 was small compared to its other recent eruptions and far below that of major eruptions Krakatoa and Tambora.

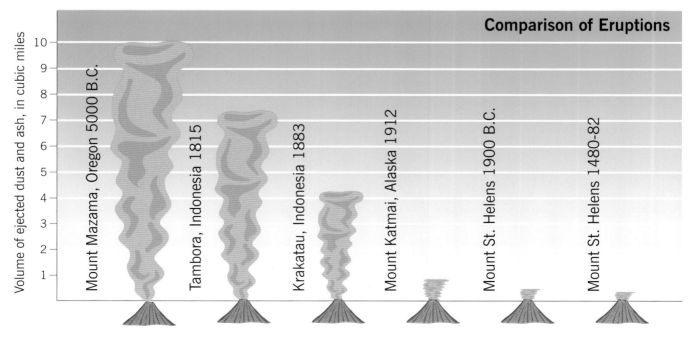

Comparison of Eruptions

Volume of ejected dust and ash, in cubic miles

Mount Mazama, Oregon 5000 B.C.
Tambora, Indonesia 1815
Krakatau, Indonesia 1883
Mount Katmai, Alaska 1912
Mount St. Helens 1900 B.C.
Mount St. Helens 1480-82

Cold Waves

"First, there is the power of the Wind, constantly exerted over the globe...an almost incalculable power at our disposal..."
Henry David Thoreau

While it is impossible to see the wind itself, the effects of wind can often be seen by watching trees and their leafy branches. Here whole palms sway in strong winds.

In earlier chapters, you've read about many types of winds – hurricanes, tornadoes and winter storms. This chapter is going to look at winds other than those associated with these stormy events and the effects of those winds (e.g. storm surges and seiches). Both surges and seiches have two components, the push onshore and then the return water flow. Here, we'll be looking at cold and warm winds on both large and small scales that are not always driven by storms.

Some background

Wind (the movement of air) doesn't just happen. It is linked to pressure and/or temperature changes. Since pressure changes are also linked to temperature changes (through a series of physical relationships involving temperature, pressure and density),

Overturned carriages lie off the tracks after a passenger train was blown over by strong winds on March 1, 2007 in the outskirts of Turpan of Xinjiang Uygur Autonomous Region, northwest China. The Ministry of Railways reported the train 5807, running from Urumqi to Aksu, was hit by gusts so strong that 11 of its 19 carriages were knocked off the rails leaving three dead and more than 30 injured.

wind is integrally tied to solar energy input and how it interacts with the earth-atmosphere-ocean system. This includes, but is not limited to, uneven heating of the Earth's surface by solar radiation and uneven cooling through terrestrial radiation to space.

Wind is a normal part of our daily existence. When winds blow gently, we seldom notice them. However, when the wind strengthens, or particularly if it becomes strong and/or gusty, we quickly become aware of it. If winds become too strong, then feelings of danger can surface.

Making pressure systems

Atmosphere pressure is simply the weight or mass of air above a point on the Earth's surface. We can directly measure that weight with a barometer. Within the weight or pressure measurement are hidden effects of air density, temperature and moisture. The pressure also describes how much air accumulates or is removed from the atmospheric column above. Accumulation involves the convergence or wind currents; divergence involves the spreading apart of winds.

Cold air is more dense (its molecules are more tightly packed) than warm air, so many people think that high-pressure areas have to be cold and storms warm. That works perfectly for arctic highs and tropical cyclones, respectively. But between these latitudinal extremes there are traveling middle latitude cyclones (see Chapter 4) and subtropical high-pressure systems. One such high-pressure system is the "Bermuda High" that parks itself off the US east coast for lengthy periods each summer.

However, temperature alone is not what makes highs and lows. Lows involve air coming together (converging) near the Earth's surface and diverging at high altitudes. This is most easily seen in hurricanes where air literally spirals cyclonically inward at low levels

Mankind has learned that wind is not purely a destructive force and that its power can be harnessed to generate electricity. Wind farms, such as this one in the United States, are being constructed in many parts of the world.

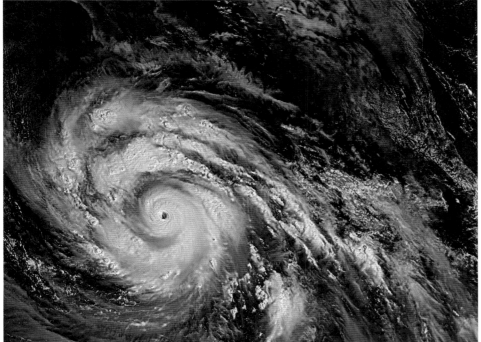

Hurricane Linda, off the west coast of Mexico, during early afternoon on September 12, 1997. Initially, one might see cyclonic inflow into the hurricane. Actually, there is a very well defined anticyclonic outflow atop the storm as shown by the high altitude cirrus clouds and thunderstorm anvils.

Fanning wind pattern associated with a December 25, 2002, east coast winter storm.

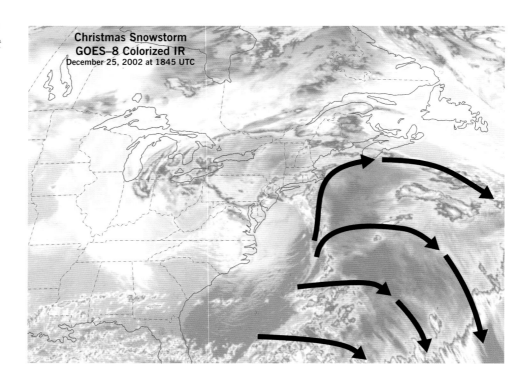

Christmas Snowstorm
GOES–8 Colorized IR
December 25, 2002 at 1845 UTC

and anticyclonically outward at high levels (see page 60). For wave cyclones (and even severe or tornadic thunderstorms), the divergence aloft happens in a fanning type of pattern.

In high-pressure systems, air converges aloft and sinks, eventually diverging near the ground. Lacking clouds, this pattern is harder to visualize. Yet, the clear skies often found in high-pressure systems are linked to sinking air that warms by atmospheric compression. Also, moist air is actually less dense than dry air. This is because water vapor weighs less than air molecules like oxygen, nitrogen and carbon dioxide.

View of highs and lows and cloud cover around the globe.

Whatever the cause(s), the net result is what is key to our everyday lives. We see the manifestation of these interactions each day on television and Internet weather maps. Hs (for high pressure) and Ls (for low pressure) dot US and UK maps. In some European countries, As (anticyclones) and Bs (base, as "low" in music) are used. Because of how weather processes take place, these high and low pressure centers can be tracked as they move across the map. This makes it easy for us to see where the weather system has been and where it is going.

But Hs and Ls are not solid objects. So, what we are viewing as motion is really an atmospheric transformation and readjustment process. As air accumulates or leaves places, air pressure changes and highs and lows redevelop. They do not move like logs floating in a river. The key to these processes is how winds affect the pressure systems and how the systems affect winds. There is definitely a feedback process at work.

High pressure systems, which typify air masses, have preferred source regions. These can be either land- or water-based and have moist or dry attributes. These are described in on page 40.

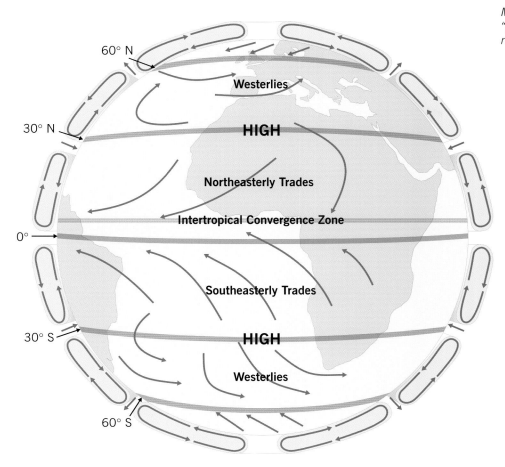

60° N

Westerlies

30° N

HIGH

Northeasterly Trades

Intertropical Convergence Zone

0°

Southeasterly Trades

HIGH

Westerlies

30° S

60° S

Global pressure and wind patterns

The same types of forces help to create our global wind pattern. Although the technical explanation can become a bit complex, consider how latitude alone affects rising and sinking air just from a temperature consideration. Air near the tropics is warm and rises; air near the Poles is cold and sinks. Without a rotating Earth, winds would blow from north to south near the ground and from south to north at high altitudes. This would cause a constant high-pressure area to be at the Poles and a matching low-pressure area near the Equator.

With a rotating Earth and the Coriolis effect (see page 236), winds flowing from the north or south would quickly become easterly and westerly winds, respectively. Hence, there would be large regions in middle latitudes without a latitudinal heat transfer mechanism. That's where middle-latitude cyclones enter the picture. These overpower the expected north-south heat transfer process. The result is an average pattern of global wind and pressure zones that we call the "general circulation". This doesn't mean we'll see this on any particular weather map (although we often see pieces of it), but it's there on the average.

Creating winds

While most storms and high-pressure systems affect larger areas (we'll even consider severe weather as a larger scale event since severe weather situations affect fairly large areas), there are more localized wind patterns. Many of these winds lack large-scale

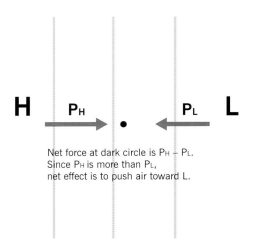

H P_H • P_L L

Net force at dark circle is $P_H - P_L$.
Since P_H is more than P_L,
net effect is to push air toward L.

The opposing forces of pressure systems and winds.

storm systems as their cause. Rather, they are driven by pressure gradients (or the change in pressure over distance). The more the pressure varies over the same distance, the stronger the wind blows.

Consider a point on a weather map. To the left there is a high and to the right a low. Although pressure is the weight of air above you, the pressure can also push sideways against the air next to it. If you swim underwater, you feel water pressure all around and you have to exert a force to overcome it in order to swim. Like water, air is a fluid.

Let's go back to the weather map. If you move a short distance toward high pressure you'll find the air pushing toward your original location with a Pressure of PH (where H stands for high). Repeat the process by moving toward low pressure and you'll find a pressure of PL.

Without rotation, air moving from the North Pole to the Equator would move southward along a longitude line (image (a)). With rotation, the expected end point "target" has rotated to a new position. To an observer at the North Pole, the apparent path of the air curves to the right. If the situation were shown for the South Pole, the apparent motion would be to the left.

Next, imagine that these two pressures represent football or rugby players lined up at the line of scrummage. Once the ball is snapped, the players push against each other for position. The stronger or more forceful player pushes the other player out of the way.

Air acts in much the same way. The stronger pressure overpowers the lower pressure and the net pressure difference starts the air in motion. As a result, air flows from high pressure to low pressure.

Coriolis and frictional forces

There are only two other things needed in order to understand larger-scale wind patterns. The first involves the effect of the Earth's rotation.

At the Equator, the Earth spins at about 1,000 mph (161 km/h). At the pole, however, there is no rotational motion. In between, the speed of rotation depends upon latitude and involves the distance between the Earth's invisible axis (see figure with seasons re axis on page 221) and the point of the earth's surface.

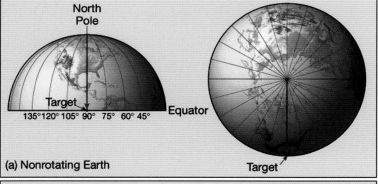

(a) Nonrotating Earth

This means that once air starts to move, it will arrive at another point on the Earth's surface after that point has also undergone a movement. Both this point and original point of the moving air typically involve different movements. So, meteorologists have to compensate for this relative motion. The Coriolis Force, a fictitious representation, does just that.

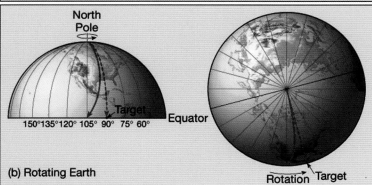

(b) Rotating Earth

Without friction, the Coriolis Force would turn the winds so they blew parallel to the isobars (lines of equal pressure). This is what happens at higher altitudes where friction is negligible. Near the ground where air is affected by buildings, trees, mountains and other obstacles, friction slows down the winds. As a result, winds blow across the isobars at an angle. The angle is directly related to the amount of friction, so the more friction, the more winds blow across isobars. Over large water bodies, friction is less and winds blow more parallel to the isobaric pattern.

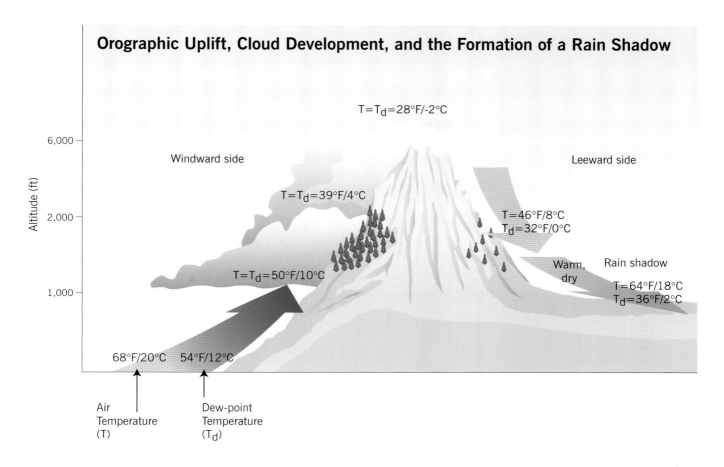

Orographic Uplift, Cloud Development, and the Formation of a Rain Shadow

Altitude (ft)

6,000

Windward side

$T=T_d=28°F/-2°C$

Leeward side

$T=T_d=39°F/4°C$

2,000

$T=46°F/8°C$
$T_d=32°F/0°C$

$T=T_d=50°F/10°C$

Warm, dry

Rain shadow

$T=64°F/18°C$
$T_d=36°F/2°C$

1,000

68°F/20°C   54°F/12°C

Air Temperature (T)

Dew-point Temperature (T_d)

This creates a large-scale diverging wind flow around highs and a large-scale converging wind flow around lows. Because outflow and inflow are less over oceans, highs and lows remain longer over water bodies.

Mountain and gap winds

In the atmosphere, wind speed typically increases with increasing altitude (at least up to about 10-12 miles (16-19 km) above the Earth's surface). It's at this high altitude that jet stream winds (at speeds up to 150 mph (241 km/h) or more in winter and maybe up to 100 mph (161km/h) in summer) occur. Not surprisingly, winds are typically stronger the higher up one goes into the mountains.

As winds blow across mountain ranges, the winds are also channeled through a narrower vertical depth and, hence, blow more strongly over and around mountain peaks or ridges. This latter effect is not much different than what happens as air blows across airplane wings.

Due to the shape of the wing and the wings presence, in general, air flows fastest across the upper surface of the wing. It is also what happens to water flow when you cover part of the opening on a garden hose. Constrict the opening and water comes out faster. This is known as the "Bernoulli Effect".

Add the presence of a large-scale pressure gradient and horizontal channeling of winds through mountain passes or gaps and wind speeds can be even stronger. The resulting so-called "gap winds" earn their name because they blow the strongest through and near gaps in mountains.

An illustration of the Chinook effect, showing how the flow over a mountain range dries out and warms an airmass.

Chinook winds

Consider what happens when strong westerly winds blow across the US Rocky Mountains. Faced with both a tall obstacle and with occasional gaps in the mountains, winds blow fastest near the tops of peaks and ridges and through higher altitude passes. In the winter, when winds tend to be strongest anyway, very strong winds can occur along, and just east of, the Rocky Mountains from Colorado northward into Canada. Winds are typically strongest when winds intersect the mountain range at a right angle and near where the jet stream crosses the mountains.

These winds blow across and then downward once they cross the highest terrain. Since sinking air comes under higher pressure as it moves toward the Earth, it becomes compressed; compression warms the air and, in turn, lowers the relative humidity of the air. The result is a warm and dry wind. If snow cover exists across the US High Plains, these winds, known as Chinooks, often decimate it through quick melting or even sublimation (where the snow actually transforms from a solid to a gas without melting in between). Chinook is a word taken from Native American Indian language that has come to mean, "snow eater".

It's not uncommon for Chinook winds to blow at 50-70 mph (80-112 km/h) with higher gusts. On occasion, winds can blow even faster. Boulder, Colorado, may be the Chinook capital of the world because no other comparably-sized or larger city experiences so many Chinook events.

According to Brian Rachford, formerly at the University of Colorado (Boulder campus), "…at least one location in or near Boulder is pounded by gusts in excess of 100 mph almost every year. Gusts have been measured as high as 147 mph. A severe windstorm in January 1982, comparable to the landfall of a Category 2 to 3 hurricane, damaged nearly half of all buildings in Boulder."

In a study of strong wind events between 1969 and 2002, Boulder reported 175 days with winds greater than or equal to 70 mph (112 km/h). Almost half of these (86 events) occurred in December and January; only 10 percent of the events occurred in the six warmest months (May to October).

Santa Ana winds

Another type of warm, dry, downslope wind is the Santa Ana. This fall to winter season wind has its beginnings when a large, cold, high-pressure system builds into or develops over the inter-mountain western United States. Winds blow outward from the high and often collide with mountains across southern California. The mountains initially block the winds westward motion. However, as the depth of the cold air increases, winds start to make their way through mountain passes. As they do, they blow downhill, quickly warming as they go. Much like their Chinook cousin, Santa Ana winds can blow at high speeds and bring compressionally-heated, dry air into a region.

Since Santa Ana winds typically come after a six-month dry period (see Los Angeles rainfall data in Chapter 12), they can easily fan the flames of any forest fires. Some of the worst fire storms in Southern California have occurred while Santa Ana conditions were present.

The föehn is the warm dry wind of central Europe. It forms as southerly winds cross the Swiss Alps and descend on their way into Germany. On a lesser scale (because the mountains have lower altitude peaks and ridges), winds crossing the Appalachians also exhibit a warming effect on many major east coast cities.

Chinook winds, sometimes called foehn winds, push the glacial sediment toward the sea. These intense, warm winds blow down mountain valleys, often taking the glacial silt with them. Because their primary movement is downward rather than lateral, these downslope winds can actually change direction with the mountain valleys they travel. A close look at this picture of Alaska in November 2006 shows the dust storm turning corners, following the Copper River Valley on its way to the sea.

Clouds and Chinook-like winds

Unless clouds are associated with these wind events, they are hard to visualize. However, when clouds are present, certain cloud features offer tell-tale evidence that a Chinook-like wind is in progress.

The Chinook arc cloud is perhaps the most dramatic. This cloud, with its base 10,000-20,000 feet (3,100-6,100 meters) above sea level, aligns itself along and downwind (east) of the Rocky Mountains. The cloud can extend for hundreds of miles and can sometimes have multiple layers within it. Although formed by high winds, the cloud appears to be standing still.

Mountain wave clouds in Maryland aligned parallel to the ridges of the Appalachian Mountain. The wind is blowing from the northwest, out from the image, perpendicular to the bands of clouds.

This stationary attribute is key to what forms the Chinook arc in the first place. As winds flow over the mountain, they are forced upwards. This push is sometimes too much for atmospheric stability to maintain, so the air sinks in an attempt to reach an equilibrium level. In doing so, it may overshoot that level. Again trying to stabilize itself, the air rises. It may take several up and down motions to finally return to its preferred level.

The resulting wave pattern is filled with areas of rising and sinking air. Where air rises, clouds form; where air sinks, air is compressed, warms and clouds evaporate. With a wave that remains nearly

An example of rotor type cloud (from an airplane at an altitude about 20,000 feet) associated with mountain-induced turbulence.

stationary, the locations where air rises and sinks are "locked in". So the overall cloud doesn't move, it just reforms in place.

While the Chinook arc is dramatic, the overall wave pattern (with many parallel lines of clouds) can extend for hundreds of miles downwind from the mountain ridges.

Cold downslope winds

There can also be cold, downslope gap winds. Perhaps the most famous of these are the Mistral (Rhone valley of France) and the Tehuantepecer (southwest coast of Mexico). In each of these situations, the air starts out over land so cold that, even with warming, the air arrives chilly. In the case of the Tehuantepecer, the cold air mass can start out as far away as Alaska or Siberia. Each of these winds primarily affects coastal water areas and thus poses significant risks to boating and shipping interests.

Again, a strong pressure gradient coupled with winds blowing perpendicular to the mountains sets the stage for a strong downslope gap wind.

In the case of the Tehuantepecer, cold air must pile up against the north-facing slopes of the Sierra Madre del Sur mountains in southeastern Mexico. Once the cold air is sufficiently deep (around 2,000 feet (6,100 meters)), it can start flowing through the mountain gap. Once it reaches the Gulf of Tehuantepec, winds can blow at almost hurricane force. Since the winds occur at tropical latitudes, the US National Hurricane Center is charged with issuing special marine forecasts and warnings for these events.

For a Tehuantepecer to form, cold air must first spill southward through the central United States. When this occurs, the arrival of the cold air can be dramatic. Often such cold air masses arrive with strong northerly winds and clear skies (the lower clouds associated with the colder air coming into the region many after the cold front passes). This type of frontal passage is known as a "Blue Norther". Even with clouds, it is called a "Norther".

Turbulence

Gap winds or even just winds blowing over mountain or ridge tops can create atmospheric wave patterns strong enough to affect airplanes. If the up and down motions are strong enough, or if the wave pattern breaks down to chaotic air motion (like a wave crashing on beach), airplanes can experience "turbulence". Turbulence can occur in other settings (e.g. winds in and near thunderstorms and along portions of the jet stream). Sometimes turbulence can occur even without clouds present; this is known as clear air turbulence.

The turbulence pattern in gap winds is relatively easy to recognize because of the wave cloud pattern. Turbulence in these settings can occur from low levels (within a few thousand feet of the ground) to high altitudes (30,000 feet (9,140 meters) or more). Studies conducted by NOAA's Satellite Research Office have shown that the greater the spacing between cloud bands, the more severe the turbulence.

Turbulence at high altitudes is also likely when cloud patterns (even when distant from mountains) exhibit a similar banding perpendicular to the wind flow. Such banding is often present near and Equatorward of high speed jet streams and in the outflow pattern of tropical cyclones.

Urban "canyon" winds

The same type of channeled wind occurs in cities or other places where tall buildings create "concrete canyons". Just as with gap winds, the winds are channeled into narrower passages and the wind speed increases.

In situations in which the prevailing winds are already strong, canyon winds can cause windows on newer, mostly glass, buildings to work loose and either blow in or blow out. This is clearly dangerous to people walking below.

Regardless, canyon winds cause the demise of many umbrellas, the loss of many hats and make it difficult for women wearing dresses or skirts. These high winds can also stress trees planted along urban streets.

Wind protection – land-based

When high winds strike, there is little that can be done to stop their effects. In some agricultural settings, wind breaks (rows of tall trees) can be planted to deflect winds upward and over an area. If spaced properly, the wind can be blocked at least partially across several fields.

Around homes, wind breaks, or even just tree-covered areas, can lessen winds. I used to live in Maryland and the neighborhood was full of 30 year old, or older, trees. Even on windy days (when tree tops were swaying) winds were light near the ground. Then we moved to a new neighborhood, where small trees were planted after homes were built. It was clearly much windier in the new neighborhood than in the old.

In addition to just blocking the wind, trees help to keep neighborhoods cooler in summer and warmer in winter.

Windiest places

There are many ways to define the "windiest" place. NOAA's National Climatic Data Center has used

In addition to serving as property markers, trees – such as these in the southeast UK countryside – provide wind breaks.

THE TOP 10 WINDIEST US CITIES

Pos	City	Average wind speed	
		km/h	mph
1	DODGE CITY, KS	22.4	13.9
2	AMARILLO, TX	21.7	13.5
3	CHEYENNE, WY	20.8	12.9
4	ROCHESTER, MN	20.6	12.8
5	CASPER, WY	20.4	12.7
6	GOODLAND, KS	20.1	12.5
6	GREAT FALLS, MT	20.1	12.5
8	BOSTON, MA	20	12.4
8	LUBBOCK, TX	20	12.4
10	CLAYTON, NM	19.6	12.2
10	FARGO, ND	19.6	12.2
10	NEW YORK (LAGUARDIA AP),NY	19.6	12.2
10	OKLAHOMA CITY, OK	19.6	12.2
10	WICHITA, KS	19.6	12.2

The windiest US cities according to NOAA's National Climatic Data Center are found primarily in the high plains region. Cities in the northeast region can also be windy, mainly due to the passage of middle latitude cyclones and strong high-pressure centers. Values shown are average wind speeds (miles per hour) using data through 2002.

average wind speed to classify the top windiest places in the contiguous United States. A global wind classification does not exist.

Excluding mountainous or hilly locations and Alaska, the windiest places in the US over the course of a full year are almost always in the central US, or the Plains States. Here, winds blow consistently strong (although usually strongest in winter and spring). While winds may occasionally blow strongly in some places (e.g. Boulder, Colorado's renowned Chinook winds), winds year-round lack the force to propel many locales into the windiest category.

The windiest mountain locations in the US are at Mount Washington, New Hampshire (elevation 6,288 feet (1917 meters)) and Blue Hill Observatory (just to the southwest of Boston, Mass., at an elevation of 635 feet (194 meters)).

In no month is Mount Washington's average wind speed below 24.7 mph (40 km/h) and in no month during winter (December to February) is the average wind speed less than 44.3 mph (71 km/h). Blue Hill's calmest month (August) has an average wind speed of 12.6 mph (20 km/h), while winds during the winter months blow consistently above 16.7 mph (27 km/h).

Although South Dakota is not the windiest place, it's pretty windy there. In fact, all four cities listed in South Dakota have average annual wind speeds of 11 mph (17 km/h) or more and in no month is the average wind speed less than 9 nine mph (14.5 km/h). These values are consistent with other Great Plains cities.

The fact that the area from the Mississippi River westward to the Rocky Mountains is so windy it makes the region ideally suited for the development of wind farming. Other locales, such as the Hawaiian Islands and south Florida (trade wind zone) and the northeast US (with so many storm passages and post-storm windy days), are also prime wind energy development areas (see also Chapter 19).

Flagged trees
Although they appear to be bent over due to high winds, flagged trees (trees that lean and have most of their vegetation on one side so they look like a blowing flag) are

Flagged trees on the west coast of Capri, Italy. The upwind side, stripped of vegetation, lies closest to the salty waters of the Mediterranean Sea. Salt spray helps to "prune" the tree of its ocean-facing vegetation.

caused by more persistent winds (i.e. winds with a prevailing direction). I've seen flagged trees in mountain areas, but also along the east and west coasts of the United States. Most recently (and shown opposite) I've seen flagged trees on the Island of Capri in Italy.

Flagged trees grow downwind from the blowing wind because existing vegetation shields the newest growth from the effects of the wind. In coastal areas, this can involve "salt pruning,", where salt spray dries out the leaves on the upwind side of the trees. In other non-salty locales such growing patterns protect tender vegetation from chilly winds or airborne particles (like dust or sand).

Ocean currents and ocean temperatures

Pressure patterns and winds are caused in part by the temperature of the underlying land or ocean surface. Resulting wind patterns can then affect the temperature of that surface. The effect is often greater over water where winds can move unhindered over the surface. The most notable effect is the creation of large-scale, wind-driven ocean circulations.

If the wind blows away more surface water than moves in horizontally to take its place, the diverging pattern requires that water from below be brought to the surface to fill the

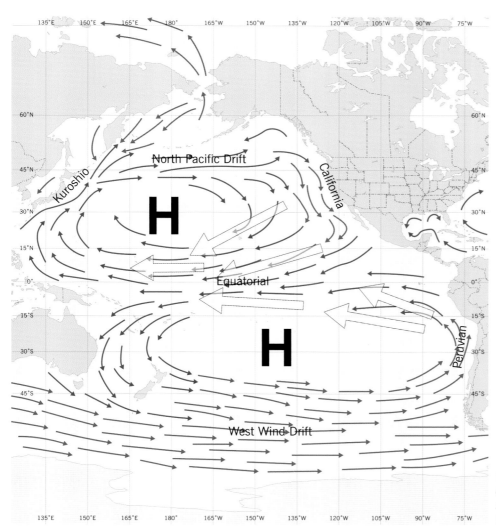

Winds (large white arrows) drive ocean currents. This illustration shows the typical current pattern in the Pacific Ocean

SEA BREEZES AND MONSOONS

Sea breezes (and their companions, lake, valley and mountain breezes) are not "high" wind events. Their circulations, like other winds, are driven by the unequal heating of the Earth's surface. Consider the following situation that plays itself over many coastal areas frequently during the warmer months of the year.

As the sun rises (and the apparent rising is based on the Earth's rotation), its solar energy is absorbed and reflected by the Earth's surface. Ground, farmland, areas with trees and other surfaces, absorb heat more easily than the adjacent water. One reason is that sunlight passes deeper into water (warming a larger volume rather than just a surface). Another is that water, once heated can move and circulate (Earth cannot). Finally, water has a high specific heat; this means that to raise its temperature by one degree requires a large amount of heat energy.

As a result, the land near the water warms more quickly. In turn, the air above the land warms more quickly by contact heating (known as conduction).

Typical sea breeze cloudiness along Florida's southwest coast, noon, July 31, 2005. A line of towering cumulus and cumulonimbus (thunderstorm) clouds parallels and is just inland from the coastline. Along the coast and offshore, skies are cloud-free.

Now there are two air masses (small ones) adjacent to each other along the coast. The warmer air mass wants to rise and the colder one wants to spread out and fill in behind it (much like a cold front). In turn, air sinks over the water and moves from land to water at some height above the ground. Where the air rises, clouds and possibly thunderstorms mark the location of the so-called "sea breeze front".

The same type of pattern occurs with mountains and valleys. With morning solar heating, sunlight strikes the eastern sides of mountains more directly than it does the valleys below. This causes air to start to rise over mountains first. As the air rises from the mountain slopes, air from the valley (even though it is cooler) rises to fill in the partial void. As a result, in this "valley breeze" pattern, clouds and thunderstorms will form over mountains.

While the sea breeze circulation pattern takes place along coasts, it is even more dramatic over peninsulas where wind flows from water to land on opposite coasts. If the sea breezes themselves, and/or the prevailing wind pattern, is strong enough then the sea breezes may meet. This converging wind pattern near the ground enhances the upward vertical motion and helps to create very intense thunderstorms.

Now imagine the same scenario on a large sub-continent, like India. Here, during the warmer months, there is an almost constant sea breeze pattern atop a valley wind regime. The result is very intense showers and thunderstorms over a large inland area.

The pattern reverses itself at night with weaker offshore winds near coasts and cooler winds blowing downhill from mountains. These wind patterns are known as land breezes and mountain breezes, respectively.

In the cooler months of the year, even with solar heating, large land areas are often chillier than adjacent waters (which retain their heat longer for the same reasons that they warm more slowly). When such breezes occur over smaller, long water bodies (like the US Great Lakes), the convergence pattern down the length of the lake can resemble that of peninsula thunderstorms in the warmer season.

During the cooler months of the year, winds blow from inland locations on the Indian sub-continent (again supplemented by a mountain effect, this time a colder one) toward the waters of the Bay of Bengal and the Indian Ocean. In these situations, with sinking air over land, skies are often cloud- and rain-free.

On the Indian sub-continent (and in other areas), the summertime onshore wind pattern with its heavy rains has come to be called the "monsoon" (from the Arabic word mausim meaning season). But, monsoon does not mean heavy rain. Rather, the word means a seasonal change in wind direction. Most places that experience a monsoon

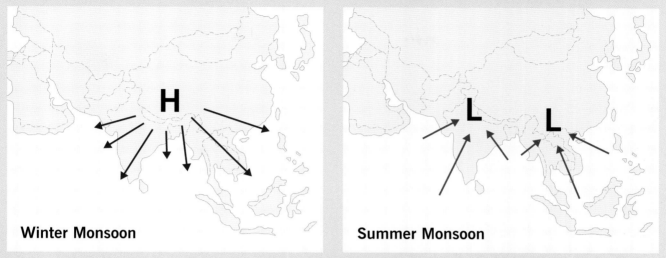

Winter Monsoon   **Summer Monsoon**

Winter (dry) and summer (wet) monsoons comprise the annual monsoon cycle in southeast Asia.

(wind) circulation also have accompanying changes in temperature, humidity, cloudiness and precipitation. In India, there is a summer monsoon and a winter monsoon.

In addition to the Indian monsoons, places like Arizona and Florida in the United States, Spain, northern Australia and parts of central Africa experience monsoonal circulations. The circulations can be driven by land-water variations and also the seasonal north-south shift in large-sale pressure and wind patterns.

When cold fronts interact with the sea breeze circulation and/or prevailing wind patterns, the associated thunderstorm pattern can be affected. If cooler winds interact with the front, they can weaken thunderstorms. This often happens as cold fronts pass over, and near, the cooler Great Lakes. In Florida, many intense to severe storms occur along Florida's east coast where the sea breeze and prevailing trade wind easterlies help to intensify frontal convergence.

Finally, the shape of the coastline can affect local wind patterns and thunderstorm occurrence. Consider the shape of the coastline in the Florida panhandle. During the day, sea breezes converge to southwest of Tallahassee, Florida. Here in the Apalachicola National Forest, there are more than 100 thunderstorm days each year. Just to the east of that area, sea breezes divergence, lessening low-level convergence and the chance for thunderstorms.

Daytime

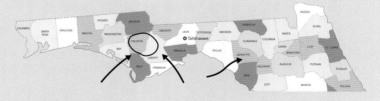

Favored locations for thunderstorm formation based on time of day and shape of coastline.

At night the pattern changes and thunderstorms are more likely to form over Apalachee Bay.

As a college student at Florida State University (in Tallahassee), I recall watching thunderstorms to the southeast in the morning and to the southwest in the afternoon almost every day from late spring into early fall. It wasn't until I understood the sea and land breeze circulations that this cycle made sense.

Nighttime

Japanese tourists brave the severe gales at Paris's Trocadero as high winds once again hit the capital in February 1990. Gusts of up to 130 km/h were forecast for the northern part of the country.

gap. This process is known as "upwelling" and typically brings colder, nutrient-rich, bottom water up. It often happens when winds blow offshore along a coastline. If you've ever been to a beach when winds are blowing strongly from land to water, you probably experienced colder than expected water temperatures. Conversely, onshore winds make water pile up on the ocean surface and this leads to "downwelling".

This process is also key to understanding the El Niño and La Niña cycle. Here, off the northwest coast of South America in the tropical eastern Pacific, upwelling (La Niña) and downwelling (El Niño) are driven by changes in sea level winds. The upwelling and downwelling affect air temperature, precipitation, pressure and wind patterns. Again, there is a significant interaction and feedback process at work.

Climatic impacts

Wind patterns affect where we live and vacation. People like to inhabit coastal areas because of refreshing daytime breezes and the cooler water. Not surprisingly, the coastal population (permanent and vacationing) has soared. In many low- and middle-latitude areas, this has significantly increased our exposure to hurricanes. In additional to higher potential property losses, it requires more massive evacuation efforts. Mountain areas are experiencing a similar swelling of population again because of cooler summertime temperatures.

Such regions also receive cooling thunderstorm showers and outflow winds. But the circulation patterns just described also provide for the potential of new energy sources. Even without high winds, monsoonal circulations tend to have more reliable winds, something that is ideal for wind power harvesting.

Persistent winds, such as those of the High Plains, also favor wind farming. In fact, many new wind farms in the US are being developed over such prairie areas because there are no obstacles to the wind flow and there are few people around to complain about the large wind turbine towers.

The Netherlands recognized the value of wind power hundreds of years ago and now other nations are now realizing its value. You can read more about wind energy in Chapter 19.

Chinooks definitely affect vegetation and water supplies. Although people claim that Chinook events increase the incidence of migraine headaches (pressure, moisture and wind effects), research does not support these claims.

Santa Ana winds are a major contributing factor in southern California forest fires. However, unless something triggers the fire, the winds only affect its spread and the drying of vegetation.

Finally, we again address any changes in winds or wind patterns. Are monsoons getting worse or are their impacts worse? Are they stronger or weaker than in the past? Are high wind events more common? Are El Niño and la Niña events more common or severe than in the past?

There is no evidence that indicates that anything, other than natural variability, is at work physically. What is at work is an ongoing population shift to coastal and mountain locales and the fact that there are simply more people and structures at risk to high winds (including hurricanes and winter coastal storms and the associated impacts of monsoons, including flooding rains).

Today, scientists have a growing number of data sources at their disposal and this is helping to uncover new relationships and helping to explain what we have experienced in the past.

High winds are not always regarded as a nuisance and are indeed welcomed by many people, including these sailors at a regatta in the Canaries.

17 AIR QUALITY

"The people have a right to clean air, pure water and to the preservation of the natural, scenic, historic and esthetic values of the environment."
Constitution of the Commonwealth of Pennsylvania, Natural Resources and the Public Estate Article 1, Section 27

The concept of air quality isn't much different than that of water quality. We want our drinking water to be free of impurities and we want it to taste good. Similarly, we want our air to be clean and we want it to smell good. While we can clean up water "relatively" easily, cleaning the air is another matter.

The recommended daily water intake for adults is around eight glasses, but we could live for days without drinking water if we had to. However, we breathe over 3,000 gallons of air each day and we take air in every few seconds. Obviously, clean air is something we want and need.

It may look spectacular, but contrails like this one add incredible amounts of water vapor to the atmosphere.

Setting the stage

There are lots of sources of pollution in our atmosphere and each source brings with it specific complications (and sometimes benefits) to our existence. For example, some gases (such as carbon dioxide, methane, greenhouse, and water vapor) contribute to an enhanced "Greenhouse Effect." While the Greenhouse Effect is an important aspect of our existence on Earth (we wouldn't be here without it), the enhanced effect may be tipping the scales too far toward global warming or unintended climate change (see Chapter 19).

Emissions from factories such as this one are contributing to the air pollution.

Los Angeles, trapped between coastal mountains and cool Pacific Ocean coastal waters, often experiences a low-level inversion. Pollutants, mostly from automobiles, become trapped under that inversion, shrouding the city in a layer of smog (smoke and fog combined).

Cities that are along high altitude airplane routes often have skies "littered" with contrails. If airplane routes intersect, the contrails can cross, as shown here. Once formed, the contrails drift with the upper level winds.

While it may be a habit both adored and despised by billions of people worldwide, cigarettes are also a pollutant.

Ozone at high altitude helps to save the planet and its human inhabitants from harmful ultraviolet rays from space. However, some gases (such as carbon dioxide and sulfur dioxide) contribute to acid rain.

Particulate matter (volcanic, fire, and industrial ash, dust and even salt spray) affects visibility (our ability to just see the natural world and how airplane pilots, car drivers, and others navigate) and how well we can use our eyes and lungs. Salt spray can cause metal objects to rust.

The emission of radioactive particles from nuclear power plant accidents (such as the Chernobyl Nuclear Power Plant in the Ukraine region of Russia on April 26, 1986) and the release of various chemicals from transportation and factory accidents often makes local, and sometimes national, headlines because of its danger to human life.

Even pollen is a pollutant (of sorts). It causes many of us to sneeze, experience watering eyes and suffer other discomforts.

Airplane exhaust emissions are often enjoyed in the form of artistic condensation trails (or contrails). Few consider its pollution or a factor in how weather and climate may be changing. Many of us enjoy colorful sunrises and sunsets, but don't appreciate the pollutant contribution (scattering and reflection of light) to its vivid coloration. As some of us puff our cigarettes, there is little thought about air quality, even though the pollutants released can be deadly and the smell can be quite offensive.

Indoor air quality also needs to be considered, because it can often be as bad for us or worse than outdoor air. This is due to the ease with which pollutants can accumulate (lack of ventilation) and how homes are built and how home materials and furniture are made. Homes are prone to the accumulation of dust, dander (dog and cat sheddings), smoke, mold, radon and formaldehyde. Unfortunately, except for noting how these pollutants get concentrated, that topic is beyond the scope of this book.

However, to the untrained, it is typically only visible pollutants (such as factory smoke and diesel car exhaust) that tell us the air is being polluted.

Pollen may be essential to the existence of plants but it is the bane of a person who suffers from hayfever. In spring, during periods without rain, it is not unusual for cars to be coated in pollen.

Many authorities are experimenting with methods of measuring air quality. Here is one such station in London.

Monitoring air quality

In the United States, the Environmental Protection Agency (EPA) and corresponding state agencies monitor air quality. The European Union has regional directives concerning air quality while individual nations and cities carry them out. In Australia, it's the Environmental Protection Authority.

The US EPA provides an Internet-based listing of links to both US and international air quality monitoring sites (see Resources). Here one can find daily pollution readings for many larger cities. Although all the sites do not provide the same detail that is provided in the US, values for both particulate and gas pollutants are often color-coded for easy comparison to standard levels.

Monitoring includes particulate matter (usually expressed as PM_{10} or $PM_{2.5}$) that indicates the upper limit of the particle size measured in microns. while 2.5 indicates particles smaller than 2.5 microns/micrometers (or 0.000098 inches/0.0025mm) in diameter. For comparison, a typical human hair is about 75 microns in diameter. The air quality measurement is typically given as parts per million parts of air (ppm). Values above a certain level indicate varying degrees of pollution. In mid-2007, the US EPA began accepting public comment about lowering the unacceptable threshold level for ozone.

Natural pollution

Perhaps the biggest obstacle to understanding pollution is that we find it hard to accept that some pollution is natural. We've been taught that all pollution is bad; that humans cause it; and that we have to clean up anything that pollutes.

Trees, for example, emit volatile organic compounds (VOCs) that help create a natural atmospheric haze. The bluish color of the Blue Ridge Mountains (part of the eastern Appalachian Mountains from Pennsylvania to Georgia) is linked to these VOCs.

Trees, grasses, and other plants produce pollen. Humans have become so sensitive to this so-called "irritant" that most of us sneeze (to varying degrees) when confronted by it. In

The Blue Ridge Mountains in North Carolina overlooking Looking Glass Rock. The distant mountains get their name from naturally occurring atmospheric pollution.

addition, pollen can coat our cars with a greenish haze and add to the dusty nature of our homes.

Volcanoes emit large amounts of ash and hazardous gases. Sometimes these encircle the globe causing significant changes in weather over periods of months to years. Forest fires, many naturally caused, spread smoke, ash, and atmospheric chemicals far from their source.

Dust, especially from major dust storms, can also be carried large distances. I recently spoke with a meteorologist from Washington, DC's Howard University who had just returned from an Atlantic research cruise. She noted that part way across the Atlantic they encountered an African dust cloud. Almost in an instant, the whiteness of the ship disappeared and locusts landed on the deck (see also Chapter 11).

Although it doesn't require high winds, waves crashing on a beach produce large amounts of salt spray. If you live near or visit an oceanic beach, be prepared for its ongoing assault.

Meteoric impacts may have spread clouds of dust and other particles across parts of, or the whole, Earth over geologic time. However, it is, perhaps, the daily introduction of about 40 tons of meteoric dust to our atmosphere that will have a larger long-term effect. According to Dr. Andrew Klekociuk from the Australian Antarctic Division in Tasmania, the dust, depending upon particle size, can affect how the Earth is cooled or warmed, as well as how atmospheric cloud formation processes occur.

Recently, a team of scientists at the Indian Institute of Tropical Meteorology, Pune, India noted that, "The November 2001 and 2002 Leonid storms, and the 2003 November outburst, caused significant enhancements of dust from just above the mesopause to the lower stratosphere. The present study shows the formation of meteoric dust layers at mesospheric levels and their subsequent descent to lower altitudes. The enhanced stratospheric layers are observed four to eight days after the peak meteor activity." The mesosphere is one of several rings of atmosphere that encircle the Earth. It is located at an altitude between 30 and 50 miles (50 and 80km) above the Earth's surface.

In short, it isn't just humans that create air pollution. However, we are major culprits and our impact on the environment (air, water, and land) is significant.

In this chapter we'll look at but a few of the factors that contribute to the quality of air. To fully comprehend this complicated and important topic, you may have to read much more about it.

Particulate matter

Perhaps, the easiest aspect of air quality to grasp involves particulate matter. That's because we can see it. Almost anything that gives off particulate matter as a pollutant also adds polluting gases to the atmosphere. Again, some of this pollution is natural and some man-made.

Forest fires were discussed in detail in Chapter 13. Biomass burning (the burning of agricultural waste) provides widespread smoke across Central and South America, Africa and southeast Asia. Depending upon wind direction (which is controlled by the

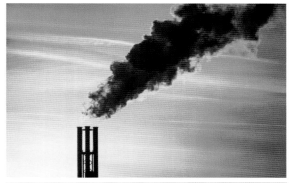

Particulate matter comes from many sources. Incinerators and smoke stacks used to be the biggest culprits, but technological advances have helped lessen their contribution to atmopsheric pollution. Instead, farming, military activities (including war), fires of all types and even events such as the World Trade Center attack and building collapse contribute large amounts of particulate matter to our skies.

This hi-resolution polar orbiting satellite image shows smoke pouring from hundreds of fires burning across parts of Brazil, Bolivia, Paraguay and Argentina. The red marks indicate locations where satellite data indicates active fires. Along the top portion of the scene is the Amazon Rainforest. In this area, fires are aligned along new roads and at the edges of existing clearings, indicating that they are caused by people clearing or managing existing agricultural land.

presence of prevailing wind patterns, transient fronts, and other factors), the resulting plume can travel for hundreds, sometimes thousands of miles.

NASA and NOAA rely on weather satellites and a special Automated Biomass Burning Algorithm to detect the source of biomass burns.

During the spring of 2007, I discovered firsthand what a distant forest fire could bring. In addition to creating reduced visibility and bringing smoky odors, there was a significant fall of particles from the sky. I didn't notice this fallout immediately, but while driving, I eventually had to clean my car's windshield. That's when I realized that my wiper blades had collected large amounts of ash. The windshield streaking made it impossible to drive safely. I had to pull over, wipe the wiper blades and hand clean my windshield in order to re-establish proper wiper capabilities.

Mount St Helens' eruption, May 18, 1980. Not all volcanoes have a smoke and gas cloud that is this voluminous. Others have even greater clouds.

Volcanic eruptions can produce similar particulate fallout. Any solid matter ejected into the air by a volcano is known by the collective term, tephra. This includes larger fragments (as large as a foot (0.3 meters) or more across) and ash (particles no larger than about 0.08 inches (2 mm) in diameter). Typically, the larger particles precipitate nearer the eruption with finer particles carried longer distances.

Volcanic ash can bury plants, clog waterways, and/or simply create a cloud that hangs over an area for hours to days and impairs breathing. It takes a long time to clean up the dust, especially if local winds, car and truck traffic, and other activities keep kicking it back up into the air.

Consider Mount St Helens, the Washington state volcano that erupted on May 18, 1980. During its nine hours of vigorous eruptive activity, it spewed millions of tons of ash into the atmosphere. US Geological Survey scientists estimate that during that time about 540 million tons of ash fell over an area of more than 22,000 square miles (57,000sq km). Before compaction by rainfall and human activities, this translates into a total volume of about 0.3 cubic mile (1.9×10^9 m³) or an area the size of a football field piled about 150 miles (240km) high with fluffy ash.

I also frequently visit beaches. Even on non-windy days, it is easy to see a cloud of "something" far up many of the beaches. That something is a cloud of the combined effect of salt spray up and down the entire length of beach. It's hard to see, however, until the amount of spray in the line of sight becomes sufficiently thick to reduce visibility.

At times, it is easy to see pollution coming from cars and trucks. Most of the time, this involves diesel vehicles. However, sometimes, it can be cars with engine problems that spew forth an oil-burning cloud.

Gases

Almost anything in our lives today involves gas emissions. Typically, we can't see these and, if we are far enough removed from the source, we can't necessarily smell them either. Yet, our skies are now filled with them.

We know water and carbon dioxide (two Greenhouse gases) are being emitted when we see almost any type of cloud emanating from a source. Clouds suggest that something is being burned and most things that burn give off those two compounds. This includes contrails or the smoke from a cigarette (see page 249).

Other gases are harder to assess. In order to understand these gaseous pollutants, specialized air quality measurements are needed. Many countries around the world now routinely monitor air quality (see the picture of the monitoring station on page 250). This includes both particulates and gaseous components.

What a difference two days can make! These images from north-central Maryland show dramatically different skies. The first image was prior to a cold frontal passage. The second image, two days later, shows much improved air quality.

These gases and others pollute our skies and affect our well-being in many different ways. For example, carbon monoxide (CO), a by-product of the incomplete burning of fossil fuels or plant material, competes with oxygen (O_2) to bind with hemoglobin in our blood. Hemoglobin brings oxygen to our body cells. When faced with a choice of bonding with CO or O_2, the hemoglobin would bond with the CO first.

Carbon dioxide can also combine with water to form a weaker acid known as carbonic acid. This is similar to the acid found in soft drinks.

Sulfur oxides (known as SO_x where X can be 1, 2, or 3 oxygen molecules) combine with water to form sulfuric acid; this acid makes rainwater much more acidic than carbonic acid.

Hydrocarbons (and this includes the myriad of compounds composed of carbon and hydrogen) react with nitrogen oxides (NO_x) in the presence of sunlight to form photochemical smog. Smog can be a major irritant to our breathing system (including lungs and mucus membranes), eyes, and skin. Some of these gases come from cleaning materials and solvents; most come from the burning of fossil fuels.

NO_x comes from the burning of fossil fuels. In addition to its role in creating smog, it, too, competes with O_2 in bonding with hemoglobin.

Ground level ozone (the "bad kind") is produced as part of photochemical smog. Among its other effects, it impairs respiratory functions.

Acid rain

Rainwater is normally slightly acidic (pH* of about 6.5) due to the combined effects of CO_2 and H_2O. With the interaction of other acid-forming gases, rainwater can have pH readings as low as 4. This is some 10-times more acidic than our morning coffee (with a pH of about 5). The figure shows pH rainwater readings across the US in 2005. While the western states have higher rainwater pH values, about half the US has pH values below 5.

While acid rain can mar marble statues and gravestones (marble is a form of calcium carbonate, a basic material), it can also shorten the life span of concrete and other building stones (which are built with calcium carbonate ingredients). Acid rain can kill trees, plants and fish. In fact, the combined effects of acid rain, acid snow and acid fog have led to the demise of large stands of trees in the Appalachian Mountains.

While the acid precipitation doesn't kill trees outright, it weakens their root systems, making them more susceptible to the effects of drought and insects. Dr Harvard Ayers, a professor of anthropology and sustainable development at Appalachian State University, Boone, North Carolina, likens the effect to that of AIDS (Acquired Immune Deficiency syndrome) in humans.

The role of weather

If weather patterns were always moving and winds were always blowing, many of these atmospheric pollutants would be dispersed. They'd still be in our air, but they wouldn't become concentrated. It is when air pollutants reach significant threshold values that their effects become significant or life-threatening.

Atmospheric inversions are perhaps the most significant factor in concentrating pollutants. An inversion occurs when a layer of warmer air caps a layer of cooler air. This happens on most evenings as the ground cools off faster than the air a few hundred feet above the ground. However, as sunlight warms the ground the next day, the inversion breaks and vertical air transport is restored.

When large-scale weather patterns (typically high pressure systems) generate an inversion, several things happen. First, air near the ground becomes trapped and vertical motions are limited. Often, winds at the Earth's surface and throughout the lower atmosphere are weak. This limits horizontal transport as well. With inversions like these, skies are often mostly cloud-free. These are the same type of conditions that accompany heat waves (Chapter 14) but they occur year-round.

Air pollution disasters

London suffered a major smog disaster during December 1952. With a cold high pressure system over the island nation, the combined effects of a strong inversion and widespread use of home heating systems (coal-based at the time), created the worst smog event in

Marble markers like this one in Rome, Italy, or other imprinted marble slabs (including cemetery stones), are easily eroded by the effects of acid rain.

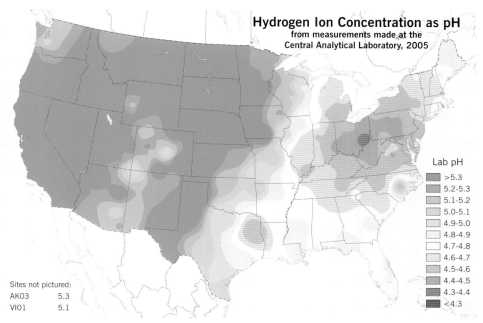

Hydrogen Ion Concentration as pH
from measurements made at the
Central Analytical Laboratory, 2005

Lab pH

■	>5.3
■	5.2-5.3
■	5.1-5.2
□	5.0-5.1
□	4.9-5.0
□	4.8-4.9
□	4.7-4.8
□	4.6-4.7
□	4.5-4.6
□	4.4-4.5
■	4.3-4.4
■	<4.3

Sites not pictured:
AK03   5.3
VI01   5.1

This map shows acid rain values for the United States in 2005. Not surprisingly, the lowest pH values are near and eastward (downwind) from the more industrial regions of the country.

Dead Fraser fir trees on the crest of the Black Mountains in Mount Mitchell State Park, North Carolina.

the UK's history. Deaths (more than 12,000 for the week based on updated statistics) were more than five-times than what would have been expected; untold numbers suffered respiratory illnesses.

During the event, the visibility in central London remained below 1,640 feet (500 meters) continuously for 114 hours and below 164 feet (50 meters) continuously for 48 hours. This brought most transportation to a standstill. Both particulate and SO_2 values exceeded current standards by a factor of at least 10. They also far exceeded the period of duration limits for daily and hourly violations.

The worst US pollution disaster occurred in Donora, Pennsylvania, a relatively small community nestled in the Monongahela River Valley of southwestern Pennsylvania. Between October 26 and 31, 1948, an industrial pollution cloud, trapped by a strong temperature inversion, bathed the town in a cloud of fluorine, sulfur dioxide and carbon monoxide gas and metal dust. Twenty people died and some 7,000 suffered from respiratory illness. Others were debilitated for life.

Image of an inversion from airplane altitude with the clearly defined layer separating concentrated low-level haze from less polluted skies above.

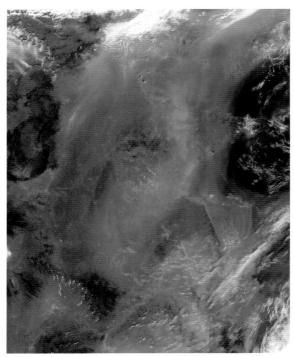

Haze cloud over eastern China on September 10, 2005 as seen by NASA's Moderate Resolution Imaging Spectroradiometer (MODIS) flying onboard the Terra satellite. In this image, haze covers China from the Bo Hai Bay coastline in the east to the mountains in the west. Events like this one are common in rapidly industrializing China due to its heavy reliance on coal as an energy source. Geographical factors (mountains to the west block dispersion), and frequent temperature inversions also contribute to the pollution situation. Some estimates suggest that hundreds of thousands of Chinese die from air quality related illnesses annually.

Some countries in southeast Asia are following in the steps that led to the two air pollution disasters in the US and the UK noted above. With unrelenting growth, an ongoing reliance on coal as a heating fuel and increasing numbers of cars, places like China are seeing more polluted days and higher pollution values.

In fact, according to a recent report issued by the European Satellite Agency, Beijing and its environs in northeast China were named as having the world's highest levels of NO2, a key ingredient in smog. NO2 originates from power plants, heavy industry and vehicle emissions. According to the World Bank, 80 percent of the world's 20 most polluted cities are in China.

Although China has pledged to have a "green" Olympics in 2008, meeting the World Health Organization's average ambient air-quality standard and having "pollution-free" skies for 80 percent of the year, it is becoming more and more likely that the nation won't be able to meet that commitment.

Other air pollution events

A gas leak occurred from a Union Carbide pesticide plant near Bhopal, India, on December 3, 1984. The pollutant, a denser than air chemical, quickly settled on the town of Bhopal.

Occurring in the early morning hours, during inversion conditions, the cloud was trapped near the ground and a carried southeastward by prevailing winds. Compounding the situation were locally induced lake and heat island circulations that further concentrated the gas. Thousands died within a few days of the incident; tens of thousands suffered side effects; and many are still suffering today.

Following the Iraq war of early 1991, retreating Iraqi soldiers set fire to oil facilities. Widespread pollution clouds were produced from these fires, that could clearly be seen from satellites above.

Cleaning up the air

Following the London Smog disaster in 1952, the UK passed legislation in 1956 and 1968 that addressed domestic and industrial sources of pollution. When meshed with changes in the urban landscape (slum clearance and urban renewal) and the widespread use of central heating in homes and offices, air quality in the UK has improved dramatically.

The US began its push for cleaner air with the passage of the Clean Air Act in 1955 (subsequently updated in 1963, 1970 and 1990). In mid 2007, the US Congress and EPA were working to update ozone level standards. Again, the push for cleaner air was the result of an air quality disaster.

However, California (and other states) led the charge for cleaner air, recognizing the problem sooner. Based on these efforts, the Federal government stepped in to address the problem nationally.

The first recognized episodes of smog occurred in Los Angeles in the summer of 1943. Visibility was only three blocks and people suffered from smarting eyes, respiratory discomfort, nausea, and vomiting. In 1945, the City of Los Angeles began its air pollution control program, establishing the Bureau of Smoke Control in its health department. Two years later, California Governor, Earl Warren, signed into law the Air Pollution Control Act, authorizing the creation of an Air Pollution Control District in every county of the state.

Australia, not being as pollution prone as the UK or the US, didn't establish national air quality standards until 1998. International pollution legislation is sporadic, with some nations stronger proponents than others.

Changes in lifestyles in the West have made a difference. For example, automobile gasoline is now lead-free; home building material no longer contains formaldehyde or asbestos; many larger cities have established a strong mass transit presence; mass transit is offered free or at a reduced cost on bad air quality days; and factories and homes now use less polluting fuels. Many appliances have now been designed to be more energy efficient. Most importantly, people are now air quality savvy. Many do things that lessen atmospheric pollution.

Perhaps the most innovative solution to reducing air pollution is the trading of air quality (SO_2 and NO_x) allowances. Much like trading stocks and futures, companies can buy and sell air quality allowances. These can be used to offset excessive emissions; can generate revenue if companies meet their standards; and can even be bought and retired so they can no longer be used to offset future excessive emissions.

The ozone hole

Although not a pollution problem at the ground as such, let's not forget the ozone hole in the stratosphere over Antarctica (and now showing up in the Northern Hemisphere, as well). It can have significant effects on our well-being. As some scientists now believe, it can also affect long-term climate change.

The ozone hole is a thinning of the high altitude (roughly 10-30 miles (16-48km) above the Earth's surface) ozone layer due to the presence of CFC's (ChloroFluoroCarbons). Ozone is an unstable oxygen molecule composed of 3 atoms of oxygen (O3), not two, as is most atmospheric oxygen (O_2). Chlorine, fluorine, and bromine gases, and also nitrogen oxides in the presence of sunlight, act to break up ozone. In doing so, they then revert back to their original state. Hence a few molecules of CFCs can destroy many molecules of ozone.

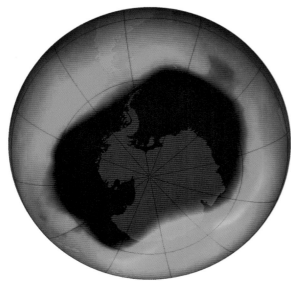

Ozone image based on data obtained from NASA's Aura satellite for September 24, 2006. The ozone hole, the region with the lowest values of ozone, shows up clearly as a large blue area. The size of the hole on this day matched its previous peak aerial coverage.

CFCs are found in refrigeration systems, air conditioners, solvents, and other industrial products. Nitrogen oxides are a by-product of combustion processes, including those involving car, lawn mower and aircraft engines.

Contrary to public perception, the ozone hole is not actually a hole. Rather it is a topographic representation of a depression in the field of total ozone depth. If there were a way to compress all of the ozone in a vertical column together, the reading (measured in Dobson Units or DU) would define the optical depth of the ozone. A DU is defined to be 0.01 mm thickness at standard atmospheric temperature and pressure (0 degrees Celsius and 1 atmosphere pressure).

The Antarctic ozone layer is typically more than 300 DU thick. At least this was its average value in October prior to about 1975 (see figure below). Scientists determined that a "hole" would be that area in which the ozone depth dropped below 220 DUs.

From September 21-30, 2006, the average area of the ozone hole (DU values less than or equal to 220) was the largest ever observed, at 10.6 million square miles. In the image above, the area of the hole was equal to the record single-day largest area of 11.4 million square miles (27.5sq km), reached on September 9, 2000.

While CFCs are the main culprit in creating the ozone, stratospheric temperatures also play a role. During this late September period, temperatures above the Antarctic were about 9 °F (5 °C) colder than average. Scientists believe that this added about 1.2-1.5 million square miles (3.1-3.9sq km) to the hole.

CFCs have half-lives of about between 50 and 150 years. Thus, once in the atmosphere, even future generations will be living with their effects. Still, international treaties, including the Montreal protocol in 1987 and subsequent revisions, have taken significant steps to control new releases of CFCs. While the European community has adopted even stricter measures, some countries have not taken any steps to control CFC emissions. As a result, the ozone hole matched its largest aerial extent in late 2006 and had its lowest DU reading since data has been collected.

SUMMARY OF MAJOR HUMAN-CAUSED GREENHOUSE GASES

Gas	ppmv Concentration*	% Greenhouse Contribution	%/yr Rate of Increase	Half-life (yrs)	Relative Greenhouse Effect	
					per kg	per mol
CO_2	355	55-60	0.4-0.5	150	1	1
CH_4	1.7	15-20	0.7-0.9	7-10	70	25
N_2O	0.31	5	0.2	150	200	200
O_2	0.01-0.05	8	0.5	0.01	1,800	2,000
$CFCl_3$	0.28 ppbv	4	4	65	4,000	12,000
CF_2Cl_2	0.48 ppbv	8	4	120	6,000	15,000

* concentrations are on a volumetric basis (mixing ratios)

Air Quality

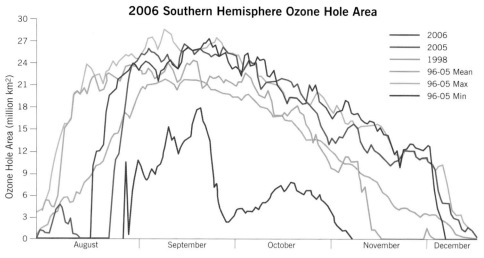

2006 Southern Hemisphere Ozone Hole Area

Ozone Hole Area (million km²)

Legend:
— 2006
— 2005
— 1998
— 96-05 Mean
— 96-05 Max
— 96-05 Min

August | September | October | November | December

The size of hole varies seasonally. In Southern Hemisphere winter (June to September), ozone levels decrease and the ozone hole grows. Ozone is not replenished until well into Spring when latitudinal mixing and sunlight return to the South Polar region.

Regardless of the size or magnitude of the ozone hole, UV radiation increases with more direct sun angle. Thus, UV is higher within two hours of local solar noon and during the warmer months of the year. UV values, excluding the effect of cloud cover, are typically highest within 23.5 degrees of the Equator (the region of the Earth with the highest sun angle).

Skin cancer

With less ozone present to block UV radiation from the sun, more UV can pass through the atmosphere and reach the Earth. Since we humans around the globe love to "catch the rays", and UV values have been climbing due to ozone loss, it isn't surprising that there has been a marked increase in skin cancer. The WHO estimates that as many as 60,000 people a year worldwide die from too much sun, mostly from malignant skin cancer. Of these deaths, 48,000 are from melanoma, and 12,000 are from other skin

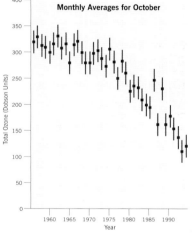

Monthly Averages for October

Total Ozone (Dobson Units)

1960 1965 1970 1975 1980 1985 1990
Year

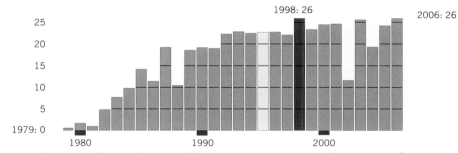

Average (Sept. 7 - Oct. 13) ozone hole area (millions of km²)

1998: 26   2006: 26

1979: 0

1980   1990   2000

This graph above shows the measured average October total ozone above the Halley Bay station in Antarctica. Note the sudden change in the curve after about 1975. By 1994, the total ozone in October was less than half its value during the 1970s, 20 years previously. Values have continued to decline into the 21st century.

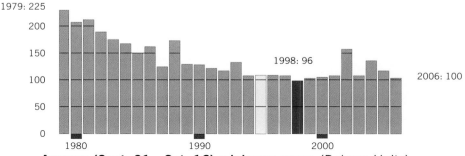

1979: 225

1998: 96   2006: 100

1980   1990   2000

Average (Sept. 21 - Oct. 16) minimum ozone (Dobson Units)

Ozone hole area and minimum values for each summer season 1979-2006 in the Southern Hemisphere (1995 data interpolated). Since international cooperation began on CFC emissions, both the size of the hole and the lowest ozone values have stabilized, although some records for size and ozone thickness continue to be broken. Values for 1998 reflect the largest ozone hole and the lowest ozone readings.

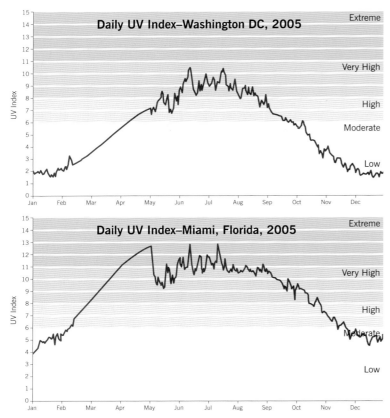

Daily UV Index–Washington DC, 2005

Daily UV Index–Miami, Florida, 2005

Annual UV index cycle (measured in terms of maximum daily UV index) is a function of latitude, length of day, and cloud cover. Miami, Florida, is closer to the Equator than Washington, DC, and has higher UV values almost daily. The sudden drop-off in May at Miami is linked to the arrival of the rainy season and the associated increase in cloudiness. Background intensity categories match those of the UN World Health Organization's Global UV Index categories.

cancers. About 90 percent of these cancers are caused by ultraviolet light from the sun.

Australia has the highest incidence of skin cancer in the world. More than 1,400 Australians die from skin cancer each year. Roughly 1 in 2 Australians will get skin cancer in their lifetime.

According to Health Canada, skin cancer has been increasing in Canada at a fairly constant rate over the past 30 years. In 2005, there were roughly 78,000 new cases of basal and squamous cell carcinomas and about 4,400 new cases of malignant melanomas reported.

Countries closest to the ozone holes (Australia and Canada) are probably the world's leaders in understanding and acting on UV safety. However, organizations in the United States are now taking a much stronger stand on sun safety, as well.

Sun safety efforts target understanding the meteorological factors that contribute to skin cancer occurrence – time of day, season, and type and amount of cloudiness. They also address lifestyle choices, including desire for a tan and application and SPF (Sun Protection Factor) value of sunscreen, and information about the types of skin cancer and how to recognize them. That some 80 percent of skin cancer occurs above the neck provides some very targeted safety measures – wear sunscreen (skin and lips) and a hat.

Sun safety information targets the very young, as well. Canada and the US both address the fact that about 80 percent of our lifetime exposure to the sun occurs prior to our reaching our 18th birthday. Many schools are aware of this and have covered playground areas, especially at elementary schools. Even on sunny days, parasols (not umbrellas) are being used to protect from the elements. There is now even UV protection clothing and swim wear.

The Sun Safety Alliance, a non-profit coalition dedicated to the task of reducing the incidence of skin cancer in America, has already developed an annual sun safety week. Many organizations and dermatologists, especially in the United States, offer special skin cancer screenings.

Sun safety, however, often overlooks eye safety. While many sunglasses (prescription and over the counter) offer UV protection, we tend to think that having sunglasses protects our eyes. Yet, sunglasses trick our eyes into opening wider to gather needed light, increasing the risk of UV-related eye problems (the most common being macular degeneration and cataracts).

Climate implications

Some of the gases and pollutants described in this chapter contribute to climate change. However, most do not. Rather, they create unhealthly or uncomfortable conditions. All, in one way or another have a meteorological component that enhances or diminishes their presence.

Parasols and hats, offering protection from the sun, are commonly seen in places like Japan. Look at the shadows to verify that the weather was sunny on the day on which this image was taken.

Contrails may have an affect on climate as they act to reflect incoming energy but do not trap Earth's radiation well. Volcanic dust, in large enough amounts, can help to cool the Earth.

Inversions, whether summer or winter season, and associated light winds trap pollutants near the Earth's surface. Stronger wind situations, including frontal passages, help to disperse the pollutants. This doesn't get rid of them, but it does lessen their concentrations.

Some weather conditions (such as droughts and lightning-caused fires) contribute to pollution onset. Other physical phenomena (e.g. sun angle, latitude, mountain location) increase pollutant levels either by photochemical interactions or trapping pollutants in valleys or regions near mountains.

While some countries (mostly developed ones) have taken strong actions to lower air pollution, developing countries have polluted more. Since, as one Chinese official noted, "pollution does not need a visa", pollution from one country can easily affect its neighbors or other countries thousands of miles away.

"Volcanic ash is hard, does not dissolve in water, can be extremely small, is extremely abrasive (similar to finely crushed window glass), is mildly corrosive and is electrically conductive, especially when wet."

US Geological Survey

Mount St Helens pictured on May 19, 1982. Note the rising steam cloud (like cumulus), the possible anvil cloud above the peak and the classic shape of volcanic crater and smaller crater within.

My first exposure to volcanoes was on the Big Island of Hawaii. Driving an open air Jeep across the volcanic terrain, smelling the sulfur and seeing the volcanic haze made my wife and I think that the dinosaurs would be just over the next rise. My second experience involved flying past Mount Rainier (near Seattle, Washington) and driving

The lava field on The Big Island, Hawaii, home to one of the most active volcanoes in the world. Kilauea continues to pour lava into the sea, thereby increasing the size of the island.

The eruption of Mount Vesuvius (left) in 79 AD is, without a doubt, one of the most renowned and catastrophic volcanic eruptions in history. The events of August 24 occurred so quickly that the inhabitants of Pompeii didn't even have time to flee for their lives as illustrated by the recreated remains of this cowering resident (below). The human shape was obtained by using plaster to fill in decomposed forms left in the hardened volcanic ash.

to Mount St Helens near Portland, Oregon. Our visit happened 24 years after the volcano erupted. It lacked the prehistoric setting. Rather, we saw both desolation and rebirth.

Then we visited Alaska and saw Mount McKinley and, in the distance, a minor eruption from nearby Mount Spurr. Our last volcanic excursion was to Pompeii, Italy, where Mt. Vesuvius loomed 20 miles (32 km) to the north. Try as we might, it was hard to envision a volcanic cloud engulfing the city and literally freezing it in time. Yet that was what occurred on August 24, 79 AD.

What are volcanoes?

Volcanoes are mountains that are quite unlike any other mountains on the Earth. They are not pushed upward like the Himalayas or crumpled and folded like the Appalachians. Rather, they are built by the accumulation of their own eruptive products – molten rock (lava), bombs (crusted over ash flows), and tephra (airborne ash). A volcano (unless it is in a highly eroded state) is most commonly cone-shaped with a vent or opening at its peak that connects with reservoirs of molten rock below the surface of the Earth. That opening allows lava, ash, and gases to escape.

According to the US Geological Survey (USGS), volcanoes both destroy and create. The catastrophic eruption of Mount St Helens on May 18, 1980, made clear the awesome destructive power of a volcano. First, a debris avalanche started moving

Volcanic eruptions (above) are one of the most fascinating and intriguing spectacles of the natural world but the lava (below) destroys everything in its path.

downhill at speeds 110-155 mph (177-250 km/h) – that is equivalent to speeds associated with an EF 2 to EF 3 tornado (see page 118) or a category 3 to 5 hurricane.

This avalanche was overtaken by a powerful lateral surge of heated ash and gases. Initially traveling at 220 mph (354 km/h), the blast accelerated to speeds of 670 mph (1,078 km/h), just under the speed of sound. Still this was more than 100 mph (160 km/h) faster than the speed at which most commercial aircraft fly.

Upon exiting the volcano, the ash cloud was picked up by winds aloft and carried mostly eastward and then southeastward at about 60 mph (96km/h).

Over a time span far longer than human memory and record, volcanoes have played a key role in forming and modifying the planet upon which we live. More than 80 percent of the Earth's surface – above and below sea level – has volcanic roots. Gaseous emissions from volcanic vents over hundreds of millions of years helped form the Earth's earliest oceans and atmosphere, which supplied the ingredients vital to evolve and sustain life. Yet, volcanic gases have also changed weather and climate, sometimes leading to mass biological extinctions. Over geologic eons, countless volcanic eruptions have produced mountains, plateaus, and plains. Thanks to subsequent erosion and weathering these have been sculpted into majestic landscapes and formed fertile soils.

Ironically, these volcanic soils and inviting panoramas have attracted, and continue to attract, people to live on and visit the flanks of volcanoes. Thus, as population density increases in regions of active or potentially active volcanoes, we humans become more at risk from these potentially geologic monsters. As the USGS once noted, people living in the shadow of volcanoes must live in harmony with them and expect, and should plan for, periodic violent unleashings of pent-up volcanic energy.

While volcanoes are close to home for us on Earth, let's not forget that volcanoes exist, or have existed, on other planets (e.g. Venus, Mars) and their moons (e.g. Neptune's Triton). In early 2007, NASA's New Horizons space probe captured an image of the volcano Tvashtar (on Jupiter's moon Io) sending a plume high into the moon's atmosphere. Since Io is so highly volcanic, it's surface is covered with lava flows, lava lakes, and giant calderas covering a

This series of pictures shows Mount St Helens before (left), during (center) and after (bottom) it erupted in 1980.

Plume from Tvashtar volcano on Jupiter's moon Io as seen from the New Horizons spacecraft (about 1.5 million miles from the moon) on February 26, 2007. In this processed image, which NASA calls, "the best view yet", it is easy to see an enormous 180 mile (290 kilometer) high volcanic plume near Io's north pole (11 to 12 o'clock position). The image also shows a much smaller, symmetrical fountain plume, about 40 miles (60 kilometers) high, from the Prometheus volcano in the 9 o'clock position. The top of a third volcanic plume, from the volcano Masubi, erupts high enough to catch the setting Sun on the night side near the bottom of the image, appearing as an irregular bright patch against Io's Jupiter-lit surface (6 o'clock position). It is also possible to see several Everest-sized mountains highlighted by the setting Sun along the terminator, the line between day and night.

sulfurous landscape. It can be liked to a volcanic area on Erath, only much more extensive.

An in-depth treatment of volcanoes is far beyond the scope of this book. However, Robert Tilling of the USGS has posted an easy to read introduction to volcanoes in an online book at http://pubs.usgs.gov/gip/volc/. There is a series of images (USGS) showing the evolution of the blast.

Ash

Volcanoes are geologic forces. Yet, they interact with weather and climate in many ways. First, volcanoes can disperse huge amounts of volcanic ash into the atmosphere. This ash has immediate local air quality implications and longer-term impacts from ongoing ash reinjected back into the atmosphere by human activities and wind. While heavier ash particles fall out closer to the volcano, finer particles can be carried downwind for hundreds, possibly thousands, of miles, and affect air quality anywhere enroute. Ash particles can seriously affect respiratory system function and can be deadly for those with respiratory illnesses. Opposite are dispersion maps for two volcanic eruptions.

Ash isn't only carried by the winds. As the particles start to fall from the ash cloud, they may first resemble mammatus clouds (pouches on the underside of a thunderstorm anvil); then they create precipitation or fall streaks across the sky, making the scene resemble that of thunderstorms. By the time the process is completed, the sky becomes completely filled with ash, and may remain so for several days.

Ash can also affect transportation, especially that involving aircraft. Since the dust is small, hard, and extremely abrasive, it can cause excessive wear on aircraft engines and even cause them to stall. Aircraft windshields, impacting the dust at flight speeds of hundreds of miles an hour, can become excessively scratched, making it difficult for pilots to see.

Consider the two 747 aircraft that encountered ash from the eruption of the Galunggung volcano in Indonesia (1982). Lacking any type of warning system, and not realizing the ash cloud was in their flight path (both planes were more than 100 miles from the volcano), one aircraft entered the ash cloud and lost power from all four engines. The craft dropped from 36,000 feet (11,000 meters) to 12,500 feet (1,810 meters) before all four engines were restarted. The airplane landed safely despite major

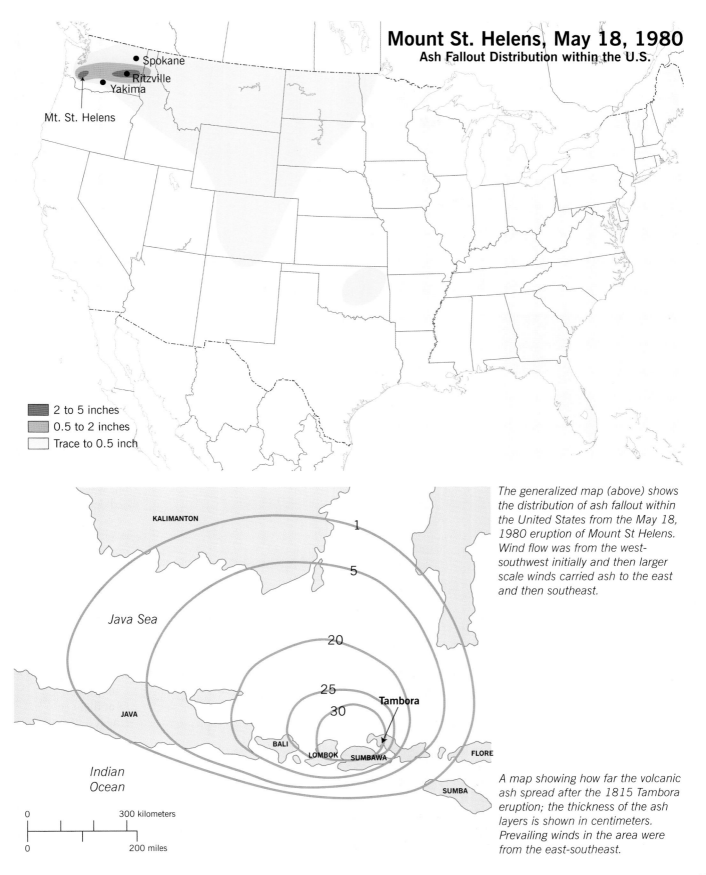

Mount St. Helens, May 18, 1980
Ash Fallout Distribution within the U.S.

Spokane

Ritzville

Yakima

Mt. St. Helens

- 2 to 5 inches
- 0.5 to 2 inches
- Trace to 0.5 inch

KALIMANTON

1

5

Java Sea

20

25

30   Tambora

JAVA

BALI

LOMBOK   SUMBAWA   FLORE

Indian
Ocean

SUMBA

| 0 | | | 300 kilometers |

| 0 | | | 200 miles |

The generalized map (above) shows the distribution of ash fallout within the United States from the May 18, 1980 eruption of Mount St Helens. Wind flow was from the west-southwest initially and then larger scale winds carried ash to the east and then southeast.

A map showing how far the volcanic ash spread after the 1815 Tambora eruption; the thickness of the ash layers is shown in centimeters. Prevailing winds in the area were from the east-southeast.

Volcanic ash billows skyward during the 1991 eruption of Mount Pinatubo in the Philippines. Cumulus clouds dot the sky in the foreground. To the untrained, the volcanic cloud might look like a giant thunderstorm.

A volcanic eruption on the island of New Zealand in January 1974.

engine damage. Before returning to service, all four engines had to be replaced. A few days later, a similar situation occurred. Although pilots should have been aware of the cloud by this time, another 747 flew into the ash cloud and also suffered significant engine damage.

Although rare, an average of three to four jet aircraft per year encounter volcanic ash with associated aircraft damage. However, the risk for disaster is great, especially with so many large aircraft flying across volcano prone areas of the Pacific Ocean basin.

Thanks to new technologies, training, and an International Airways Volcano Watch (IAVW) program, the number of incidents is decreasing. The IAVW program involves nine centers operated by different countries that collaborate on providing ash cloud information to aircraft. Their goal, plain and simple is "to keep volcanic ash and aircraft completely separated."

On the ground, dust can get into automobiles, air conditioning systems, and appliances. Ash can be heavy, too. When piled up on roofs (especially with added rain water), it can mimic the weighting of snow loading. Unless removed, it can strain the roof causing collapse.

Gases

In addition to ash, the volcanic cloud may contain aerosols of methane, carbon dioxide, and sulfates. These gases are unto themselves potentially deadly. However, the gases can create highly acidic rainfall, affect respiratory systems, and contribute to long-term atmospheric warming or cooling. For larger eruptions, gases and dust can remain in the stratosphere for years. There, some of the gases can cause a decrease in stratospheric ozone.

Gases can escape a volcano quickly or over time. This depends, in part, on the type of eruption. When a volcano erupts explosively, gases escape into the air quickly. If the volcano involves a lava flow, gases can escape quickly (thick lava, pressure buildup and release) or slowly (thinner lava that allows gases to seep into the atmosphere).

Hawaii's Mauna Loa volcano and the island's newest volcano Kilauea (south side of the Big Island of Hawaii) have both types of lava. While writing this book, Kilauea's

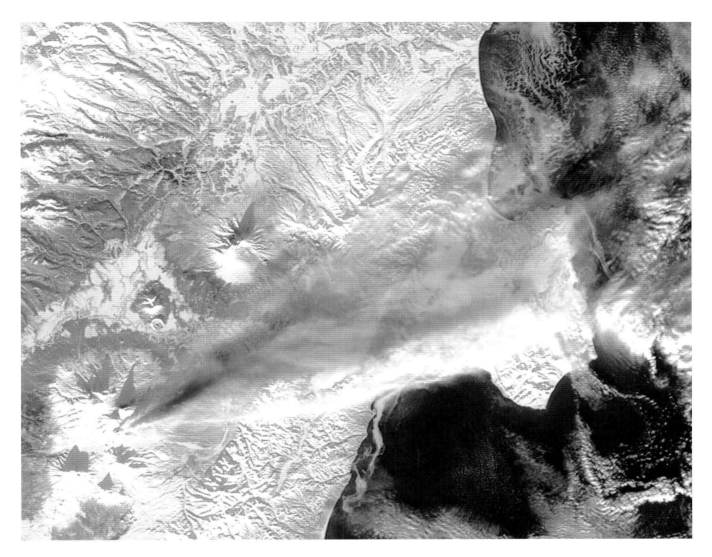

ongoing eruption produced high enough levels of sulfur dioxide that the National Park Service had to temporarily close the Hawaii Volcanoes National Park until the air cleared.

While volcanoes may be the main source of gaseous geologic emissions, hot springs and geysers can also emit sulfur and other polluting gases. When the gases and smoke combine, "vog" (volcanic smog) results.

This MODIS image shows an ash plume from the January 13, 2004, eruption of Bezymianny Volcano on Russia's Kamchatka Peninsula. According to the Alaska Volcano Observatory, the plume eventually reached a height of 3.75 miles (6 kilometers).

A volcanic explosivity index

Some scientists recently proposed the Volcanic Explosivity Index (VEI) to attempt to standardize the relative size of an explosive eruption. The VEI is based on the volume of ejected matter and other factors and ranges from 0 to 8. A VEI of 0 denotes a nonexplosive eruption, regardless of volume of erupted products. Eruptions designated a VEI of 5 or higher (which occur only once every 20 years on average) are considered "very large" explosive events. The May 1980 eruption of Mount St Helens just reached a VEI rating of 5. Although its lateral blast was powerful, it had little magma output.

Scientists have computed VEIs for more than 5,000 eruptions that occurred in the last 10,000 years. None of these eruptions have reached the maximum VEI of 8. For

For weeks volcanic ash covered the landscape around Mount St Helens and for several hundred miles downwind to the east. Noticeable ash fell in 11 states. The total volume of ash (before its compaction by rainfall) was approximately 0.26 cubic miles (1.01 cubic kilometers), or enough ash to cover a football field to a depth of 150 miles (240 kilometers). In this photograph, a helicopter stirs up ash while trying to land in the devastated area.

example, the eruption of the Vesuvius volcano in 79 AD, which destroyed Pompeii and Herculaneum, was only rated a VEI of 5. Since 1500, there have only been 22 eruptions with VEI 5 or greater. The greatest is the 1815 Tambora eruption (VEI 7) (see page opposite). There have been four VEI 6 eruptions, including Karate in 1883, and 17 VEI 5s, including Mount St Helens in 1980 and El Chichon, Mexico, in 1982. Scientists estimate that the enormous caldera-forming eruptions that occurred more than 10,000 years ago would have earned VEI 8 ratings.

A tourist takes a snapshot of lava on Mount Etna in Italy.

The snowcapped Popocatepetl, an active volcano in Mexico.

The number of casualties and the extent of destruction have also been used to compare the "bigness" of volcanic eruptions. Although there is an overall relationship between VEI and the number of casualties, that relationship can be tenuous based on how the volcano erupts and geographical and population density aspects. Consider that some of the most destructive eruptions have not been "very large".

For example, mudflows triggered by the November 1985 eruption of Nevado del Ruiz (Colombia) killed more than 25,000 people, making it the worst volcanic disaster in the 20th century since the catastrophe at Mont Pelee in 1902. Yet, the eruption was very small, producing only about three percent of the volume of ash ejected during the May 1980 eruption of Mount St Helens.

The eruption of Mount St Helens had a higher VEI rating than five of the deadliest eruptions in the history of mankind (see figure above), but only 57 people died due to the eruption and its related effects. Part of the reason involved a hazard warning system

DEADLIEST VOLCANIC ERUPTIONS OF THE PAST 500 YEARS AND THEIR VEI RATINGS
Volcanic Explosivity Index (VEI) of the deadliest eruptions since 1500

ERUPTION	YEAR	VEI	CASUALTIES
Nevado del Ruiz, Colombia	1985	3	25,000
Mont Pelee, Martinique	1902	4	30,000
Krakatau, Indonesia	1883	6	36,000
Tambora, Indonesia	1815	7	92,000
Unzen, Japan	1792	3	15,000
Lakagigar (Laki), Iceland	1783	4	9,000
Kelut, Indonesia	1586	4	10,000

An eruption in Piton de la Fournaise on Reunion Island in August 2003. Reunion Island is a small volcanic island in the western Indian Ocean

and the establishment of a zone of restricted access. Although, the area affected was sparsely populated to begin with.

Volcanoes and weather

Once formed, the volcano itself can influence local weather patterns. Most immediate, following an eruption, is the potential for melting snow and landslides to create local floods. If there is a significant tree fall, it is also possible to dam waterways, setting the stage for dam breaks and additional flooding.

If the volcano melts the snow and/or covers it with ash, local temperature patterns can be disrupted. The image on page 267 shows how prevailing winds carried ash from Mount St Helens mainly in one direction.

In tropical latitudes, where an inversion may often cap warm and humid air, the presence of a volcano can create "interesting" weather and rainfall patterns. Let's take a look at Hawaii.

Mauna Loa on Hawaii's Big Island is a large volcano. Although it only rises to a height of about 2.5 miles (4 kilometers) above sea level, it is still the world's tallest active volcano. That's because the volcano sits on the ocean floor some 3.1 miles (5 kilometers) below the ocean surface. When viewed this way, the volcano is more than 5.6 miles (9 kilometers) tall. Due to further computations involving the depression of the Earth's crust by this large mountain, the volcano might even be considered to be 10.6 miles (17.2 kilometers) tall.

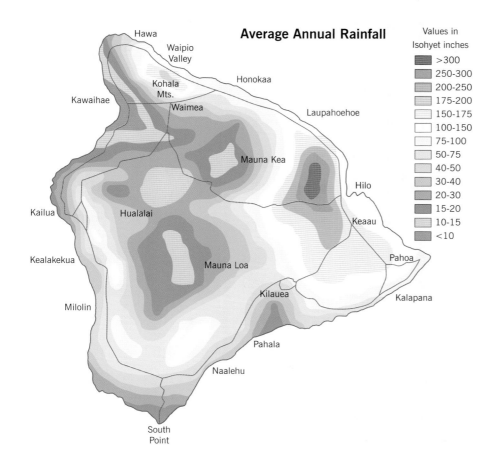

Average Annual Rainfall

Values in Isohyet inches
- \>300
- 250-300
- 200-250
- 175-200
- 150-175
- 100-150
- 75-100
- 50-75
- 40-50
- 30-40
- 20-30
- 15-20
- 10-15
- <10

Rainfall map for the Big Island of Hawaii, the youngest of the Hawaiian islands and the one with the tallest mountains. Notice that rainfall maxima for the island lies on the east coast (upslope side of the mountain in the trade wind easterly wind regime) and that heaviest rain lies well below highest terrain. In fact, rainfall near the peak is less than 4% of the peak rainfall on the island

Under typical conditions a low-level inversion (around 7,000 feet (2134 meters) above sea level) dominates the northeast trade wind weather regime. As a result, low-level cumulus and stratocumulus clouds dominate. On the upwind side of the islands, these clouds pile up against the mountains and drop their rains. As the clouds cross the islands, sinking air prevails and a rain shadow pattern exists.

The situation takes an added twist on the younger islands, Maui, and the Big Island of Hawaii. Here volcanic peaks extend higher into the atmosphere. The inversion keeps clouds and precipitation confined to the lower levels of the mountain, leaving the upper levels mostly cloud and precipitation free. This makes the peak of Mauna Kea (the taller northern peak on the island at 13,796 feet (4,205 meters) above sea level) ideally suited as an astronomical observatory site.

Located on Maui, Haleakala (an elevation of 10,023 feet (3,055 meters)) has an almost desert-like climate at its upper reaches. The terrain and weather give the landscape a Martian appearance. The National Park is noted for its sunrise and sunset scenarios and, having driven the 38 miles (11.5 km) winding road up the mountain between 3.00 am and 5.00 am, I can attest to their magnificence.

Climatic considerations

Volcanoes, once thought to have one main effect on global climate, now typify the whole climate change debate. Volcanic ash and sulfur gases may lead to global cooling while the addition of Greenhouse Gases (methane and carbon dioxide) can lead to warming. One recent study (involving scientists form Oregon State University, Rutgers

Once volcanoes become extinct, they make very good defensive strongholds. Edinburgh Castle in Scotland stands upon the basalt plug of such a volcano that is estimated to have risen some 340 million years ago.

University, and the University of Denmark) showed volcanoes not only caused atmospheric impacts, but also oceanic changes. Using geologic tracing techniques (including chemical and aging analysis), the scientists linked significant global oceanic warming (as much as 9-11 °F (5-6 °C)) to the occurrence of major volcanic activity around 55-60 million years ago. In addition, the scientists suggested that there was also a dramatic change in ocean chemistry due to the volcanic activity. Oceans became more acidic and numerous deep-sea species became extinct.

Other studies suggest that sulfur aerosols (not the volcanic ash) are the cause of global cooling. After studying three historical explosive eruptions of different sizes, and with different amounts of ejected ash in Indonesia, Tambora (1815), Krakatau, also known as Krakatoa (1883), and Agung (1963), scientists found that decreases in surface temperatures after the eruptions were of similar magnitude (32-34 °F (0.18-1.3 °C)).

After the 1783 eruption of the Laki volcano in Iceland, the cooling over Iceland, Europe, and the eastern US was dramatic. The eruption, lasting eight to nine months, sent an estimated 80 million tons of sulfuric acid aerosol into this atmosphere. This was

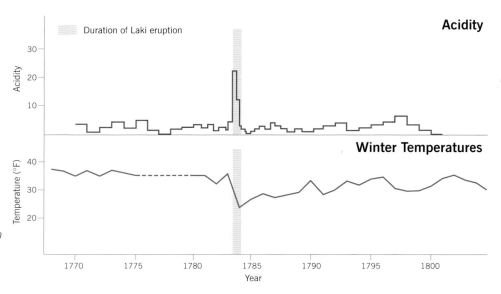

Graphs showing the effects of the 1783 eruption of the Laki volcano in Iceland. More acidic rainfall and much cooler temperatures followed the eruption.

four times more than the El Chichon and 80 times more than the Mount St Helens eruptions. The result was a bluish sulfur aerosol haze across Iceland. The haze drifted eastward, covering Europe during the winter of 1783-84. During the "blue haze" period, most summer crops in Iceland were destroyed (by acid rain), 75 percent of the island's lifestock was lost (eating fluorine contaminated grasses); and 24 percent of the population died (famine). The winter of 1783-84 across Europe was deemed, " unusually severe" by University of Rhode Island vulcanologist Haraldur Sigurdsson, who studied the event.

The event can also be linked to a very cold winter in the eastern United States with temperatures almost 9 °F (5 °C) lower than a 225 year average and overall Northern Hemisphere temperatures were nearly 2 °F (1 °C) colder than average. The event also caused a spike in acidity values over Greenland.

Another study, conducted by scientists at New York University and the University of Hawaii, indicate that the eruption of Toba, on the Indonesian island of Sumatra, around 73,500 years ago, may be linked to the start of the last ice age (in which global temperatures dropped some 6-9 °F (3-5 °C) in just a few years. This eruption, the largest in two million years, dispersed more than one billion tons of volcanic ash and sulfur gases high into the stratosphere and may have been triggered by an already-begun cooling period, including the lowering of sea levels. The scientists speculate that the loss of ocean water atop the volcano's base may have also allowed it to erupt explosively.

Scientists now believe that the cooling due to the sulfur aerosols is linked to its longer airborne lifetime and how the sulfur gas absorbs and reflects solar and terrestrial energy. The most significant impacts result from the conversion of sulfur dioxide (SO_2) to sulfuric acid (H_2SO_4), which condenses rapidly in the stratosphere to form fine sulfate aerosols. The aerosols reflect more solar energy back to space, thus cooling the Earth's lower atmosphere or troposphere. However, the sulfur aerosols absorb heat radiated up from the Earth, thereby warming the stratosphere. In addition, sulfate aerosols interact with CFCs in the stratosphere to cause further declines in the ozone layer.

The Arenal Volcano in Costa Rica vents its fury.

19 CLIMATE CHANGE
PAST, PRESENT AND FUTURE

"Nothing endures but change."
Heraclitus, Greek philosopher (540-480 BC)

Back in the 1970s, some scientists were concerned that we'd soon be in another ice age. Now, some 30 years later, we are focused on global warming. If we go back 400 years or so, much of the Northern Hemisphere was just starting to exit from a mini-ice age. Look to geological and other indirect evidence from several billion years ago and you will find that the Earth was more heated than it is now.

All of these reflect different scales in the Earth's weather record. The long-term patterns and changes (measured in hundreds of thousands to billions of years) and shorter time frames (hundreds to thousands of years) reflect true climate. Then we have the 30-year average of weather conditions that we often refer to as climate. And it is this time frame (to maybe a hundred years or so) that the world's population is looking at when it comes to global warming or climate change concerns.

Listen to the news and you'll hear terms like "worst drought in 25 years" or "highest temperatures in the last 100 years". Yet, in the geological frame of reference, this latter scale is like a hiccup or noise. It may bear little resemblance to the real climatic trends.

Four examples of different climates that abound on Earth (viewing counterclockwise): the lush La Selva River in a Costa Rican rainforest; desert dunes in Morocco; Acacia trees on the Serengeti plains; explorers braving the Arctic.

Anything in a time period less than 30 years, for lack of better classification, falls under the guise of weather. This includes decadal shifts and also day-to-day or hour-to-hour changes. I've even heard statements like, "if you don't like the weather, wait a minute", applied to weather in such far-removed places like the San Francisco area, Chicago, Boston and Washington, DC.

In fact, Robert Heinlein's famous quote, "Climate is what you expect and weather is what you get", couldn't be truer. We expect the "average" conditions that we hear on television and other weather reports, but, more often than not, we get conditions that are either close to, or far removed from, these averages.

Still, short-period changes are easier to see than long-term climate shifts. Cold fronts pass by, thunderstorms come and go while seasons change. However, the year-to-year variations, and changes over even longer time scales, are what drives our overall planetary condition. In each of these data sets there are short- and long-term variations. When these patterns are mixed together, the result can sometimes be like waves interacting on a beach. There may be lots happening, but the patterns are harder to see. Thus, understanding these complex patterns and how they inter-relate is the key to understanding whether we are in real global warming mode or merely in a period of warming. Such studies may also help ascertain whether the changes are "natural" or driven by human actions.

The main waves at a beach are often those produced by the swell, while the long-period waves are produced by winds and storms in other places. Travelling large ocean distances, these waves, known as swell, eventually reach coastlines. When they do, they interact with the underwater topography and "break", causing waves like those shown here. As these waves crash ashore, water rushes back down the beach toward the ocean, interacting with the incoming waves (foreground). In addition, there can be locally wind-driven waves and waves caused by boaters. Collectively, these can make a sometimes chaotic wave pattern, even though each component is easy to describe.

Weather pattern changes

In the 1930s the central US was drought-stricken. In mid-2007, other parts of the US were in drought (e.g. southern California and parts of the southeast) while the Midwest and Texas were reeling from too much rain. In fact, numerous daily, monthly, annual rainfall and stream flow records were broken in Texas, Oklahoma and Kansas in June 2007.

The western states and the southeastern region were in drought and reported extremely low stream flow levels. Although some claim these extremes are linked to global warming, they are more the result of a stagnant weather pattern. On many occasions,

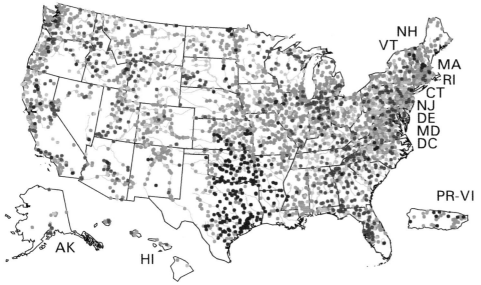

Streamflow conditions across the US on July 1, 2007. Locations shaded in blues and blacks indicate above average values; reds and oranges below average. Notice how close the two extremes are across the lower Mississippi River Valley.

High
≥90th percentile
75th-89th percentile
25th-74th percentile
10th-24th percentile
<10th percentile
Low
Not ranked

The colored dots on this map depict streamflow conditions as a percentile, which is computed from the period of record for the current day of the year. Only stations with at least 30 years of record are used.

weather patterns persist in one place for weeks (even months). At other times, it is the effect of snow cover versus non-snow cover that locks in storm track patterns. Snowy regimes tend to beget more snow. As my wife notes, "the pattern will persist until something causes it to change."

Although persistency and process feedback may dominate weather news, the same effects can occur on climatic scales. If snow begets more snow, then excessive snow (without sufficient melting in between snowy seasons) can help to create glaciers. While glaciers take long periods to create, even just one season's change in snow cover can affect the upcoming weather over one or more seasons.

Similar feedback and interactions occur with other types of weather and weather patterns can also "migrate" from one area to a nearby one. Compare the US drought index maps, in Chapter 12, between March 2005 and March 2007 for an example. In Australia, the current drought is now described as the worst in 1,000 years, a long period by most standards but far below geological standards. Yet, during the drought, in late June 2007, the worst flooding in 37 years occurred in southeastern Victoria State (the state in which Melbourne is located). Rainfall amounts were not excessive – in the order of one to two inches (25-50 mm) – but the rain fell in a short time period and it fell on baked, hardened ground. Runoff was excessive.

Tornadoes and hurricanes

When it comes to hurricanes and tornadoes, there are similar periodic, geographical groupings of events. According to a research team from the University of South Carolina's Department of Geography, the center of tornado hazard incidence has remained in southeast Missouri since the 1950s. However, the location has exhibited a cycloidal pattern with a systematic shift toward the southeast. This can reflect the higher incidence of hazard (death, injury, or damage from tornadoes) as population has shifted to the southeast.

However, during this period, there has been a marked increase in tornado safety education and a proliferation of storm spotters and chasers throughout the central US. These human "eyes of the community", when coupled with new Doppler radar detection capabilities, could have helped the National Weather Service issue at least some warnings in a more timely manner. Furthermore, TV stations and the National Oceanic and Atmospheric Administration (NOAA) Weather Radio transmit warnings faster than 30-40 years ago. The result is that some tornadoes might not cause injury or death (although damage may still occur). The statistics have considerable error, but they do provide another way to look at tornado incidence across the US.

There was a high incidence of major hurricanes along the east coast of the US during the 1950s – it was a series of hurricanes that affected the New York City area where I was growing up at the age of nine that pushed me into a weather career.

That was the only such decade in 200 years in which this occurred. There were three decades in which the Gulf Coast was the target zone for hurricanes. However, look at the transition between the 1950s and the 1960s and you'll see that clearly something was different in the steering wind patterns that took these weather systems to different landfall locations.

In fact, there are many global and local variables that may have some effect on hurricanes. These include wind shear, sea surface temperatures in the Atlantic and even the track of storms. If the storm track moves hurricanes over some of the larger Caribbean islands, especially those that are mountainous, or peninsulas, such as Florida and the Yucutan, the storm can be weakened to the point of being less dangerous to later landfalling locales.

Consider Wilma in the fall of 2005. She stalled over the Yucutan Peninsula, weakening from a category 5 to a category 1 before landfalling in southwest Florida. If Wilma had not been affected by the Yucutan, but had remained mostly over water, she would have affected some part of Florida much more severely.

El Niño and la Niña

El Niño and La Niña* – the dramatic warming and cooling, respectively, of ocean waters off the coast of Peru and Ecuador – are often linked to the number of hurricanes in the Atlantic Ocean. That's because when El Niño arrives (warmer than average water temperatures), it results in warmer air above the warmer waters.

This leads to lower sea-level pressures; increased showery weather; and increased westerly winds at higher altitudes that start in the region and extend across the tropical Northern Hemisphere Atlantic. With low-level easterly trade winds already in place, any tropical depressions, tropical storms and hurricanes are, generally, carried toward the west. The increased westerly winds at higher altitudes help to shear, or blow, the tops off of these tropical systems prematurely, lessening their ability to grow into stronger storms. The reverse occurs when La Niña arrives.

In fact, according to numerous researchers, the chances of a hurricane striking the United States are greater, and the chances of an intense hurricane striking the United States are much greater, during non-El Niño years. In a joint Florida State University-NOAA study, researchers noted that there was a more than 4:1 ratio of intense hurricanes landfalling in the United States in La Niña years compared to El Niño years. If one combined neutral and La Niña years, the ratio jumped to almost 12:1.

Equally significant, is where the hurricanes strike. In La Niña years, storms make landfall further south, affecting Central America. In neutral years, landfall is more along the Gulf Coast. In El Niño years, landfall is further north (including the US east coast). The researchers also found much higher chances of at least two landfalling US hurricanes during La Niña and neutral years.

Unfortunately, there are many caveats to these findings. First, there is more than one definition of El Niño and La Niña in use – though both center on having sea surface temperatures at least 0.9 °F (0.5 °C) warmer (El Niño) or colder (La Niña) for at least a five-month period. The definition used is significant when determining which temperature regime defines the peak Atlantic Ocean hurricane season. By one definition, 1992 would be an El Niño year; by the other a neutral year. 1992 was a relatively tranquil hurricane season in the Atlantic Ocean basin, but it had one major landfalling hurricane – Andrew. Until Katrina, Andrew was the costliest US hurricane and remains one of the strongest (category 5) to strike the US.

ENSO

The El Niño and La Niña events are actually part of a larger circulation pattern that involves the entire Southern Hemisphere and for this reason is actually known as ENSO (En Niño-Southern Oscillation). ENSO is the periodic shifting of sea level pressures

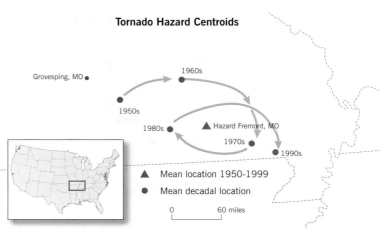

Tornado Hazard Centroids

Centroid of decadal tornado incidence across the United States has remained in southeast Missouri for the past 50 years.

* Normally cool waters are present in the eastern Pacific near Peru and Ecuador. The anomalous situation, El Niño, occurs about every two to seven years and may last for one to two years. Every event is different. Significant cooling is known as La Niña. The names, linked to the Christ child, come from the warming events that occur around the time of Christmas.

El Niño and La Niña sea surface temperature patterns are quite different. Here the two are compared. The most striking variations are across the central and eastern Pacific Ocean.

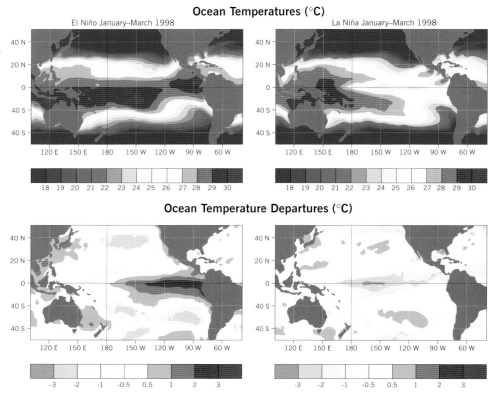

Ocean Temperatures (°C)

El Niño January–March 1998   La Niña January–March 1998

Ocean Temperature Departures (°C)

across the tropical Pacific Ocean. It is typically measured by comparing the variation in pressures from average values at Tahiti (French Polynesia) and Darwin (Australia). The Southern Oscillation Index (SOI) can be easily tracked since these two locations have long-term records of pressure.

ENSO also affects seasonal rainfall and temperature patterns, not just in the southern hemisphere, but globally. When it comes to droughts, ENSO is almost always involved. ENSO does this through changes in surface and upper air wind and pressure patterns and resulting changes in sea surface temperatures by either favoring upwelling or downwelling of ocean waters.

Upwelling occurs when winds in a local or regional area tend to move water away from a place. This creates a lowering of the sea level surface. To replace the partial void of water, bottom waters are brought upward. These bottom waters are typically cooler than surface waters. The result is a lowering of the sea surface temperature. If winds act to pile water up in a place, surface waters are forced downward (downwelling).

This process is a major part of the ENSO (both in the eastern and western tropical Pacific Ocean). It also happens along coastlines when winds blow offshore (upwelling) and onshore (downwelling). This can cause significant daily variations in sea surface temperatures, especially along the east coast of continents in middle latitudes. Even hurricanes can cause up- and downwelling as they move across ocean areas.

Local wind changes can also disrupt normal ocean biological processes. For example, lacking cool, nutrient-rich, upwelled, surface waters, fishing in the area near Ecuador and Peru decreases during an El Niño period. (See Resources for an interactive listing of global impacts.)

Were ENSO the only pattern that controlled global weather patterns, things would be easy. However, there are other patterns to consider.

North Atlantic oscillation

One of the most profound is the North Atlantic oscillation (NAO). As with SOI, this involves a pressure variation across a large geographical area. This pressure comparison is between north-south locations (not east-west locations).

The Climatic Research Unit in the UK centers on the NAO for winter temperature and rainfall patterns. Here, a strong NAO means the Icelandic low pressure is more intense and the pressure gradient between Iceland and the Azores is large. This means a stronger westerly wind flow south of the Icelandic low (see illustration on page 282). Within this westerly flow, there is a related storm track from the north Atlantic into western Europe. The result is more precipitation and warmer temperatures across northwestern Europe. Related to this is a wetter and warmer pattern across the southeast United States. With storminess to the north in Europe, precipitation is below average across much of the Mediterranean; in North America, below average precipitation tends to occur in eastern and western Canada.

With a neutral NAO, the pressure gradient between the Azores and Iceland is lower and weather patterns reverse. More storminess affects the Mediterranean; less affects northwest Europe.

The United States looks instead at the NAO as a predictor of hurricane behavior. A high or strong NAO has an associated strong high pressure system near the Azores. This means a weaker Bermuda high pressure system and a favored tropical cyclone pathway along the east coast of the US.

When the NAO weakens, the Bermuda high becomes stronger and hurricanes track more toward the west (affecting the Florida and the Gulf Coast).

Even more oscillations

There are also Arctic, Antarctic, North Pacific and a tropical Madden-Julian Oscillations. These are known, respectively as AO, AAO, NPO and MJO. The AO

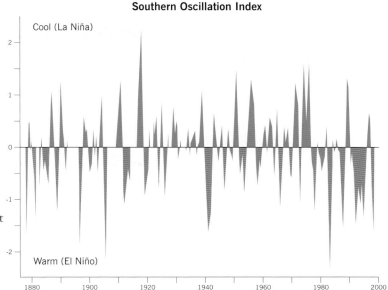

This SOI graphic from Oceanographer Dr. William Kessler (PMEL and University of Washington) shows the periodicity of the El Niño and La Niña cycles. Longer term assessments through proxies – or indirect, climatic data in the form of tree ring analysis, sediment or ice cores, coral reef samples and even historical accounts from early settlers – are typically used to extend this record back in time.

Precipitation and temperature anomalies during El Niño in a Northern Hemisphere summer (below left) and winter (below). Orange areas indicate locations where drought conditions often develop during El Niño situations.

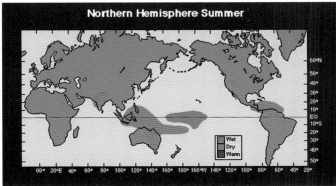

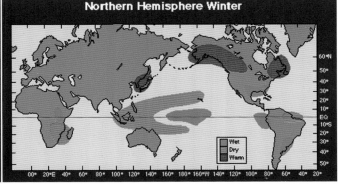

involves pressures north of the Arctic Circle and resulting variations in storminess at latitudes between 40 and 65 degrees north. The AAO is similar, but for the southern hemisphere. The AO's "positive phase" has relatively low pressure over the polar region and high pressure at middle latitudes (about 45 degrees north). In this phase, frigid winter air remains locked at higher latitudes and does not extend as far into the middle of North America as it would otherwise. This keeps much of the United States east of the Rocky Mountains warmer than average, but leaves Greenland and Newfoundland colder than usual. Again, when the pressure pattern changes, so does the associated weather.

According to the US National Snow and Ice Data Center, since the 1970s the AO has tended to stay in the "positive phase", causing lower than average air pressure over the arctic and higher than average temperatures in much of the United States and northern Eurasia.

The North Pacific Oscillation (NPO) also known as Pacific Decadal Oscillation (PDO) is a long-lived El Niño-like pattern of Pacific climate variability. Similar to El Niño in context, the NPO exists on scales roughly 10-25 times longer. Several research studies have found evidence for just two full NPO cycles in the past century. Cool PDO regimes prevailed from 1890-1924 and again from 1947-76, while "warm" PDO regimes dominated from 1925-46 and from 1977 through to (at least) the mid-1990s.

Major changes in northeast Pacific marine ecosystems have been correlated with phase changes in the PDO; warm eras have seen enhanced coastal ocean biological productivity in Alaska and inhibited productivity off the west coast of the contiguous United States, while cold PDO eras have seen the opposite north-south pattern of marine ecosystem productivity.

Observational NPO data has been correlated with southern California Douglas fir and California-Mexico Jeffrey Pine tree ring assessments (one form of "proxy" data) in order to extend the NPO record. In their study, Franco Biondi, Department of Geography, University of Nevada, and his team discovered similar time scale variability in the PDO and significant climate shifts around 1750, 1905 and 1950.

There is also a Madden-Julian Oscillation (MJO) that incorporates the combined effects of average low-level and upper-level winds and outgoing long-wave radiation measurements (satellite-based) in near-equatorial regions.

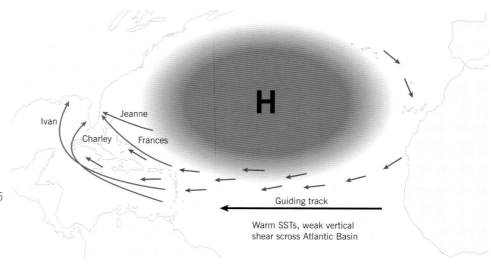

This map shows strong and neutral NAOs in 2004. The 2004 and 2005 hurricane seasons were influenced by a more neutral NAO pattern with storms being directed into Florida and the eastern Gulf of Mexico.

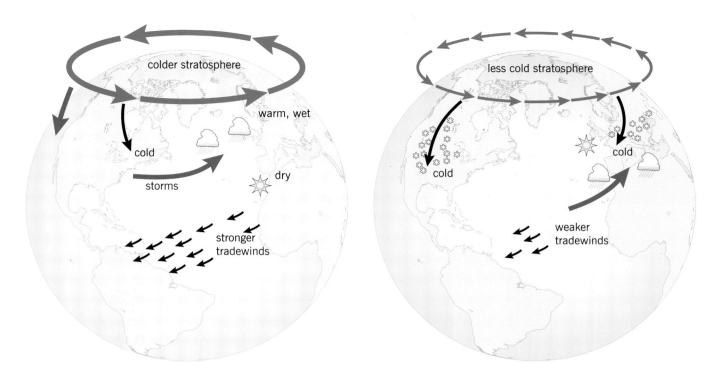

The Positive Phase of the Arctic Oscillation (left) is accompanied by stormier weather in northwest Europe. Under the Negative Phase of the Arctic Oscillation (right) storminess shifts to the Mediterranean.

All of these, and other patterns, can be grouped under the term "teleconnections." Teleconnections involve relationships among weather and oceanographic patterns (and associated weather regimes) around the world. Put in simple terms, these relationships show that when it's warm in one place, it is cold somewhere else. Out of balance weather patterns tend to be compensated for by their opposites. We see this all the time on daily weather maps.

It is harder to get a handle on the causative factors of the pattern shifts. For example, why does the polar low intensify in some years and not in others? Why do easterly winds weaken to create the El Niño pattern? How inter-connected are the various patterns themselves? Can one or two teleconnections provide the information or must there be numerous indices or evaluations to understand how weather is related to larger scale events? Can new indices be computed that incorporate several indices?

Recent climate change

Phil Jones at the UK's Climatic Research Unit, has summarized the weather pattern for the past 1,000 years. His "Millennial Temperature Record" highlights information obtained from numerous paleoclimatology researchers. Paleoclimatology involves studying ancient climates through the use of "proxies" (indirect climatic evidence rather than direct observations). Proxies can include fossils found in sedimentary (layered) rocks, information from coral reefs, trees and ice cores. Ice cores provide the longest historical information and include carbon dioxide (CO_2), methane (CH_4), other greenhouse gases and dust concentrations. In some situations, historical documents and native folklore provide information about short-term past weather or climate.

Proxies cannot pinpoint temperatures in space and time, but can at least offer evidence of overall patterns (e.g. annual temperature and sometimes rainfall information), principally based on the growing season of many of the proxies. Since most of the proxy data (95 percent) is found in the northern hemisphere, most reconstructions are for that hemisphere.

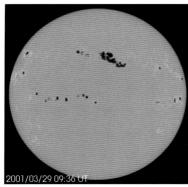

The largest sunspot in 10 years occurred on the Sun on March 29, 2001. Sunspots, like this one, are darker areas on the visible surface of the Sun caused by a concentration of distorted magnetic fields. The strong magnetic field slows down the flow of heat from the Sun's interior and keeps sunspots slightly cooler (darker) than their surroundings. The number of sunspots increases and decreases as the Sun's 11-year cycle of stormy activity rises and falls. Violent solar activity (including flares and coronal mass ejections) is believed to be caused by the release of magnetic energy. This sunspot produced four flares, one of which was classified in the "most intense" category; the other three flares were in the second most intense class.

There seems to strong agreement among findings by different researchers. However, many researchers have used common data sets, so the comparisons are not fully independent. Some findings involve averaging some of the available data sets, while others employ more robust techniques of data analysis. Understanding the way data sets are used is a key part of understanding research findings.

Still, there is little doubt that there has been a dramatic rise in temperatures during the 20th Century, making it the warmest period of the millennium. The 1990s were the warmest decade and 1998 the warmest year (note that since this data set was developed, many years between 2000 and 2006 have earned the title of warmest year of the last 100). Still, there was a comparable dramatic rise in the 1600-1750 period. While data uncertainties were greater earlier in the millennium, overall trends are fairly easy to detect.

Reconstructions of this sort are important because proxy evidence provides our only means of estimating the levels of natural variability before meteorological instruments were used in the past 100-200 years. While proxies are useful in this time frame, they are even more useful when more ancient climates are studied.

Jones and others suggest that natural changes caused by the sun and volcanoes dominate the pre-1850 part of the record and only human activities appear to adequately explain the rise in temperature during the 20th Century. These changes are typically linked to the use of fossil fuels and similar environmental impacts. However, by considering only human population changes (not even our alleged impacts on greenhouse gases), there seems to be very dramatic correlation with the most recent warming period (since about 1800).

On the other hand, smaller variations since the late 1880s can be linked to solar influences. Rodney Viereck, of NOAA's Space Environment Center, has compared sunspot data to global temperatures and his findings suggest that solar factors may be playing a larger role than previously thought in recent (and past) climate change. As ocean temperatures rise, so do air temperatures through direct heat transfer (conduction), evaporation and enhanced convection and there is clear-cut evidence that the sun has contributed to the global warming of the 20th Century. It's estimated that as much as one third of global warming may be the result of an increase in solar energy due to greater sunspot activity. Because this shows up in recent data, then there is probably a similar link to climate change in the past.

Long-term climate change

The figure opposite shows temperature change during the past 1,000 years and appears quite ominous. There's what appears to be an unprecedented recent rise in temperature that shrieks of "doom and gloom". Now look at the proxy records that date back 100 million years.

Temperatures from millions of years ago are much higher than they are today. Even with a proposed doubling of CO_2 values (through the human presence), it would still be cooler than it was in the distant geologic past. However, numbers alone don't tell the whole story.

University of Nebraska at Lincoln geoscience researchers Tracy Frank and Christopher Fielding and their research team, didn't just look at the temperatures, but also the variation in temperatures. Writing in early 2007 for *Science* magazine, Frank and Fielding noted, "In a multi-pronged study of the period from 305 million to 265 million years ago…found evidence of several major climate shifts, alternating between extreme

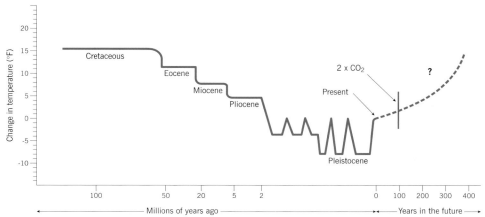

cold, or glacial, periods and periods of relative warmth. In each instance, an increase in atmospheric carbon dioxide (CO_2) accompanied a shift to a warmer climate and a decrease in atmospheric CO_2 accompanied a shift to a glacial climate."

The research team also found atmospheric CO_2 varied dramatically. During the 40-million year period, CO_2 values in the atmosphere varied from 280 parts per million (PPM) to 3,500 parts per million, with consideration variability in between. CO_2 values are often expressed as a value compared to the larger air volume. A value of 280 PPM means 280 molecules of CO_2 compared to a million air molecules.

Current CO_2 levels, roughly 380 PPM, are the highest in at least 650,000 years. This value is some 50 percent higher than at the start of the Industrial Revolution in the early 19th Century. Values are also rising at a faster rate than anything found in the geologic record.

Fielding notes that it's getting warm now at a value of CO_2 around 380 PPM. Yet, we have never seen CO_2 values anywhere close to 3,500 PPM. Fielding's concern is that if CO_2 values continue to rise (and temperatures warm accordingly), the Earth could warm dramatically.

Cycles and more cycles

"Zooming in" on limited time periods makes it possible to see greater variation. This suggests many cycles and their causes, are superimposed on the long-term record. Consider just the past 18,000 years. The most recent 6,000 years or so, in which much of our history resides, was a period of relative thermal calm. Prior to that, starting around 10,000 years ago, things were much cooler.

While CO_2 is often touted as the greenhouse gas of importance, several greenhouse gases have been moving in tandem through recent geologic times. Any or all of these can be easily matched to the warming of global temperatures.

Finally, looking at almost a million-year period, two things stand out. First, temperatures almost always remained below that observed around 1900 (zero baseline). This suggests that our planet was basically colder throughout nearly all of geologic time than it was in 1900 and certainly was colder than it is today!

Second, there is a pronounced periodicity of up and down swings in temperatures. These can be related to wide array of "natural causes" including sunspot variability; meteor impacts; changes in sunspot activity; changes in the Earth's orbit (wobble and tilt of its

The evolution of our current continental land mass configuration during the past 225 million years based on plate tectonics.

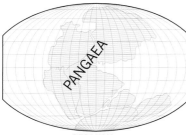

PERMIAN
225 million years ago

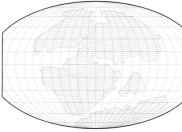

JURASSIC
135 million years ago

LAURASIA

TETHYS SEA

GONDWANALAND

TRIASSIC
200 million years ago

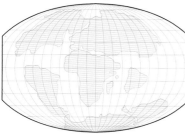

CRETACEOUS
65 million years ago

NORTH AMERICA · EUROPE · ASIA · AFRICA · SOUTH AMERICA · AUSTRALIA · ANTARCTICA

PRESENT DAY

axis); the movement of continents (plate tectonics); shifts in wind and ocean currents due to glacial advance and retreat; and so on. There are several theories currently circulating that relate meteor impacts to possible mass extinctions, global cooling and even the breakup of the supercontinent, Pangea. Vulcanism would not account for these longer period temperature transitions.

Spheres of our lives

Although we live on Earth, we really exist in many worlds. There is the geosphere (or lithosphere), which includes the Earth and rocks, minerals, sands, volcanoes and landforms. The hydrosphere defines almost of the water processes and the cryosphere encompasses the ice portion. The atmosphere, of course, handles all weather and climate. And let's also include the many spheres actually surrounding our Earth – including the stratosphere, troposphere and magnetosphere – that serve, layer-by-layer, as an interface with space. And the new field of study (known as space weather) addresses the myriad of extraterrestrial forces that control weather and climate on Earth and other planets.

Muliti-displinary perspectives

For many years, it's been thought that scientific advances require specialization. In some areas (such as nuclear engineering), this is true. But, more often, a single disciplinary approach to understanding or solving a problem simply doesn't work. If there is one US Federal agency that recognized this, it would have to be NASA. In order to undertake any type of space mission, especially those involving human cargos, NASA had to look at a range of disciplines. There was the physics of flight and re-entry, the meteorology of launch and return to Earth, the biological and chemistry components of the crew (including water, oxygen, waste, space clothing and muscle function), impacts of space radiation and the solar wind, transforming food stuffs into easily stored versions, communications, photography and much more. It was not just "let's launch a space mission".

A similar approach can be applied to the framework of extreme weather and climate. There are many factors at work with the sun perhaps the biggest driver of our weather and climate. This includes short and long period changes in solar energy, the daily solar cycle and so on. Mixed with this is the tilt of the Earth's axis, our daily rotation, and the seasonal variation in distance from the sun.

Weather blends with geography, oceanography, chemistry and people. Each of these disciplines in turn interacts with others. Weather and climate could be described using the "first law of ecology": everything is connected to everything else.

Geologic climatic conclusions

In studying climate change, several things become clear. There is an incredible amount of proxy data available now that is helping us understand our planet's history. That data is incomplete and when coupled with the wealth of data, it sometimes provides conflicting evidence about climates of the past. Furthermore, our planet doesn't look like it did in the past. Continents have moved, ocean currents have changed, mountains have either grown or been eroded and meteors (some of massive size) have impacted our planet. These in turn have contributed to changes in local, regional and sometimes global climatic zones. What was once a forest area may now be a desert and what was once under water may now be a mountain.

Nothing provides greater evidence to this shift in land masses than the locations in which fossils fuels are found. Coal deposits require the long-term submergence of large wooded (forested) areas into swampy or ocean areas. Eventually peat forms. If the peat is covered by sand, soil, or other deposits, it can become capped by a layer of sedimentary rocks. Over time, the peat would become compressed and have its water squeezed out. Eventually, meaning millions of years, it would become coal (and possibly oil or natural gas). Coastal deposits are found under many varied land areas, some far removed from oceans (see, for example, Chapter 13 where you can read about an Illinois coal mine filled with plant fossils).

Oil and gas were formed primarily as marine organisms died and became buried by sediment. If large amounts of sediment material piled atop buried organisms, the compression could produce high heat or pressure yielding more natural gas than oil; if there were less heat and more pressure, oil would be more likely.

Oil and natural gas are found under water, but also in deserts and under tundra regions. Without a shift in continents, this geographical distribution would not be possible.

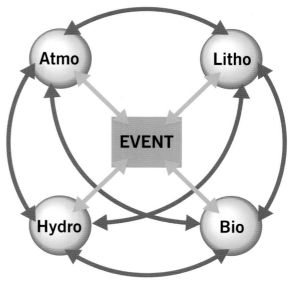

As ice ages have come and gone, sea level has fallen and risen respectively. This has resulted in some river canyons becoming submerged, while still keeping their topographic characteristics. Consider the Hudson Canyon that extends onto the Continental Shelf (more gently sloping underwater region adjacent to continents) from the New York City area. It looks like a V-shaped river valley, but it could not have formed underwater. One explanation is that it was once above sea level and that it was carved out by moving surface water, much like any other river valley.

Our world can be thought of as involving interactions across many disciplines or "spheres". This image shows the linkages among just four of them. Arrows show linkages between pairs of spheres. The relationships can involve links among more than two spheres and can be extended to include other spheres.

Other evidence of changes in sea level can be inferred from how the "falls" of some rivers (like the Potomac near Washington, DC) have migrated upstream. The falls mark the transition between the mountains and the coastal plain. Yet the falls near Washington, DC are much further upstream than expected. If sea level had lowered due to ice age conditions, then the elevation difference between the coastal plain and sea level would have increased, increasing the slope of the river and fostering increased erosion. Now that sea levels have risen, erosion has slowed.

Climatic controls

It should be easy to understand weather and climate control. Each and every one of us is affected by, and also controls, climate and weather on a daily basis. Perhaps it's the clothes we wear that either trap heat or allow it escape. Maybe we run the air conditioning in our home or our car. We may put up storm windows or reinsulate our attic.

A USGS survey in 2002 showed that water levels have risen around 400 feet during the last 21,000 years.

The Earth carries out similar climate control activities. Changes in cloud cover or pollution can either trap heat or block sunlight from reaching the Earth. An insulating cover of snow will keep plants and burrowing animals warmer, but the upper surface of the snow reflects sunlight and radiates energy strongly away to space. Cooling sea breezes may temper afternoon heat in coastal zones, while strong downslope winds in southern California may increase temperatures. In winter, winds blowing from warmer waters may cause snow to transform to rain in some Chicago suburbs; at other times, cold air blowing over the warmer Great Lakes may cause incredible lake effect snows.

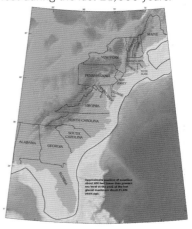

As noted in Chapter 1, the Earth also regulates itself by moving excess heat from tropical latitudes toward the poles and bringing colder air from the poles southward. Similar balancing takes place in moisture. While much of this balancing is effected by atmospheric storms and winds, ocean currents play a significant role, as well. With a

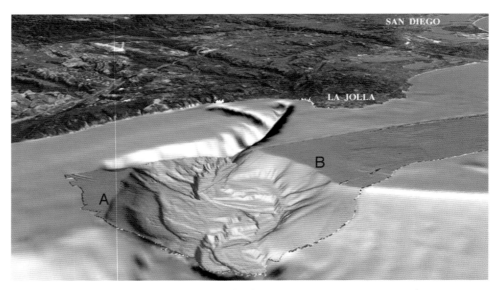

An example of an undersea canyon. Most likely, this canyon was formed above sea level through classic erosion processes and is now below sea level due to rising ocean levels.

circulation pattern that resembles a large high-pressure system, ocean gyres bring warmer waters to the eastern sides of mid-latitude oceans (latitudes of 45 degrees and higher) and cooler waters southward (between 25 and 45 degrees). This is one reason that the weather is milder in the UK than it is in eastern Canada (though they have similar latitude locations).

Even with its weather and climatic extremes, Earth typically returns to equilibrium. It may quickly leave that delicate balance, but overall the sum and average of these extremes creates an average planetary weather and climate system that remains the most hospitable one for human life.

Earth's average temperature is around 60 °F (15.5 °C). That's not much different than the temperature you would experience in an underground cave or cavern (55 °F (13 °C)). Average global rainfall and snowfall, winds and cloud cover can also be computed and compared to daily conditions.

Some tropical and middle latitude areas experience several hurricanes/typhoons/cyclones a year. In some years, many storms affect an area (e.g. Florida had seven hurricanes in 2004-05; the Philippines were struck by four typhoons in 2006). In other years, there is a dearth of activity (e.g. Florida had no hurricanes and only two tropical storms in 2006). Other parts of the tropics lack tropical storms and hurricanes and receive much less rainfall. Some tropical locales are far from what we think the tropics should resemble and are more like a desert.

Earlier in this chapter, I referred to "teleconnections". This linkage between weather in one place and another is best typified on seasonal and long-term scales, but there is also a connection between weather in one place and that of another, each day. You can see this by looking at daily weather maps such as the one on page 289.

Here, to the east of high pressure systems in middle latitudes, colder air is carried Equatorward; on the west of side of the high pressure system, Equatorward air is moved poleward. It is easy to see how the temperature patterns reflect this relationship every time a high pressure system advances across the weather map.

Low pressure systems operate in a similar fashion, although with a reversed circulation, the movement of temperatures is reversed.

Surface Weather Map at 7:00 a.m. EST

Although scientists tend to refer to "teleconnections" on a global scale, similar relationships are present on weather maps each day.

The balancing continues with droughts in one place and floods nearby, with extreme warmth in one place and cooler temperatures in another. And, as shown earlier in this chapter, the balancing can occur on relatively small scales. Sometimes that can involve a few states; at other times it can involve one part of a state; at other times, the balancing can even be on the county level.

Between June 1 and July 12, 2007, Miami, Florida recorded 20.36 inches (51.7 cm) of rain. This was nearly 10 inches (25.5 cm) more than average for the six-week period. About 100 miles (161 km) to the west, in Naples, Florida, rainfall for the same period was 4.90 inches (12.4 cm) or about 6.5 inches (16.5 cm) below average. Rainfall for the period at both locations would normally be around 11 inches (28 cm).

Even within Collier County (home to Naples), there were large variations in rainfall. Day after day rain fell inland, but not near the coast.

Human factors

We humans do affect the weather. We do this in many ways including changing land cover, changing drainage, clearing snow cover faster than it would normally melt, adding greenhouse gases and much more. Some of these affect weather directly; others have potential long-term weather or climate impacts. For example, years ago, we released aerosols from spray cans into the atmosphere. At the time, no one realized that the aerosols contained CFCs. We've since learned the unintended consequences of this action. We'll be dealing with a diminished ozone hole and the associated higher skin cancer risk that comes from more UV rays entering the atmosphere for centuries to come.

Similar types of impacts came from the release of sulfur, carbon dioxide and other gases that decreased the pH of rain and snow. Once steps were taken to reduce these gases, pH values stopped their decline. Still, rainfall across the entire US remains far more acidic that it should be.

Sometimes firefighters will utilize any and every means at their disposal to combat wildfires. Here, a World War II era B-17 bomber goes in for a raid; only this time they're dropping water on fires in southern California.

Much of the global warming scenario has keyed on carbon dioxide. Yet, through the decomposition of natural materials (e.g. wood, plant, human and animal waste) one of the by-products is methane (another major greenhouse gas). Recognizing this, some landfills are now capturing and selling methane.

What is especially significant is that sometimes we make changes to protect a location from a weather hazard and, in turn, we create a problem for someone nearby. Consider the case of flooding. To protect a place (either along a river or a coast), someone builds levees or a sea wall. That structure protects the place for which it was intended. But the water would have naturally flooded that place. Now it can't. So water levels nearby have to rise. When this happens nearby places may become flooded. Quickly, there is a cry to "save us, too". Structures continually get built until the landscape is so protected that natural processes are stopped from occurring.

Along the Mississippi River in the United States, so many structural improvements have been made that the river no longer floods extensively and no longer erodes as it once did (see Chapter 10 for more details). Flood plains no longer get flooded and have nutrient rich soils added to them. Decreased sediment loads in the river no longer feed the delta region south of New Orleans and so it is now being lost to normal processes.

Other structure improvements and building prevent normal sand transport along barrier islands. This allows storms to erode beaches and requires massive sand replenishment activities to protect coastal properties.

Forests and grasslands are a natural carbon storage and temperature – rainfall-regulating system. These take carbon out of the air and store it in vegetation. Trees are long-term carbon storage systems. Forests and any vegetated land area allow water to enter the ground more easily, lessening runoff and possibly flooding. Replacing these with "developed" land areas changes the landscape into one filled with impervious surfaces. With the loss of plants (and their natural evaporative cooling process) and their replacement with darker surfaces (e.g. asphalt), temperatures in developed areas are normally hotter.

As noted in Chapter 13, over the past 100 years, when we fought forest fires, thinking all fires were bad, we actually allowed forest fuels to build up. This has set the stage for more and more destructive fires.

Climate changes

With a perspective that spans far more than our most recent past (few years to few decades) and which addresses Earth, oceans and space, it starts to become easier to see how complex and ever-changing the weather and climate machine can be. It is easy to see how a multi-disciplinary team approach is needed to more fully assess what happened in the past. However, it is far more difficult to explain all of the causes and effects.

It is easy to want to see things in black and white – shades of gray make it more difficult for people to understand complex science. But, the black and white approach does not always work. For example, the Inter-governmental Panel on Climate Change's (IPCC) most recent report states emphatically that humans are the cause (95 percent certainty) of our current global warming period and that global warming is the cause of all the storminess and weather extremes that are occurring. I have considered the evidence myself, read and talked with other scientists and ascertained for myself what a logical conclusion could be. And in researching this book, I have uncovered considerable information that suggests that other factors are playing major roles, as well.

MISINTERPRETING FACTS

Throughout this book, you've seen how media reports weather records and extreme events and how careful one has to be using statistics. Sometimes we don't always know the details behind the graphs we see and we sometimes have to separate the real story from what is reported.

Consider the following which made the news in early July 2007. It isn't a climate change situation, but it does highlight media reporting and hidden stories. It was reported that several joggers and one person mowing a lawn had been struck with lightning while using Apple™ iPods. The news story also noted that there were also 13 documented cases in which cell phone users had been struck. The "news" keyed on the fact the iPods and the cell phones where electric and this conducted the electricity, making the injuries far worse.

The real story involves basic lightning safety. Had these individuals not been outdoors in a thundery environment, they never would have been struck in the first place.

This doesn't mean that we humans are not contributing. Rather, I have made it clear that we humans have done much to "mess up" our planet (development, erosion, loss of trees and fossil fuel use). Sometimes our best intentions have had unintended negative consequences (e.g. stopping forest fires set the stage for worse fires).

Based on increased amounts of CO_2 in the atmosphere, the 2007 IPCC report (with references to associated predictive climate models) talk about wholesale shifts in climates. This includes having some climates vanishing and new ones forming. Clearly, some species can migrate and others can increase or decrease their range. However, it is much more difficult for tree and large plant species in natural settings to adjust their range in short time periods.

We humans, however, can move tree species for home or community purposes. As temperatures warm and rainfall patterns change, some trees have become more suited for different locales. For example, I have seen palm trees planted in London and in coastal North Carolina.

Because of human intervention, the Arbor Day Foundation has created new plant hardiness zone maps. These show where tree species can be safely grown. Given the recent amount of global warming, many species have shifted poleward by one or two hardiness zones.

These changes can be good or bad, depending on your perspective. For example, grain-growing areas can expand into more northern latitudes. However, southern parts of prairie areas may soon become too hot and dry. Grape growing regions will shift, with some areas becoming better qualified to produce wines; other areas may lose their advantage. In the UK, grape growing – part of the warm period about 1,000 years ago – has returned. Insects and airborne diseases will have new venues in which to thrive. Periodic winter freezes might not kill ticks, fleas and similar pests. How will the world look when palm trees survive in Washington, DC and London, England?

Although we can work to stop global warming and climate change, the question to ask is, "should we?" We may prefer a specific climate or we may dislike another. But do we have the right to decide which climates stay and which ones go?

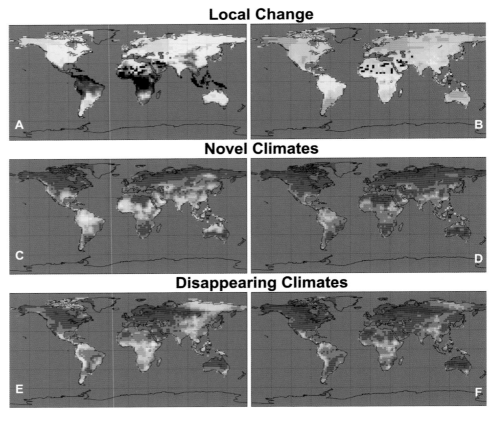

Local Change

Novel Climates

Disappearing Climates

Solutions

I could write a whole book about solutions to the myriad of weather- and climate-related problems that are plaguing the Earth right now (especially global warming). However, the book would be dated before it was written – so much is happening in this arena, so quickly, that it makes sense here to only touch upon the topic.

With recent political and economic factors, "green" has finally become a mainstream word. Everyone from individuals to state, local and federal agencies are either already on the bandwagon or are getting ready to jump on. Hybrid cars, mini-fluorescent light bulbs, solar and wind power, energy-efficient windows, programmable thermostats, roof-top gardens and much more have joined the green energy suite.

Unfortunately, just because it reads "green" doesn't mean that it comes without a price. Hybrid cars may save lots in fuel costs, but they come with a hefty upfront cost premium. I calculated this using my average driving expenses and if I were to buy a hybrid vehicle, it would take me about 8-10 years to recover the increased cost (given current gas prices). Hydrogen cars tout they don't pollute, yet water vapor is their main by-product. The last time I checked, water vapor was a greenhouse gas.

Biofuels are supposedly the cure-all for petroleum needs. Yet these come with a cost in terms of fertilizer needs (some of which must be supplied via petrochemical processes), soil use impacts and – perhaps the most significant – the cost increases that spill over into cattle, poultry and other food production as crops are diverted to biofuel production. Notwithstanding these, scientists and industry leaders are working to find ways around the trade-offs.

Biodiesel is becoming more widely available as an alternative to traditional fuels.

The media's role in climate change

Like it or not, many reporters and their media home stations are no longer simply reporting news. They decide what is newsworthy (it's always been that way), but now they typically take a position. Often, in print media, the position is "liberal", if one has to assign a label. The emphasis is on equality, equal rights and often anti-establishment. At other times, the reporters simply accept what governmental agencies provide and rewrite press releases. And we know that following entertainment stories often trump important news stories. Radio talk shows tend to take on more conservative stances.

Today, "fair and balanced" and investigative reporting are typically missing from most media outlets. And once the media (as a group) defines which bandwagon they are going to follow, there are few, if any, stragglers.

Nothing could be closer to the truth than in the area of "environmental reporting". Based on an informal survey (I read the newspapers and watch TV religiously, both at home and wherever I travel), I don't think I've ever seen an environmental story that doesn't save an endangered species and I've only seen a couple of stories that are not warning of the dire consequences of global warming.

That is why I have tried to show where evidence supports global warming (at least in the short-term) and where it does not. I've also tried to set the stage for you, the reader, to look into the issue and decide for yourself if global warming is really happening, what its cause(s) might be and what consequences (good and bad) might result from it.
So, let's take a specific look at some things that have recently occurred which unbalance our understanding of global weather and climate.

The Weather Channel™ typically reports only one side of global warming – it is happening and things are bad. Solutions are always given. And even when the weather

Below let: Four generators are pictured here using methane from a landfill site to produce electricity.

Below: A hybrid car drives in the car pool lane in Orange County, California.

In the NASA Aqua image (May 9, 2007), the storm circulation off the east coast of Florida appears like a tropical system; but unfavorable winds and cold temperatures aloft suggest that Andrea may not be tropical at all.

is good, the Weather Channel™ almost always pulls the plug and warns that UV radiation may kill us!

Two cases of mistaken identity? Subtropical Storm Andrea and Tropical Storm Barry

In early May, 2007, a middle latitude low pressure system developed off the Carolina coast. The system had its origins as an upper level low pressure system that had moved southward from New England. The system was referred to initially as an extra-tropical low ("extra" as a prefix means "outside of"). So this low was originally outside of the tropics.

As the low sat off the Carolina coast, it sat over the warm waters of the Gulf Stream (a ribbon of warm water with its origins in the Gulf of Mexico – see map and satellite image). This helped to start to transform the low pressure from a cold core low to one with warmer characteristics. Still, the upper wind and temperature pattern was not conducive to having a named tropical system even though, on occasion, some extra-tropical lows do make the switch to tropical.

The speed at which the National Hurricane Center named the system and described it as "sub-tropical", surprised even veteran meteorologists such as myself. I don't know the reason for the naming of this storm, but I can go back in time to similar situations (on radio talk shows) in which meteorologists from a national private weather service pushed for naming a cold core low (a low with a cold center; tropical systems have warm centers) as a tropical system because the government was "hiding information from the public". Rather than counter this push, the government went along with creating a case of mistaken identity.

A similar scenario occurred with Tropical Storm Barry. I watched satellite images throughout the event and it was hard to discern the needed warm core, favorable upper winds and cloud patterning necessary to apply the tropical label.

Back in the 1970s when I personally experienced this situation, naming another storm was no big deal. Now, in the heat of global warming, these situations have added two named storms to the seasonal totals. Although the National Hurricane Center noted that such early storminess is not unusual, these two events will surely be used, in some circles, to bolster claims that global warming is causing storms to form earlier and for these storms to be more intense.

Aside from fueling global warming proponents, the process actually confused the public. I live in Florida where hurricanes are the watchword. I was asked by dozens of people about the naming of the system. Questions included, "what is a sub-tropical low anyway?" and "was that really a tropical storm?"

Finally, there are some interesting impacts to both the insurance and real estate industries. Will the addition of another storm help to raise insurance rates and/or will a storm near Florida cause a decrease in home sales or negatively impact tourism?

In short, there may be some very significant unintended consequences even from such a simple action as naming a tropical low-pressure system.

Final thoughts

Imagine what you'd think if you read the following headline some 20 years from now, "After 45 years of marked warming, scientists have now concluded that a major cooling trend has begun. With solar output dropping and glaciers rebuilding, it is clear why snow

So-called Tropical Storm Barry (June 1, 2007) may have also been non-tropical.

Flooding has become more widespread in the UK during the first few years of the 21st Century. This picture shows Tewkesbury, Gloucestershire, where flood waters forced the local water treatment plant to shut down thereby depriving 150,000 homes of fresh water. Many areas of the UK were hit by the worst floods in 60 years following torrential rainfall in July 2007.

has fallen as far south as Houston, Texas and Rome, Italy on the first week of summer. Nations have started to hoard oil supplies in the face of a global winter…"

That headline is not real and it is not even a prediction. However, it does fit the theme of this book. I wanted to explain weather and climate in understandable terms and in short-, long- and geologic-term perspectives and to show how weather and climate are inter-linked with other natural forces and including human existence. This cuts both ways – weather and climate affect us and we, in turn, affect weather and climate. While scientists (I am one) and the media (I have been a TV meteorologist) provide important information and perspective in our lives, they are not "gods". We can make mistakes, arrive at incorrect conclusions and even get involved in political and social messaging (muddying the waters).

Yes, humans have affected the functioning of our planet in many ways: increased carbon emissions, transformed landscapes by creating excessive impervious surfaces, removed trees and managed water resources poorly. But overall, changes to long-term climate have occurred on scales and magnitudes far greater than our most recent history.

I also wanted to stoke some controversy and foster greater research and understanding on the part of all global citizens and encourage readers not to blindly follow mainstream thinking. I hope readers will gather as much information as possible and then digest it in context with their experiences and with a geological time scale in mind. Because scientists keep providing new information about our planet's past, we must continue to keep reading and learning.

With so many storms and other forms of extreme weather and climate change in the news, keeping abreast of these will become easier, as long as we have an open mind to information and ideas. And because so many of these affect us either directly or indirectly, we have to consider how we live our lives.

This is because, whether you are a global warming proponent or not, many of the "solutions" to global warming are simply good ideas. What can be wrong about saving energy, developing alternative fuels, preserving our forestlands, keeping our planet's air and waters pollution-free and having reliable water supplies?

Let me close with a quote from the late Arkansas Senator J William Fulbright who once noted, "We must dare to think 'unthinkable' thoughts. We must learn to explore all the options and possibilities that confront us in a complex and rapidly changing world." Although he wasn't writing about global warming or climate change at the time, his words couldn't be more appropriate to today's meteorological and climatological issues.

While political and economic forces suggest that people use mass transit (where available), convenience and other factors may win out. This traffic jam in Los Angeles, California, demonstrates that people are prepared to drive to their destination regardless of cost or the consequences of "global warming".

PICTURE CREDITS

158 bottom NASA/GSFC/METI/ERSDAC/JAROS and US/Japan ASTER Science Team;
159 top two images A B Miller and G R Brakenridge, Dartmouth Flood Observatory;
159 bottom © Donald P Schwert, North Dakota State University;
160 UCAR;
161 top © Bettmann/Corbis;
162 top © BlueRed Press Ltd;
162 bottom NOAA/NOS/NGS;
163 top © Stephen Finn/Fotolia;
163 bottom this file is licensed under the Creative Commons Attribution ShareAlike 2.5 License;
164 bottom © BlueRed Press Ltd adapted from NOAA;
165 top © BlueRed Press Ltd adapted from NOAA;
166 bottom© BlueRed Press Ltd adapted from NWS;
167 top © Christa Eder/Fotolia;
167 bottom © iStockphoto.com/Michael Graham;
168 Top © iStockphoto.com/Emil Schreiner;
168 bottom NASA;
170 top NASA;
170 middle © BlueRed Press Ltd;
171 top © BlueRed Press Ltd;
171 above left © Ho Weihao/Fotolia;
171 above right © Alexis Lesaffre/Fotolia;
171 below © Dmitry Pichugin/Fotolia;
172 top © Gavriel Jecan/Corbis;
172 bottom left and right H Michael Mogil;
173 top © iStockphoto.com/Tinik;
173 bottom H Michael Mogil;
174 top NASA;
174 bottom © BlueRed Press Ltd;
175 top left NASA;
176 top NASA/Goddard Space Flight Center;
177 top © BlueRed Press Ltd;
177 bottom USGS;
178 top and bottom SeaWiFS Project, NASA/Goddard Space Flight Center, and ORBIMAGE;
179 top left, right and left MODIS Rapid Response Team, NASA-Goddard Space Flight Center;
179 bottom right Cpl Alicia M Garcia, US Marine Corps;
180 left three images MODIS Rapid Response Team, NASA-Goddard Space Flight Center;
181 top left five images NASA;
181 right two images NASA;
182 bottom left © iStockphoto.com/Alexander Hafemann;
182 top left © iStockphoto.com/Chad Purser;
182 top right © iStockphoto.com/Jeremy Voisey;
182 bottom left © iStockphoto.com/Phill Hunt;
182 bottom right © iStockphoto.com/Sakakawea7;
183 top © BlueRed Press Ltd;
183 bottom US Environmental Protection Agency;
184 top © iStockphoto.com/James Phelps;
185 top NOAA/BRP;
185 right three images © BlueRed Press Ltd;
186 top Plymouth State University;
186 bottom three images NOAA/NCDC;
187 top four images Graphic from Living with Drought/ Bureau of Meteorology/copyright Commonwealth of Australia, reproduced by permission;
189 top right Mike Mogil;
189 top left USGS;
189 bottom © iStockphoto.com/Kris Hanke;
190 top USGS;
190 bottom four images USGS;
191 top H Michael Mogil;
191 bottom © iStockphoto.com/Martin McCarthy;
193 top © Monique Pouzet/Fotolia;
194 top © BlueRed Press Ltd;
195 top three images NOAA;
195 bottom H Michael Mogil;
196 top left © Leon Run/Fotolia;
196 top right H Michael Mogil;
198 top H Michael Mogil;
199 bottom © BlueRed Press Ltd;
200 top © iStockphoto.com/Ryan Gardner;
200 bottom © Frans Lanting/Corbis;
201 top © iStockphoto.com/fabphoto;
201bottom © iStockphoto.com/Scott Vickers;
202 top left USDA Forest Service;
202 top right © iStockphoto.com/Scott Leman;
202 bottom © BlueRed Press Ltd;
203 bottom Plymouth State University;
204 top NOAA;
205 top © BlueRed Press Ltd;
205 bottom two images NASA;
206 top NIFC;
206 bottom © Romulo Moya Peralta/Fotolia;
207 bottom ©Marti Timple/Fotolia;
208 top NASA;
209 top © iStockphoto.com/Brasil2;
209 bottom US Dept of Agriculture;
210 top © iStockphoto.com/Tim Osborne;
210 bottom left © iStockphoto.com/Sean Locke;
210 bottom right © iStockphoto.com/Donald Gruener;
211 top NOAA;
211 bottom © Chris Hondros/Getty Images;
212 top NASA;
212 bottom © iStockphoto.com/Sakakwea7;
213 top © iStockphoto.com/Irving Bartlett;
213 bottom Plymouth State University;
214 top © BlueRed Press Ltd;
215 middle NASA;
215 bottom right two images NASA;
216 top NOAA;
217 top NOAA;
218 top © Timothy A Clary/AFP/Getty Images;
219 bottom © Daniel Berehulak/Getty Images;
220 top © iStockphoto.com/Seth Loader;
220 bottom © Michael S Lewis/Corbis;
221 top NASA;
221 bottom NOAA;
222 top left © iStockphoto.com/Karl Dolenc;
222 top right © iStockphoto.com/John Snelgrove;
223 top NOAA;
223 bottom © Scott Peterson/Getty Images;
224 top © iStockphoto.com/tiburonstudios;
224 bottom NASA;
225 top Public Domain;
225 bottom © Alison Wright/Robert Harding World Imagery/Corbis;
226 top left © iStockphoto.com/Ian Francis;
226 top right © iStockphoto.com/Debra James;
226 bottom NASA;
227 top © Jessica Rinaldi/Stringer/Reuters/Corbis;
228 top USGS;
231 bottom © BlueRed Press Ltd;
232 top© Daniel Gustavsson/Fotolia;
232 bottom © China Photos/Getty Images;
233 top © deserttrends/Fotolia;
233 bottom NASA;
234 top NOAA;
234 bottom BRP/NOAA;
235 top © BlueRed Press Ltd;
236 bottom adapted from meteored.com;
237 top graphic adapted from Charles Graves/Understanding the Weather/Saint Louis University;
239 top NASA;
239 bottom H Michael Mogil;
240 top H Michael Mogil;
241 top H Michael Mogil;
242 bottom H Michael Mogil;
243 bottom © BlueRed Press Ltd;
244 top H Michael Mogil;
245 top two images © BlueRed Press Ltd;
245 top right twp images © BlueRed Press Ltd;
246 top © Charles Platiau/Reuters/Corbis;
247 bottom © iStockphoto.com/João Freitas;
248 top © Jeffrey Lai/Fotolia;
248 bottom left © iStockphoto.com/Thaddeus Robertson;
248 bottom right © iStockphoto.com/Daniel Stein;
249 top H Michael Mogil;
249 middle © Sascha Burkard/Fotolia;
249 bottom Emily Boynton;
250 top H Michael Mogil;
250 bottom© iStockphoto.com/Pamela Cowart-Rickman;
251 top © iStockphoto.com/Jaap Hart;
251 bottom © Hubert Boesl/dpa/Corbis;
252 top NASA GSC;
252 bottom USGS;
253 top two images H Michael Mogil;
254 bottom H Michael Mogil;
255 right NADP;
255 middle Public Domain;
255 bottom H Michael Mogil;
256 top NASA;
257 top H Michael Mogil;
258 top NASA;
259 top NOAA;
259 middle Cambridge University, Centre for Atmospheric Research;
259 bottom two images NOAA;
260 top two images NOAA;
261 top H Michael Mogil;
262 top USGS;
262 bottom © iStockphoto.com/gkanivets;
263 top© iStockphoto.com/Tatiana Mironenko;
263 bottom © iStockphoto.com/Danilo Ascione;
264 top © Julien Grondin/Fotolia;
264 bottom © iStockphoto.com/koch valérie;
265 top Harry Glicken/USGS;
265 middle Peter Lipman/USGS;
265 bottom Lyn Topinka/USGS;
266 top NASA;
267 top USGS;
267 bottom © BlueRed Press Ltd;
268 top USGS;
268 bottom National Geophysical Data Center;
269 top Jeff Schmaltz, MODIS Land Rapid Response Team at NASA GSFC;
270 top Lyn Topinka/USGS;
270 bottom © Tangy de Saint-Cyr/Fotolia;
271 top © iStockphoto.com/Marco Regalia;
272 top © iStockphoto.com/koch valérie;
272 bottom © iStockphoto.com/gkanivets;
273 top © BlueRed Press Ltd;
274 top © iStockphoto.com/Kenneth Gray;
274 bottom graphic adapted from Volcano World;
275 bottom © iStockphoto.com/David Stearn;
276 bottom top © iStockphoto.com/Glenn Bartley;
276 bottom middle © Franck Lachaud/Fotolia;
276 bottom © iStockphoto.com/Chris Crafter;
276 bottom right© Andrew Watson/Fotolia;
277 top H Michael Mogil;
277 bottom USGS;
279 top © BlueRed Press Ltd;
280 top four images NOAA;
281 bottom two images adapted from Dr William Kesler, PMEL, University of Washington;
282 bottom USGS;
283 top adapted from J Wallace, University of Washington;
284 top GSFC/NASA;
285 top adapted from Thomas Crowley, 1990 in the Journal of Climate;
286 side five images USGS;
287 top Copyright© 1997-2000, Wheeling Jesuit University/NASA Classroom of the Future™. All rights reserved;
287 bottom USGS;
288 top USGS;
289 top © BlueRed Press Ltd;
292 top adapted from "Projected distributions of novel and disappearing climates by 2100 AD" by John W Williams, Stephen T Jackson and John E Kutzbach;
293 top left © iStockphoto.com/James Benet;
293 top right © Pizano/Fotolia;
293 middle © iStockphoto.com/Logan Buell;
293 bottom left © iStockphoto.com/Andrew Martin Green;
293 bottom right © iStockphoto.com/T; Back cover image Collection of Wayne and Nancy Weikel, FEMA Fisheries Coordinators

REFERENCES

The following is a brief listing of web-based resources that you may find helpful in learning more about the topics explored in this book. Some of these may have been included within this book; others may have been included in the book, but not listed here. Many of the pages provide links to additional websites and direct you to sources you might not otherwise be aware of. But, please don't limit yourself to anything listed here or within the book. Instead, I encourage you to judiciously search for other information about extreme weather and climate on the Internet and in various books and periodicals and continue to learn more about the wide range of weather and related hazards that affect our planet.

CHAPTERS 1-3
http://www.nws.noaa.gov
http://www.ncdc.noaa.gov/oa/reports/weather-events.html
http://lwf.ncdc.noaa.gov/oa/climate/severeweather/extremes.html

CHAPTER 4
http://www.nohrsc.nws.gov/nsa/
http://www.ncdc.noaa.gov/oa/climate/research/snow-nesis/
http://www.erh.noaa.gov/buf/lakeeffect/indexlk.html
http://www.erh.noaa.gov/buf/spotters/totalsnow.htm
http://www.erh.noaa.gov/buf/lakeeffect/indexlk.html
http://www.ncdc.noaa.gov/ussc/

CHAPTER 5
http://www.nhc.noaa.gov/
http://australiasevereweather.com/cyclones/index.html
http://www.aoml.noaa.gov/hrd/tcfaq/tcfaqHED.html
http://www.nrlmry.navy.mil/tc_pages/tc_home.html

CHAPTERS 6-9
http://www.spc.noaa.gov
http://www.spc.noaa.gov/faq
http://www.essl.org/cgi-bin/eswd.cgi
http://australiasevereweather.com/
http://www.torro.org.uk/TORRO/index.php
http://www.tornadoproject.com/
http://www.publicaffairs.noaa.gov/storms/
https://thunderstorm.vaisala.com/tux/jsp/explorer/explorer.jsp
http://www.bangladeshtornadoes.org/
http://www.cocorahs.org/

CHAPTER 10
http://water.usgs.gov/waterwatch/
http://www.bom.gov.au/climate/austmaps/puld_menu/index.shtml
http://www.bom.gov.au/hydro/flood/
http://www.icold-cigb.net/
http://www.nws.noaa.gov/oh/hic/flood_stats/Flood_loss_time_series.htm
http://news.nationalgeographic.com/news/2001/06/0601_chinadust.html

CHAPTER 11
http://www.nps.gov/archive/whsa/Sand%20Dune%20Geology.htm
http://earthobservatory.nasa.gov/Study/Dust/
http://www.met.rdg.ac.uk/~aer/dodo/dodo.html

CHAPTER 12
http://www.drought.noaa.gov/
http://www.drought.unl.edu/dm/current.html
http://www.bom.gov.au/sat/NDVI/NDVI2.shtml
http://www.osdpd.noaa.gov/PSB/IMAGES/gvi.html
http://svs.gsfc.nasa.gov/stories/drought/index.html
http://www.edwardsaquifer.net/desalination.html

CHAPTER 13
http://www.nifc.gov/links.htm
http://www.cbc.ca/news/background/forcesofnature/forestfires.html
http://www.fire.uni-freiburg.de//current/globalfire.htm
http://firewatch.dli.wa.gov.au/landgate_firewatch_public.asp
http://tucupi.cptec.inpe.br/meio_ambiente/
http://www.cyberwest.com/cw17/fire1.html

CHAPTER 14
http://www.isse.ucar.edu/heat/
http://www.bom.gov.au/weather/wa/sevwx/perth/heatwaves.shtml
http://www.epa.gov/hiri/about/heatguidebook.html
http://environment.about.com/od/healthenvironment/a/air_quality_hea.htm

CHAPTER 15
http://lwf.ncdc.noaa.gov/oa/climate/extremes/1999/february/extremes0299.html
http://mcc.sws.uiuc.edu/climate_midwest/mwclimate_events_cold1.htm#
http://www.msnbc.msn.com/id/10994127
http://www.reliefweb.int/rw/RWB.NSF/db900SID/KHII-6L787L?OpenDocument

CHAPTER 16
http://www.mountainnature.com/Climate/Chinook.htm
http://ggweather.com/winds.html
http://wxpaos09.colorado.edu/windstorms/chinook.html
http://klima.hr/priopcenja/bura_2004_e.html
http://www.cdc.noaa.gov/Boulder/wind.html
http://archive.orr.noaa.gov/cameo/dr_aloha/terrain/terrain.html
http://www.nyc.gov/html/oem/html/news/03_11_13_wind.shtml

CHAPTER 17
http://airnow.gov
http://airnow.gov/index.cfm?action=where.world
http://ec.europa.eu/environment/air/links.htm
http://www.epa.gov/airmarkets/trading/basics.html
http://www.bom.gov.au/info/about_uv.shtml
http://www.atm.ch.cam.ac.uk/tour/dobson.html
http://www.sunsafetyalliance.org/wmspage.cfm?parm1=173
http://ozonewatch.gsfc.nasa.gov/

CHAPTER 18
http://vulcan.wr.usgs.gov/Volcanoes
http://www.volcano.si.edu/
http://volcano.und.edu/
http://www.ssd.noaa.gov/VAAC/
http://www.avo.alaska.edu/
http://www.spacedaily.com/reports/JAXA_Satellite_Watching_Indonesian_Volcano.html
http://earthobservatory.nasa.gov/Newsroom/NewImages/images.php3?img_id=17598
http://rapidfire.sci.gsfc.nasa.gov/gallery/?search=volcano
http://rsd.gsfc.nasa.gov/goes/text/goes.volcanoes.html
http://www.spaceflightnow.com/news/n0703/01io/

CHAPTER 19
http://www.ncdc.noaa.gov/oa/climate/research/cie/cie.html
http://www.nationalgeographic.com/wildworld/terrestrial.html
http://www.koshland-science-museum.org/
http://www.epa.gov/climatechange/
http://www.ipcc.ch/
http://www.unep.org/themes/climatechange/index.asp
http://www.bom.gov.au/lam/climate/levelthree/climch/climch.htm
http://www.pewclimate.org/
http://edugreen.teri.res.in/EXPLORE/climate/causes.htm
http://www.nationalcenter.org/NPA455.html
http://www.cpc.ncep.noaa.gov/products/precip/CWlink/MJO/enso.shtml
http://jisao.washington.edu/pdo/
http://ircamera.as.arizona.edu/NatSci102/NatSci102/lectures/climate.htm

REMOTE SENSING
http://earthobservatory.nasa.gov/Observatory/#
http://www.ssec.wisc.edu/data/
http://www.goes.noaa.gov/
http://rapidfire.sci.gsfc.nasa.gov/
http://www.bom.gov.au/weather/satellite/
http://www.nws.noaa.gov/radar_tab.php
http://www.weatheronline.co.uk/radar/euroeuro.htm
http://mirror.bom.gov.au/weather/radar/

OTHER REFERENCES, LINKS, ETC
http://earthobservatory.nasa.gov/Newsroom/Headlines/
http://www.wmo.int/pages/members/index_en.html
http://www.bom.gov.au/lam/wpage.shtml
http://www.jma.go.jp/jma/en/Centers/indexe_cnt.htm
http://www.weatherworks.com/links.html
http://www.disastercenter.com/disaster/TOP100C.html

There are other organizations and periodicals that provide useful information about extreme weather. These include:

Weatherwise Magazine (US): http://www.weatherwise.org/

Weather Magazine (UK):
http://www.rmets.org/publication/weather/index.php

EnviroZine (Canada – online):
http://www.ec.gc.ca/envirozine/english/home_e.cfm

Monthly Significant Weather Summaries (Australia):
http://www.bom.gov.au/inside/services_policy/public/sigwxsum/sigwmenu.shtml

Severe Weather Association (Australia):
http://www.severeweather.asn.au/public/index.php

Past Weather (US):
http://www.noaa.gov/pastweather.html

Natural Hazards Center (US):
http://www.colorado.edu/hazards/

Natural Hazard Links: http://www.rothstein.com/links/rothstein_recommended20.html

International Startegy for Disaster Reduction:
http://www.unisdr.org/

International Hazards Data Base: http://www.em-dat.net/disasters/list.php

FURTHER READING

Blame It On The Weather, David Phillips, Michael Parfit and Suzanne Chisholm, Portable Press, © 1998

Furious Earth, Ellen J Prager, McGraw Hill, ISBN 0-07-135161-2, © 2000

Hurricane Watch, Bob Sheets and Jack Williams, Vintage Books, © 2001

Skywatch East, Richard A Keen, Fulcrum Publishing, © 1992

The Weather Book, Vintage Books, Jack Williams, © 1992 (but there is an updated edition)

Weather – A Visual Guide, Bruce Buckley, Edward Hopkins and Richard Whitaker, Firefly © 2003

Extreme Weather – A Guide and Record Book, Christopher Burt, WW Norton & Company, © 2004

These four books comprise a set (but are also sold individually) from Voyageur Press:
Global Warming by Mark Maslin, World Life Library, © 2002,

Volcanoes, Peter Clarkson, World Life Library, © 2000

Tornadoes, H Michael Mogil, © 2003

Glaciers, John Gordon, World Life Library, © 2001

GLOSSARY

Adiabatic – process that occurs without gain or loss of heat.

Albedo – fraction or percentage of reflected light.

Anticyclonic circulation (anticyclone) – a high-pressure spin in the atmosphere; the spin is clockwise in the Northern Hemisphere and counter-clockwise in the Southern Hemisphere.

Anvil – the top, mostly flat, ice crystal part of a thunderstorm.

Aquifer – a layer of rock, sand or gravel that contains water' typically sandwiched between two rock layers that do not contain water.

Boundary – a region separating two masses of air with different characteristics. Includes weather fronts, thunderstorm outflow boundaries and even sea breeze fronts.

Climate change – a more correct term applied to the overall, long-term variation in Earth's temperature and rainfall.

Convergence zone – any area in which air is forced together. If the convergence occurs near the ground, air is forced to rise; if it occurs at high altitude, air is forced to sink.

Carbon Monoxide (CO) – a toxic, odorless, colorless gas composed of one atom of carbon and one atom of oxygen. It results from incomplete oxidation of carbon during combustion.

Carioles Force – a fictitious force, due to latitudinal variations the speed of Earth's rotation, which has to be considered in studying air motions.

Convection – vertical motions linked to heating and cooling. Typically associated with rising air currents that produce cumulus-type clouds.

Cumulus – flat-bottomed, puffy-topped cloud type that forms primarily due to the heating of the Earth's surface and the lower atmosphere.

Cumulonimbus – tallest stage of a cumulus cloud; forms with strong rising air currents. Brings lightning, thunder, heavy rainfall and various forms of severe weather.

Cyclonic circulation (cyclone) – a low-pressure spin in the atmosphere; the spin is counter-clockwise in the Northern Hemisphere and clockwise in the Southern Hemisphere.

Dew point – temperature to which air must be cooled in order for dew to form.

Diurnal temperature cycle – the daily rise and fall of temperature due to the day-night cycle.

Downdraft – a downward current of air, typically associated with thunderstorms.

Dry line – a boundary separating warm and dry air from warm and moist air; typically found in the High Plains of the US and in other similar regions around the world.

Dust devil – a small whirlwind not associated with cumulus clouds; often found in dry or desert regions when the ground becomes superheated.

El Niño – from the Spanish for the (Christ) child; associated with abnormally warm ocean waters off the northwest coast of South America.

Electromagnetic radiation – energy which moves in the form of an advancing electric and magnetic field disturbance; includes visible, infrared, ultraviolet and x-rays.

Evapotranspiration – the process by which plants breathe (taking in CO_2) and give off water vapor and O_2.

Flood plain – the low area, typically adjacent to a river or other body of water that is prone to flooding.

Föehn (wind) – a warm, dry, downslope wind formed as air descends the downwind side of the European Alps. Similar to the US or Canadian Chinook winds.

Front – the boundary or transition zone between two air masses.

Fujita Scale – (and new **Enhanced Fujita Scale**) a scale relating tornado intensity indirectly to structural and/or vegetative damage.

Funnel cloud – a tornado-like, rotating, appendage extending from a cumulus-type cloud but not reaching the ground.

GOES (Geostatational Observational Environmental Satellite) – a weather satellite parked in a fixed location above the Equator whose daily orbit matches that of the Earth. A series of these satellites rings the Earth providing full Earth viewing day and night.

Global warming – an ongoing increase in the Earth's atmospheric and oceanic temperatures linked to an increase in the greenhouse effect (higher values of CO_2 and other greenhouse gases).

Inversion – a reverse of the typical vertical temperature profile (applied to both atmosphere and ocean); in the atmosphere, it signifies increasing temperature with increasing altitude.

Jet stream – a zone of relatively high-speed winds concentrated within a narrow stream in the atmosphere.

Katabatic winds – in mountainous areas, the downslope flow of air driven by nighttime cooling of the Earth's surface; can also be applied to the more general flow of cold air downhill (including bora winds).

La Niña – see El Niño; named for the female child and signifies colder waters off the northwest coast of South America.

Lead (Pb) – a metallic element which can be poisonous when ingested (e.g. deteriorating old paint) or through cumulative contact. It has many positive uses, as well (e.g. storage batteries).

Lithosphere – the solid portion of the Earth.

Little Ice Age – the cold period in which mountain glaciers grew and advanced; occurred between about 1500 (or perhaps as early as 1300) and 1850.

Microburst – an intense and localized downburst.

MODIS – the Moderate Resolution Imaging Spectroradiometer instrument package (flown aboard NASA's polar-orbiting TERRA and AQUA satellites) used to obtain cloud, temperature and other measurements of the Earth's surface, atmosphere and oceans.

Nitrogen Dioxide (NO_2) – a reddish-brown gas with a pungent and irritating odor. In the air, it interacts with other gases to form nitric acid and toxic organic nitrates. It is also involved in atmospheric reactions that produce ground-level ozone.

Nocturnal cooling – the part of the diurnal temperature cycle linked to the cooling of the ground at night.

Ozone (O_3) – an atmospheric gas composed of three atoms of oxygen. Near the ground, ozone acts as a pollutant, is harmful to breathe and it can damage sensitive plants. At high altitudes, ozone absorbs ultraviolet (UV) radiation from the sun.

Particulate matter (PM) – a descriptor for small particles and liquid droplets in the atmosphere smaller than about 10 micrometers (a micrometer is one millionth of a meter or one thousandth of a millimeter. A human hair is between 20 and 150 micrometers wide) in diameter. PM can include acids (such as nitrates and sulfates), organic chemicals, metals and soil or dust particles. The smaller the particles, the easier they are to inhale and harm the human respiratory system.

Plate tectonics – theory in geology that describes how small parts of the Earth's surface (called "plates") float on and travel across the interior of the Earth.

Rainshaft – the region beneath a cloud in which falling precipitation is easy to see. Often seen as a thin or small area that masks clouds in the distance.

Smog – originally a mixture of smoke and fog; today, more commonly applied to photochemically-induced pollution filled with nitrogen oxides, ozone, carbon compounds and particulate matter.

Solar heating – the part of the diurnal temperature cycle linked to the heating of the ground by the sun.

Solar wind – plasma (a charged mass that has gas and other properties) that is continuously ejected from the sun's surface into and through interplanetary space.

Squall line – a line of active thunderstorms (either continuous or with breaks) that often brings gusty winds.

Sulfur Dioxide (SO_2) – is one of a family of sulfur oxide gases (SOx) that can easily dissolve in water (creating various acids, such as sulphuric acid). SOx gases are formed when fuel containing sulfur, such as coal and oil, is burned.

Super-cell – a long-lived thunderstorm that often consists of a single, strong updraft. Supercells spawn a large percentage of the most intense tornadoes and often bring other types of severe weather.

Super-cooled liquid water – water remaining unfrozen at temperatures below 32 °F (0 °C).

Synoptic – large-scale view of meteorological data; often meaning weather observations taken at the same time.

Tehuantepecer – a violent squally wind from the north or north-northeast in the Gulf of Tehuantepec (southwest Mexican coast) in winter. Occurs when a large, cold high-pressure system moves far southward in the Gulf of Mexico.

Tornado – a violently rotating column of air, associated with a cumulus-type cloud, that is in contact with the ground; it is often, but not always, visible as a funnel cloud.

Trade winds – region of mainly easterly winds between about 23.5 degrees N and 23.5 degrees S latitude. One of the most favored and reliable wind regions during the period in which sailing ships dominated global navigation.

Train echo effect – the repeated movement of convective precipitation across a particular area; often occurs with stationary lines of thunderstorms and lake effect snow bands.

Troposphere – the sphere (or shell) of atmosphere closest to the Earth. Roughly 8 to 12 miles thick, it houses most of what we call "weather,"

Ultraviolet radiation – electromagnetic radiation with shorter wavelengths than visible radiation but longer than x-rays; can be dangerous to humans in large doses.

Upper-level disturbance – a cyclonic circulation in the upper level wind flow pattern that is more strongly defined than near the ground.

Upwelling – the rising motion of subsurface water to the surface due to the removal of surface water by wind flow.

Urban heat island – the warmer region within an urban region caused, in large-part, by human activities.

Warm-core – a storm system that has a warmer central region, as opposed to the usual colder central region. Hurricanes and other tropical low-pressure systems almost always contain warm cores.

Wind shear – the variation in wind speed and/or direction in any direction. The term is typically applied to winds associated with microbursts and downbursts.

INDEX

Figures in italics indicate captions.